GLOBAL PATTERNS

GLOBAL PATTERNS
Climate, Vegetation, and Soils

By Wallace E. Akin

UNIVERSITY OF OKLAHOMA PRESS : NORMAN AND LONDON

Also by Wallace E. Akin

The North Central United States (Princeton, N.J., 1968) (with H. Wayne Pritchard)
Earth: The Stuff of Life, by Firman E. Bear. 2d ed., rev. (Norman, 1986)
Global Patterns: Climate, Vegetation, and Soils (Norman, 1991)

Library of Congress Cataloging-in-Publication Data

Akin, Wallace E.
Global patterns : climate, vegetation, and soils / by Wallace E. Akin.
p. cm.
Includes bibliographical references and indexes.
ISBN 0-8061-2309-5
1. Climatology. 2. Vegetation and climate. 3. Soils and climate. I. Title.
QC981.45.A38 1990
581.5′22—dc20 90-50227

The paper in this book meets the guidelines for permanence and durability of the Committee on Production Guidelines for Book Longevity of the Council on Library Resources, Inc. ∞

CONTENTS

Three great patterns dominate the earth and are of tremendous importance to man—the pattern of climate, the pattern of vegetation, and the pattern of soils. When the three patterns are laid one upon another, their boundaries coincide to a remarkable degree because climate is the fundamental dynamic force shaping the other two.

D. I. BLUMENSTOCK AND C. W. THORNTHWAITE, "Climate and the World Pattern"

PREFACE

The geographic similarity of the great world patterns of climate, vegetation, and soils has intrigued many natural scientists and has led to the development of cause-and-effect theories of relationships between them. Of the three patterns, climate inevitably was perceived as the principal dynamic component shaping the other two, vegetation and soil. Other environmental factors, such as geology and human activities, were perceived as playing only a secondary role.

In the first half of the twentieth century, overzealous attempts to force vegetation and soil patterns into a climatic framework led to overgeneralization and failure to recognize that the patterns, although similar, do not correlate perfectly. Progress in soil science and a better understanding of the complexity of soils in tropical areas have both shown the need for and made possible a reassessment of the climate-vegetation-soil relationships that once seemed so beautifully simple. There is no doubt that relationships exist between these three great global patterns—only they are more complex than they were once thought to be. Some knowledge of these dynamic relationships and their associated patterns is vital if we are to comprehend the human habitat with all its mounting environmental problems.

Of the three patterns, obviously climate is the independent variable, whereas vegetation and soil depend to some degree both on climate and on each other through various feedback mechanisms. Although vegetation and, to a lesser degree, soil may exert some influence on climate, particularly on *microclimate,* their influence on broader climatic patterns (*macroclimate*) is less easy to demonstrate.

Climate can be defined as "average weather," but this definition fails to convey the intricate variations occurring from day to day, month to month, and year to year. In fact, a statistical analysis of weather records would probably fail to reveal a completely average year or month. More accurately, climate is the *long-range pattern of weather*—a definition that allows for the march of the seasons and the tendency for climate to vary from the norm. Climatologists once viewed climate as relatively static, notwithstanding minor fluctuations; now, more and more attention focuses on actual climatic change.

The decade that has just passed offered a good example of climatic change, since it included the hottest years of the twentieth century (1981, 1983, 1987, 1988, and 1989). It now appears that the decade of the 1980s was the warmest in the instrumental record. This sweltering decade focused world attention on the *greenhouse effect* and the threat of global warming, which will affect not only temperatures but also world patterns of precipitation, human activities (especially agriculture), and eventually vegetation and soil.

World patterns of vegetation and soil relate to global climatic patterns, and changes in world vegetation over time are an important key to past climatic change. Such relationships, however, are not exact since other environmental factors, particularly geology and terrain, are important in determining distribution patterns of plant communities and soils. Human activity, in some places continuous over thousands of years, also plays a major role in determining the character of some vegetation and soil complexes.

Major changes in soil science have taken place in the last several decades. This is true especially in soil classification, with the publication in 1975 of *Soil Taxonomy* by the U.S. Department of Agriculture. The new system completely replaced the classification nomenclature that was used in the United States before the 1960s, but older terms still remain in the scientific literature, and some are retained in classification systems of other countries and the United Nations. For this reason, both the old and the new systems are reviewed in chapter 8, "Classification and World Distribution of Soils."

In part 1 we begin our review of world patterns with a summary of the evolution and dynamics of the earth's life-sustaining atmosphere. We review world climatic patterns as they exist today, consider the water budget (an integral component of climate of the "water planet"), and take a brief excursion into the climatic past and possible climatic futures. In part 2 we examine the nature and dynamics of vegetation and its global distribution, while part 3 deals with the nature and origin of soils and their global patterns. The human role in the past, present, and future evolution of these global environmental systems is a major focus throughout.

It is impossible to give credit to the many teachers and colleagues who have inspired me and added to my knowledge and appreciation of earth environments. Particularly, I am indebted to the late Thomas F. Barton and to Clarence F. Jones, my mentors in undergraduate and graduate school, for introducing me to the concept of environmental interrelationships, both in the classroom and in the field. I owe special appreciation to the staffs of the University of Copenhagen's Geographic Institute and the Danish Heath Society for acquainting me with the Danish landscape featured in the section on human impact on vegetation. To the late Walter Jeske, USDA, I owe many of the excellent illustrations from U.S. government sources. And I am grateful to H. Wayne Pritchard, retired executive vice-president of the Soil Conservation Society of America, for contributing from his extensive knowledge of the soil conservation movement and the people involved in it. I owe special thanks to my wife, Peggy, for her hours of proofreading and her invaluable contribution to style and syntax. Finally, however, none of the above contributors should be held responsible for errors of fact or interpretation; as the author, I alone bear that responsibility.

WALLACE E. AKIN

Des Moines, Iowa

Part 1

CLIMATE

This round of green, this orb of flame,
Fantastic beauty; such as lurks
In some wild poet, when he works
Without a conscience or an aim.
TENNYSON, *In Memoriam*

Chapter 1

EARTH, THE WATER PLANET

Elements and Controls of Weather and Climate

Among the nine planets orbiting the sun, Earth is unique. Although similar geologically to its nearest neighbors in space—the moon, Venus, Mars, and Mercury—its climate is markedly different. Only Earth has conditions favorable for life. We can speculate on the possibilities and probabilities of life on other planets within other planetary systems, but only on Earth is life a certainty.

Atmospheres and Climates of the Solar Planets

We now know from Soviet and American space probes that Venus, our nearest planetary neighbor, is extremely hot, with surface temperatures approaching 477° C (890° F) and atmospheric pressure on the surface almost one hundred times greater than at sea level on earth. Such temperatures are well above the melting point of lead (327.5° C, or 621.5° F). Mars is also inhospitable to life because of its thin atmosphere, high winds, and low but extremely variable temperatures. Mercury, the planet closest to the sun, is intensely hot on the sunward side, very cold on the dark side, and, like Earth's moon, it lacks an atmosphere. The giant outer planets and their satellites are extremely cold. It is only on Earth that surface temperatures mostly fall within that very narrow range between 0° C and 100° C (32°F and 212°F) in which water is liquid—an essential condition for the origin and sustenance of life.

Intriguingly, photographs of the Martian surface suggest that at one time water played an important role in producing some erosional landscapes, such as the famous Giant Canyon, three times deeper than Earth's Grand

Canyon. The origin of these water-derived features is debatable. Some planetary geologists are convinced they were formed when subsurface ice was melted by volcanic activity or meteorite impacts, whereas others believe they are evidence that in the past Mars had a more humid, warmer, and denser atmosphere in which water vapor condensed and fell as rain on the planet's surface.

On the other planets of the solar system, temperatures are either so high that water can exist only in the gaseous state (water vapor) or so low that most water is ice. It is true that water vapor can exist at temperatures below 0° C, but it passes from water vapor to ice, or from ice to water vapor, without passing through the liquid state—a process, called *sublimation,* which produces snow or frost in the Earth's atmosphere. On Mars, where there are snowcaps in the polar regions (possibly a mixture of snow and frozen carbon dioxide), the equatorial surface temperatures may sometimes be high enough to melt ice, but because of the very low barometric pressure, any water produced would evaporate rapidly and form water vapor.

Origin of the Earth's Atmosphere

How did Earth's life-supporting climate originate? Evidence suggests that Earth, Mars, and Venus all have atmospheres that originated from gases ejected from the planetary interiors during volcanic activity and that initially these atmospheres contained approximately the same mixture of gases. Why the three planetary atmospheres evolved into quite different mixtures of gases seems to have been primarily the result of the distance of each planet from the sun.

In addition to the distance factor, temperature played a critical role in the evolution of atmospheres. The key to the present temperatures on the surfaces of Venus, Earth, and Mars lies in the relative abundance of water vapor and carbon dioxide in their respective atmospheres. In the evolving atmospheres, temperatures were such that the stable phase of water was ice on Mars, water vapor on Venus, and liquid water on Earth. Most of the water vapor in Earth's early atmosphere was precipitated out as water and, thus, oceans were formed. Today, oceans cover more than 70 percent of Earth's surface, and they contain about 97 percent of all the water on the planet. Of the remaining 3 percent, about 2 percent is contained within glaciers, and about 1 percent is accounted for in the atmosphere and in the rivers, lakes, soil moisture, and groundwater on land.

Once formed, oceans had a profound effect on the evolution of Earth's atmosphere. Oxygen, which is chemically very active, initially combined with carbon to form carbon dioxide (CO_2), or with hydrogen to form water (H_2O). On Mars, most of the oxygen is still combined as CO_2 or H_2O, or it is joined chemically with other elements in the surface rocks. On Venus, water vapor was present, but since the planet is some 28 percent closer to

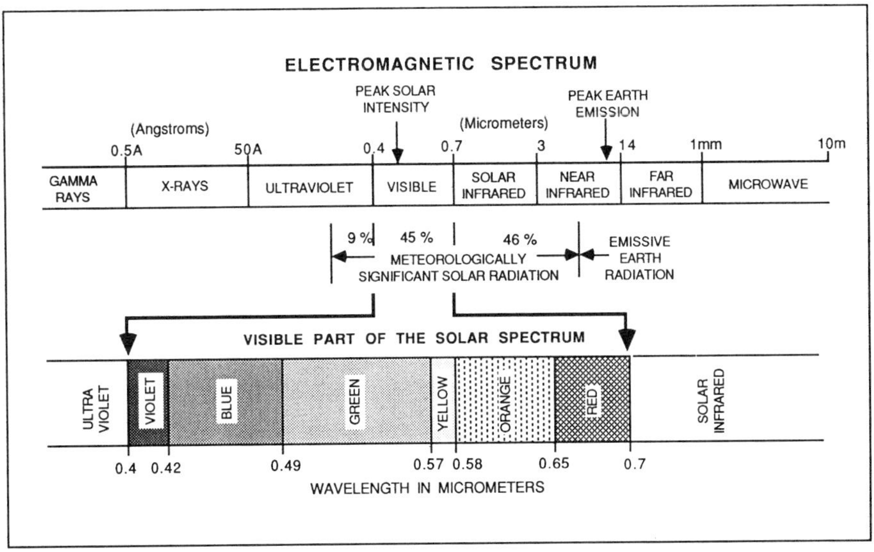

Fig. 1–1. The electromagnetic spectrum. Note that the energy received from the sun is concentrated in the near UV and visible wavelengths of the spectrum, whereas reradiation of that energy from the Earth—a much cooler radiator than the sun—is in the form of longer wave infrared radiation.

the sun than is the Earth—108 million km (67 million mi) compared to our 150 million km (93 million mi)—it was too warm for the water vapor to condense and form oceans. In the upper atmosphere of Venus, the water molecules were broken down into hydrogen and oxygen by ultraviolet radiation from the sun, the lightweight hydrogen atoms escaped to space, and the oxygen combined with carbon to form CO_2. The present Venusian atmosphere is about 95 percent CO_2, and Venus has a "runaway greenhouse effect," accounting for its extremely hot surface.

A *greenhouse effect* also is operating in Earth's atmosphere. The sun is a very hot radiator (5,727°C, or 10,340°F), and its radiant energy is concentrated in the shorter wavelengths of the electromagnetic spectrum in the near infrared (IR), visible, and ultraviolet (UV) regions (figure 1-1). Earth's atmosphere is relatively translucent to these wavelengths. When solar radiation reaches the surface, it is absorbed, the surface temperature increases, and the absorbed energy is reradiated into the atmosphere. The cooler Earth (27°C, or 81.6°F) radiates this energy as long-wave IR radiation, the atmosphere absorbs this energy, and air temperature rises—the greenhouse effect. A number of greenhouse gases constitute the "glass" in the earth's greenhouse. The most important of these are CO_2, H_2O, methane (NH_4),

TABLE 1-1 Characteristics of the Sun and Planets of the Solar System

	Diameter Km (Mi)	Average Distance from Sun Km × 10^6 (Mi × 10^6)	Average Surface Temperature C° (F°)	Main Atmospheric Components
Sun	1,392 × 10^3		5,800 (10,500)	—
Mercury	4,880 (3,032)	58 (36)	260 (500)	—
Venus	12,112 (7,526)	108 (67)	480 (900)	CO_2
Earth	12,742 (7,920)	150 (93)	15 (59)	N_2, O_2
Mars	6,800 (4,216)	228 (142)	−60 (−76)	CO_2
Jupiter	143,000 (88,700)	778 (483)	−110 (−166)	H_2, He
Saturn	121,000 (75,000)	1,427 (886)	−190 (−310)	H_2, He
Uranus	51,800 (32,187)	2,869 (1,780)	−215 (−355)	H_2, CH_4, ?
Neptune	49,000 (30,447)	4,498 (2,790)	−225 (−373)	H_2, CH_4, ?

Source: Ahrens 1985, 3.

and ozone (O_3). Venus, with its CO_2 atmosphere, has a "super greenhouse," which explains its extremely high surface temperatures (see table 1-1).

On Earth, most of the original CO_2 was removed from the atmosphere by solution in the oceans, where it combined with other elements, such as calcium, to form calcium carbonate ($CaCO_3$) and other mineral salts. Today, this surplus CO_2 is stored in the vast volumes of carbonate bedrocks composed of calcite ($CaCO_3$) and dolomite [$CaMg(CO_3)_2$]. Later, when green plants evolved, they extracted additional CO_2 from the atmosphere through *photosynthesis* (see chapter 5), a process by which CO_2 combines with water in plants to form carbohydrates [$(CH_2O)_n$]. Much of the organic matter produced in past geologic ages is stored in the fossil fuels, coal and petroleum. Human beings now are burning these fuels and rapidly releasing millions of years of CO_2 accumulation into the modern atmosphere, thereby enhancing the greenhouse effect. In chapter 4 we shall examine some of the possible climatic implications of this CO_2 enrichment.

In addition to its role in cleansing the atmosphere of the excess CO_2, photosynthesis by green plants and some bacteria is responsible for Earth's oxygen-rich atmosphere. Since photosynthesis produces an excess of free oxygen, it replenishes the atmospheric supply. Without it, chemically active free oxygen would combine quickly with other elements and would disappear from the atmosphere.

Free oxygen also contributes to the survival of life on Earth by contributing to the formation of the ozone layer. Ozone is a form of oxygen consisting of molecules formed by the joining of three atoms of oxygen (O_3), whereas most atmospheric oxygen consists of molecules formed by two atoms of oxygen (O_2). Ozone reaches its highest concentration about 25 km above the earth's surface, where it is formed by the action of solar UV radiation on O_2. Besides absorbing harmful wavelengths of UV radiation and protecting animal and plant life, the ozone layer also absorbs outgoing IR radiation and is a factor in the greenhouse effect, particularly in maintaining the heat balance in the stratosphere. Ozone is unstable, and the molecules disassociate into O_1 and O_2 at about the same rate as they form. Atmospheric scientists are concerned that certain air pollutants will deplete the ozone layer, with potentially disastrous consequences to life on earth. Should the concentration of stratospheric ozone decrease, it could have an adverse effect on crops and animals, cause higher rates of skin cancer in humans, and alter the earth's heat balance.

Changes in the Sun's Energy Output Through Geologic Time

Astronomers have presented climatologists with a seeming paradox: the sun is brighter now than it was during the early formation of the Earth's atmosphere. Astronomers estimate that 3–4 billion years ago the sun emitted 30 percent less energy than now, a level that today would produce runaway glaciation covering the entire Earth. Why did such a catastrophe not occur in an earlier period of the Earth's history? The answer most likely lies in the higher CO_2 content of the early atmosphere, which produced a more efficient greenhouse effect. As the processes noted above slowly reduced the concentration of atmospheric CO_2, the resulting decrease in the efficiency of the greenhouse effect seems to have been balanced by a gradual brightening of the sun at a rate that maintained surface temperatures well within a range favorable for life. We shall consider the long-range climatic consequences of these and other atmospheric and terrestrial changes in chapter 4.

Composition of the Earth's Modern Atmosphere

Earth's modern atmosphere is a mixture of a number of gases, the most abundant being nitrogen (N_2) and oxygen (O_2) (table 1-2). Oxygen, of course, is necessary for animal life and for many chemical reactions between the atmosphere and the earth, including weathering of rocks, decay of organic matter, and combustion. Nitrogen is relatively inert chemically and is unavailable directly to plants from the atmosphere. Of the other gases, water vapor and CO_2 are of particular significance to weather and climate: CO_2 for its role in the greenhouse effect, and water vapor because of its importance in the hydrologic cycle.

Unlike most other atmospheric gases, water vapor occurs in highly variable atmospheric concentrations. Extremely dry air may contain less than 1 percent water vapor by volume, whereas very warm, moist air may hold as much as 4 percent. Because an equal volume of water vapor weighs only about 64 percent as much as molecular nitrogen (N_2), and 56 percent as much as molecular oxygen (O_2), moist air is lighter than dry air of the same temperature. Because the source of the water vapor in the atmosphere is the

TABLE 1-2 The Earth's Atmosphere Today

Component	Molecular Composition	Percent by Volume	Approximate Molecular Mass[a]
Gases			
Nitrogen	N_2	78.08	28
Oxygen	O_2	20.94	32
Argon	Ar	0.93	40
Carbon dioxide	CO_2	0.03 (variable)	44
Neon	Ne	0.0018	20
Helium	H	0.0005	4
Methane	CH_4	0.0001	16
Ozone	O_3	0.00006	48
Hydrogen	H_2	0.00005	2
Xenon	Xe	0.000009	131
Krypton	Kr	(trace)	84
Water vapor	H_2O	0–4	18
Atmospheric Pollutants (average value)			
Carbon monoxide	CO	0.00002	28
Ozone[b]	O_3	0.000004	48
Sulfur dioxide	SO_2	0.000001	64
Nitrogen dioxide	NO_2	0.000001	46
Particles (dust, etc.)		0.00001	—

[a]Molecular mass indicates the relative weight of a gas.
[b]Ozone is a component of polluted air as well as one of the permanent atmospheric gases.

Earth's surface, particularly the oceans, most water vapor is concentrated in the lower atmosphere. Although water vapor may be relatively abundant near the surface, it accounts for less than 1 percent of the total atmospheric volume.

Vertical Structure of the Atmosphere

The atmosphere can be divided into layers based on temperature changes related to elevation; this vertical temperature gradient is called the *lapse rate*. Changes in this lapse rate, which occur at well-defined height intervals, produce a layered structure in the atmosphere, ranging from the lowest layer, the *troposphere*, upward through the *stratosphere*, and ever higher to the *mesosphere* and *thermosphere* (fig. 1-2).

Troposphere

Most weather phenomena occur in the lower 8 to 18 km, the zone designated the troposphere. This layer contains about 75 percent of the atmosphere by weight, practically all of the water vapor, and most of the atmospheric dust. Its upper boundary, the *tropopause*, reaches its greatest

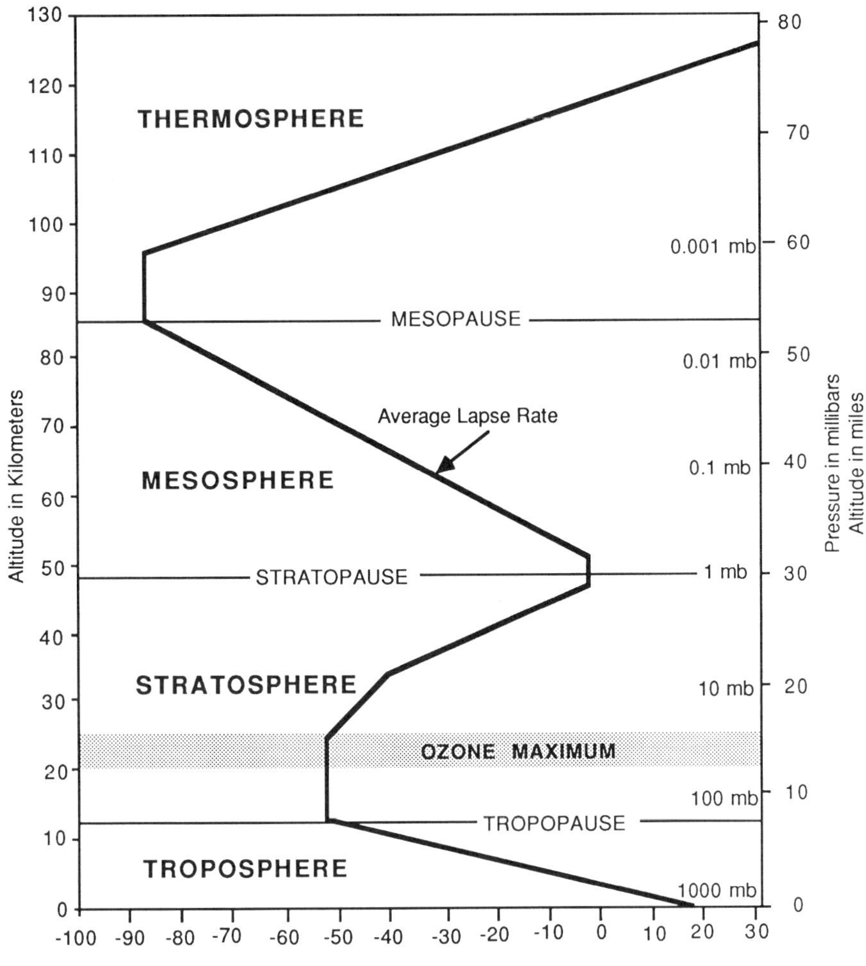

Fig. 1–2. Temperature zones in the atmosphere. Ozone (O_3) is present in all levels of that atmosphere, but reaches its greatest natural concentration in the lower stratosphere. Here, oxygen (O_2) is more plentiful and UV radiation, necessary for the formation of ozone, is strong. Above this so-called ozone layer, the concentration of oxygen decreases rapidly with altitude; below, the ozone formed in the stratosphere filters out the shorter wavelengths of UV necessary for rapid ozone formation.

height over the equator, and is closest to the surface over the poles. The average temperature shows a rather uniform decrease of about 6.5°C per km (3.5°F per 1,000 ft) from the surface upward to the tropopause. Daily and seasonal changes occur in the thickness of the troposphere, and the lapse rate may vary considerably from the mean, especially near the surface.

Under some conditions, the lapse rate may be reversed, with temperatures increasing with higher elevation, a condition called a *temperature inversion.*

The troposphere is the layer where most of the heat exchange between the atmosphere and the earth's surface takes place. Because the surface is the source of most of the heat in the troposphere, and because the air is denser, average air temperatures are higher near the surface. Heating from below causes the surface air to become more buoyant and results in vertical motion and overturning within the troposphere. Of course, the supply of heat must be sufficient to overcome the density gradient in the atmosphere so that air can rise against the acceleration of gravity. Actually the warmer, less dense air on the surface is displaced upward by cooler, denser air settling from above.

Stratosphere

Immediately above the tropopause is the stratosphere, a layer in which the lapse rate is nearly constant in its lower levels but is inverted in the upper levels, thus forming a temperature inversion. Vertical motion is reduced by this temperature inversion; consequently, the stratosphere acts as a lid on the troposphere, limiting the height to which overturning can take place. Very few clouds are present in the stratosphere, although a few giant thunderstorm clouds (*cumulonimbus*) may push their tops through the tropopause and into the lower stratosphere. The top of the stratosphere—the *stratopause*—is about 45 km (28 mi) above the surface; thus, it contains the ozone layer (fig. 1-2) and about 24 percent of the atmosphere by weight. Supersonic jet aircraft operate in the lower stratosphere because flying conditions there are nearly ideal because of lack of clouds and air turbulence.

Mesosphere and Thermosphere

The two outer layers contain only 1 percent of the atmosphere by weight. The mesosphere lies between 45 and 92 km (27–55 mi) above the earth, between the stratopause and the mesopause. Temperatures decrease with altitude in the mesosphere, reaching a low of −93° C (−137° F) at the mesopause, the coldest region of the atmosphere.

The outermost layer is called the thermosphere because of the extremely high temperatures observed in its upper regions. There, temperatures may be as high as 1,300° C (2300° F). Because of the very low air density at these altitudes, however, these high temperatures are not strictly comparable to those of the denser atmosphere at lower elevations. In the thermosphere, temperature is a measure of molecular activity, and the air actually holds very little heat, although the individual air molecules are very hot. An object exposed to this "hot" atmosphere would gain very little heat because it would have relatively few contacts with the widely spaced hot molecules; thus, vehicles such as *Skylab,* orbiting within the upper thermosphere, have no problem maintaining reasonable internal temperatures.

Heating of the Atmosphere

The sun is the source of almost all energy on Earth. It drives the atmosphere-ocean engine; this, in turn, produces weather and climate. A small amount of energy is derived from the earth's interior (*geothermal energy*), and this is produced by radioactive decay. Another energy source is the gravitational fields of the Earth, moon, and sun. These nonsolar sources supply less than .02 percent of the total atmospheric energy and have a correspondingly minor effect on the earth's atmospheric systems.

Close to 30 percent of the solar radiation reaching the earth (*insolation*) is reflected and scattered immediately back into space by clouds, atmospheric dust, the atmosphere itself, and ground and water surfaces. This reflected radiation, called the *albedo*, has no effect on the land-ocean-atmosphere system. The remaining 70 percent of the sun's energy heats the atmosphere, oceans, and land surfaces; drives the winds and ocean currents; evaporates water; and supplies energy for life through the process of photosynthesis.

The path taken by the solar energy that is retained within the land-ocean-atmosphere system is relatively complicated, as indicated in figure 1-3, a simplified diagram of the energy exchanges that take place between the atmosphere, the Earth's surface, and outer space. Because of the greenhouse effect, the insolation is not returned immediately to space, but is absorbed and reradiated back and forth between the earth and the atmosphere. Eventually, the solar energy retained by the land-ocean-atmosphere system must be balanced by outgoing radiation; otherwise, the Earth's temperature would be constantly increasing. With the present atmospheric composition, the delay in returning the solar energy to space assures an average surface temperature some 40°C (72°F) higher than would be the case if there were no atmosphere.

About 51 percent of the direct, diffuse, and scattered insolation is absorbed by the surface. The temperature of the surface rises, and it radiates the energy back into the atmosphere as infrared radiation; water vapor, CO_2, ozone, and other greenhouse gases absorb most of this. The temperature of the atmosphere rises and, in turn, it radiates some of the heat back to the surface; consequently, the surface receives energy equal to about 140 percent of the sun's radiation. Since the surface acquires a major share of the solar energy retained in the system, it becomes the major source of heat for the atmosphere. Thus, the atmosphere is warmed from below, an important factor that causes air in the troposphere to overturn and thereby to produce weather phenomena.

In addition to the direct heating of the Earth-atmosphere system by solar radiation, some 18 percent of the sun's energy evaporates water from the land and oceans. This energy is not lost but retained in the atmosphere as *latent heat of vaporization*. When condensation occurs, this energy is released

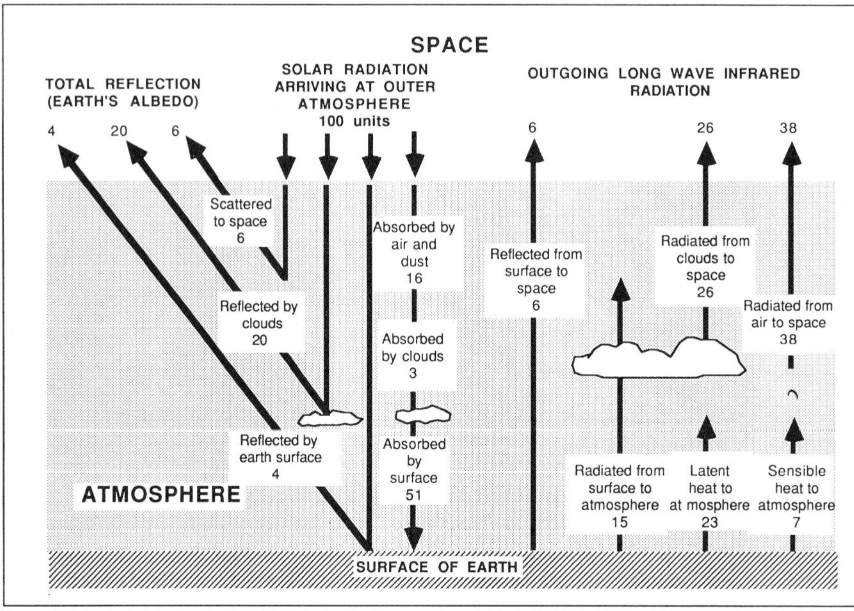

Fig. 1–3. Earth's radiation balance. About 30 percent of the solar radiation arriving at the outer atmosphere is returned unchanged to space by reflection and scattering, about 19 percent is absorbed directly by the atmosphere and clouds, and about 51 percent is absorbed by the earth's surface. This absorbed energy is returned to space as long wave radiation through the several paths indicate in the diagram. Some 6 percent of the IR radiation is reflected from the surface to space, and 45 percent is radiated by the surface and absorbed by the atmosphere, causing air temperature to rise. The air in turn radiates this IR energy to space. This is the greenhouse effect, the process by which Earth's atmosphere is heated. (Data from National Research Council 1975)

into the atmosphere as heat; often, this heat is released far from its original point of arrival.

Ultimately, if the heat balance of the Earth-atmosphere system is maintained, the same amount of energy received from the sun must be radiated or reflected from the outer atmosphere into space. If changes were to occur in any of the inputs and outputs of the system, the Earth's average temperature conditions would change, with parallel changes in the climate. Should energy received from the sun increase, or should the system's ability to transfer heat back to space decline, temperature would rise until a new balance was reached at a higher overall temperature level. Should Earth receive or retain less energy, a cooler climate would result. In chapter 4, we will discuss some possible solar and atmospheric changes that might lead to climatic fluctuations.

Variation in Solar Radiation Received at Earth's Surface

The heat budget varies from place to place, from day to day, and from season to season. A number of factors contribute to this, including the Earth's rotation on its axis, causing night and day; the inclination of the axis from vertical to the plane of the orbit, causing seasonal change; and atmospheric conditions—particularly the presence or absence of clouds—resulting in daily and hourly variations in the energy received. Some less obvious variables include distance from Earth to the sun and small fluctuations in the intensity of solar radiation.

Variations in Energy Arriving from the Sun

The average distance from the Earth to the sun is about 150 million km (93 million mi), but, because the Earth follows an elliptical orbit with the sun located at one focus of the ellipse, its distance from the sun varies throughout the year. Inhabitants of the Northern Hemisphere often are surprised to learn that they are nearest to the sun on January 3, and farthest from it on July 4. Obviously, the seasonal change in distance from Earth to sun has only a very minor effect on earth temperatures. Although the amount of solar energy arriving on Earth during the Southern Hemisphere summer (December–March) is about 7 percent greater than that received during the Northern Hemisphere summer, the uneven distribution of land and water between the two hemispheres makes it difficult to evaluate the effect of the difference. In truth, Northern Hemisphere summers are warmer and winters are colder than those south of the equator. Why? The reason is the rapid seasonal heating and cooling of the great continental land masses dominating northern latitudes.

The absorption and reflection of the incoming solar radiation by the atmosphere also plays a role in determining the amount of solar energy reaching the surface. Of primary importance is the extent and density of cloud cover, since thick clouds may reflect 80 percent or more of the insolation.

Angle of Incidence and Duration of Daylight

The progression of the seasons of the year is marked by changes in the angle of the sun above the horizon and in the duration of daylight caused by the inclination of the Earth's axis by about 23.5° from perpendicular to the plane of its orbit. As the Earth moves around the sun, its axis maintains the same orientation in space, with the North Pole always pointing toward Polaris, the North Star. Polaris, because of its great distance from our solar system, shows no observable parallax when viewed from different points on the orbit, and it remains above the North Pole, even though the distance from one side of the orbit to the other is some 300 million km (186 million mi). The axis, therefore, is always parallel to its orientation at any point on the orbit.

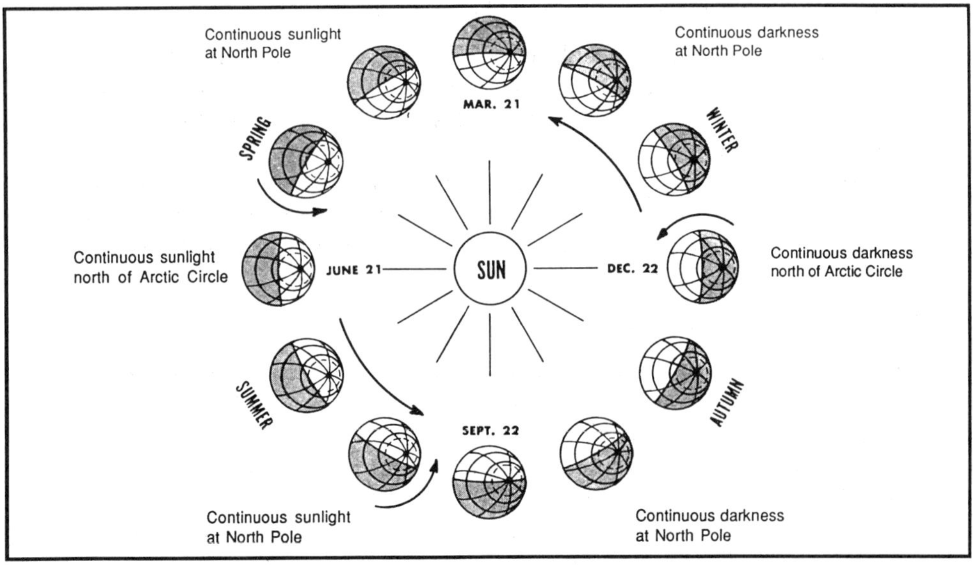

Fig. 1–4. Illumination of the earth at various seasons and positions on the orbit as viewed from over the North Pole. Since the North Pole is always inclined in the same direction throughout the year, it is inclined toward the sun for six months and away for the remainder; thus, it has six months of daylight and six months of night. Note that there is continuous darkness north of the Arctic Circle on December 22. On the same date, areas south of the Antarctic Circle have continuous daylight. (U.S. Air Force Manual 105-5, *Weather for Aircrew Trainees,* 1956.)

On June 21 the North Pole is at the point of maximum inclination toward the sun, making this date the first day of summer in the Northern Hemisphere, the day with the longest daylight period and the highest noon sun (fig. 1-4). On December 22, the situation is reversed; the North Pole is inclined away from the sun, the noon sun is low in the southern sky, and the Northern Hemisphere winter begins. (Of course, seasons are reversed in the Southern Hemisphere.) These two dates are called the summer and winter *solstices.*

Halfway between these two points on the orbit, on about September 23 and March 21, neither pole is inclined toward the sun, and the circle of illumination passes through the poles and cuts all parallels in half; thus, all points on the earth except the poles have twelve hours of daylight and twelve hours of darkness, if twilight and atmospheric refraction are ignored. These are the *equinoxes,* meaning equal nights, which mark the beginning of fall and spring. Table 1-3 gives the approximate duration of daylight at selected latitudes on the equinoxes and the solstices. Note in figure 1-4 that the circle of illumination always bisects the equator; the result is twelve hours each of daylight and darkness during all seasons of the year.

TABLE 1-3 Approximate Time Between Sunrise and Sunset for Selected Latitudes with No Allowance Made for Refraction by the Atmosphere

Latitude	June 21 (hours: min.)	December 22 (hours: min.)	March 21 and September 23 (hours: min.)
90°N (N. Pole)	24:00	00:00	12:00
66.5°N	24:00	00:00	12:00
50°N	16:09	07:51	12:00
40°N	14:51	09:09	12:00
30°N	13:56	10:04	12:00
20°N	13:13	10:47	12:00
10°N	12:35	11:25	12:00
0° (Equator)	12:00	12:00	12:00
10°S	11:25	12:35	12:00
20°S	10:47	13:13	12:00
30°S	10:04	13:56	12:00
40°S	09:09	14:51	12:00
50°S	07:51	16:09	12:00
66.5°S	00:00	24:00	12:00
90°S (S. Pole)	00:00	24:00	12:00

Note: For this table sunrise and sunset are determined from center of sun to center of sun on horizon.
Source: Harrison 1935.

In addition to the effect of longer daylight hours, the high angle of the summer sun causes more energy to be concentrated on a smaller area. If the sun is at the zenith (90° above the horizon), all the energy reaching the surface in a sunbeam with a cross-sectional area of 1 m^2 is concentrated on 1 m^2 of surface, after it penetrates one atmosphere (the total depth of the atmosphere measured vertically to the Earth's surface). On a clear day, the atmosphere transmits about 78 percent of the sun's energy to the surface at sea level (through one atmosphere). If the angle of the sun is only 45° above the horizon, the sunbeam passes through 1.4 atmospheres, and only about 50 percent of its energy reaches the surface, where it is spread over an area of 1.4 m^2; consequently, the square meter receives only about 45 percent as much energy as it would have if the sun were at the zenith.

The sun's rays are vertical only in that portion of the earth bounded by the Tropic of Cancer (23.5° N) on the north, and the Tropic of Capricorn (23.5° S) on the south—the "mathematical" or "astronomical" tropics. The seasonal variations in the angle of the sun and the duration of daylight are relatively minor within the tropics; thus, seasonal fluctuations of temperature are small. Poleward of 23.5° N and S the sun is never overhead, and one must look toward the equator to see the noon sun. Outside the tropics, seasonal variations in duration of daylight and angle of the sun increase poleward, giving rise to great contrasts between winter and summer temperatures in high latitudes.

TABLE 1-4 Albedos of Various Surfaces Expressed as Percent of Total Received Solar Radiation that is Reflected

Surface	Percent
Freshly fallen snow	80–85
Sand (light-colored dry)	25–35
Sand (light-colored wet)	20–30
Soil (dry)	15–25
Soil (wet)	5–15
Forest (broadleaf)	10–20
Forest (needleleaf)	5–15
Water (sun near horizon)[a]	50–80
Water (sun near zenith)[a]	3–5
Clouds (thin)	25–40
Clouds (thick)	60–80
Average for Earth and atmosphere	30

[a] Reflectivity is dependent on sun angle, wave action, and direction of wave movement.

Contrasting Response of Land and Water to Solar Radiation

When solar radiation reaches the surface, it may be reflected or absorbed, depending on the nature of the surface materials. Table 1-4 gives the reflective capacity (albedo) of different types of surfaces, and indicates that the net reflection and absorption of land and water are about the same. Although land and water surfaces reflect and absorb about the same amounts of solar energy, their response to the absorbed energy is not the same because of the different physical natures of the two surfaces. These physical differences—specific heat, translucency, evaporation, and mobility—are discussed below.

Specific Heat. The specific heat of water is the highest of any natural substance—that is, it requires more heat energy to raise the temperature of 1g of water by 1°C than 1g of mineral material, such as sand or soil. Since land and water are absorbing solar energy at about the same rate, water surfaces heat more slowly and, at the same time, also cool more slowly; thus, oceans are great storehouses of heat and have a moderating effect on temperature fluctuations in the atmosphere. Conversely, land masses experience rapid temperature changes, particularly from season to season, and continental temperature regimes show much more fluctuation than do oceanic ones.

Translucency. Water is translucent, whereas land surfaces are opaque. This means that solar radiation penetrates to a considerable depth in clear water, and a much greater volume is heated. On the other hand, only the top few millimeters of land surfaces absorb radiation, and the heating of materials below the surface involves a slow process of heat transfer by conduction.

Evaporation. A large amount of the incoming solar radiation evaporates water and does not contribute to an increase in surface temperature. Of course, when the water vapor eventually condenses in the atmosphere, the process releases latent heat of vaporization and warms the air. Evaporation occurs on land as well as water. It involves direct evaporation from wet soil or wet vegetation and also water *transpired* from the subsoil by way of plant roots, through plant tissues, and into the atmosphere as water vapor from openings in the leaves (see chapter 5). Although some of the radiation received on both land and water is involved in evaporation, the amount from oceans is larger.

Mobility. Movement of water, both vertically and horizontally, can transfer vast amounts of absorbed heat over great distances. Of course, the location of land masses is fixed. The great ocean currents help balance the heat surpluses of the tropics with the heat deficiencies of middle and high latitudes. An example is the moderation of the climate of Western Europe by the Gulf Stream. London has an average January temperature of 4.2°C (39.6°F), compared to −2.9°C (26.8°F) for Chicago, although London is as far north of the equator as is the southern end of Hudson Bay in Canada—much farther north than Chicago. Likewise, Keflavik, Iceland, just a few miles from the Arctic Circle but bathed by the Gulf Stream, has a January average that is above freezing (0.01°C or 32.2°F).

Water in the Atmosphere

Although the percentage of Earth's total water supply in the atmosphere at any one time is very small (about 0.001 percent), the atmosphere, nevertheless, is the most important pathway by which water is circulated from one part of the planet to another, particularly from oceans to land. In chapter 3, we will examine details of this movement of water in the *hydrologic cycle*, and also the role it plays in the Earth's water budget.

Measurement of Moisture in the Atmosphere

Quantitatively, there are several ways to designate the water vapor content of the atmosphere at any place and time; usually this atmospheric moisture is referred to as *vapor pressure*, *specific humidity*, *mixing ratio*, *relative humidity*, or *dew point temperature* each of which involves a different method for measuring the same quantity:

1. *Vapor pressure.* The pressure exerted by the atmosphere is the sum of the pressures of each of its constituent gases, so that there is a nitrogen pressure, oxygen pressure, CO_2 pressure, argon pressure, and so on, as well as a water vapor pressure. At a constant barometer reading, the vapor pressure changes only with a change in the water content of the atmosphere. Water surfaces also have a vapor pressure, which is higher than that in the atmosphere if the air is not saturated; thus, there is a vapor pressure gradient from water to air, causing water molecules to escape into the air; thus evaporation occurs.

2. *Specific humidity and mixing ratio.* These measure the weight of water

vapor in a certain weight of air, or the mass of water vapor per given mass of air, usually expressed in grams per kilogram. If water vapor is included in the total mass, it is *specific humidity*; if not, it is *mixing ratio*. Mixing ratio generally is used in place of specific humidity by meteorologists. These measurements are valuable in meteorology since these values can change only with a change in the moisture content of the air. They are independent of temperature changes, and pressure changes at a fixed elevation are insufficient to have a major effect on their value.

3. *Relative humidity.* This measurement of atmospheric moisture is most familiar as the value given in television and radio weather summaries. It measures the actual quantity of water vapor in air compared to the amount the air could hold *at a particular temperature*, expressed as a *percentage of saturation*. If air temperature changes, relative humidity changes, even though there is no change in moisture content; it rises as temperature falls, and it falls as temperature rises. For example, if air with a temperature of 10° C has a relative humidity of 50 percent, it has half as much water vapor as it could hold at that temperature; should the temperature rise to 15° C, the relative humidity would be less than 50 percent; conversely, if the temperature should fall to 8° C, relative humidity would be greater than 50 percent. If cooling were to continue, the air eventually would become saturated, and the air would be at the *dew point*.

4. *Dew point temperature.* The dew point is the temperature at which air becomes saturated; the higher the water vapor content of the air, the higher will be its dew point. Since the dew point remains constant unless water vapor is added to or removed from the air, it is a good indicator of its moisture content. If air is cooled to the dew point, the relative humidity becomes 100 percent; if air is cooled further after the dew point is reached, condensation will occur, and then the dew point will be lowered as water is removed from the air, but relative humidity will remain at 100 percent. At the surface, this may produce dew if the dew point is above freezing, or frost if it is below freezing. Cloud formation and precipitation indicate that air aloft has been cooled to the dew point; when rain is falling, the surface air temperature generally is above the dew point, and relative humidity is less than 100 percent; of course, saturation has occurred at cloud level.

Precipitation of Water from the Atmosphere

Dew, frost, fog, and cloud droplets and ice crystals may form by condensation and sublimation alone, but if these particles are to grow large enough to fall as *precipitation*, simple condensation or sublimation is not enough. Vast quantities of air must be brought to the dew point, and then they must continue to cool very rapidly if rain, snow, sleet, or hail is to form. Only one atmospheric process, *adiabatic cooling*, is sufficient to bring this about.

Adiabatic cooling occurs when rising air expands because of the rapid reduction of barometric pressure with increasing altitude. It can be demonstrated that rising air will cool at a fixed rate of about 10°C for each kilometer of vertical rise. This process is called *adiabatic* since there is no exchange of heat between the rising air mass and its surrounding environment.[1] Horizontal pressure changes are so small that the small adiabatic changes can be ignored. When air descends, warming occurs at the same rate. The term *dry adiabatic lapse rate* refers to this temperature change with changing altitude.

Condensation or sublimation occurring in the rising air releases the latent heat of vaporization, and the cooling rate falls to about 6°C for each kilometer of rise. Actually, the air continues to cool at the dry adiabatic rate of 10°C/km, but it is gaining heat at about 4°C/km; thus, the net change in temperature is about 6°C/km. This rate is called the *pseudoadiabatic lapse rate* (the terms *false*, *wet*, or *saturated adiabatic lapse rate* also are used), since it is a combination of two rates, one of which is not adiabatic.

It is obvious that for precipitation to occur, there must be some mechanism at work for air to rise and cool adiabatically. Where meteorological conditions frequently favor uplift of air, the climates are humid; where subsidence of the atmosphere is the rule, climates are dry. Some of the conditions favorable for uplift of air are: (1) *orographic uplift* along mountain ranges lying across the paths of prevailing winds; (2) *frontal uplift* along weather fronts; (3) *nonfrontal uplift* brought about by local convection; and (4) *nonfrontal uplift* associated with large-scale convergence in low pressure systems such as tropical hurricanes.

Where a mountain range lies in the path of prevailing winds, uplift may be rapid. Rainfall often is heavy on the windward side and, conversely, on the leeward side, where air is subsiding and warming adiabatically, conditions may be dry. A dry area downwind from a mountain range is called a *rain shadow*; if it is very dry, it is termed a *rain shadow desert*. An excellent example of a rain shadow is found to the east of the Sierra Nevada in California, Nevada, and Utah. The Sierra Nevada lies across the path of the prevailing westerly winds. The dry northern Great Plains are in a similar position to the east of the Rocky Mountains.

Frontal convergence and uplift occur along the boundaries of air masses of different temperatures—the *fronts* on weather maps. Cold air on one side of a front is denser than the warm air on the other; consequently, where the cold air is advancing along a *cold front*, it pushes under the warm air ahead of it, forcing the warm air to rise and cool adiabatically. Precipitation may result if moisture conditions are adequate in the warm air. Where warm air is advancing along a *warm front*, it climbs up the slope of the retreating cold air where it is cooled adiabatically, and it may produce clouds and precipitation.

[1] *Adiabatic* (Greek *adiabatos*, not passable) refers to temperature changes without a gain or loss of heat energy. In the atmosphere, when pressure drops, air expands, and some of the air's heat energy performs the work necessary for the air to occupy more space. When pressure increases, it compresses the air and work is performed on the air, causing its temperature to rise. The adiabatic process, then, is self-contained and requires no transfer of heat across the boundary of a mass of rising or subsiding air.

Local convection may develop because of intense heating of the surface by the sun, particularly during the warm part of the year. A column of warm air rises from the surface, cools adiabatically, and may reach the condensation level, which is marked by the flat base of a *cumulus* cloud. If the upper air also is unstable, the column may continue to rise and form a *cumulonimbus* cloud, producing a local thunderstorm. The weather forecast for such a period might read "scattered thunderstorms."

Nonfrontal convergence on a scale much larger than that in a local thunderstorm happens in tropical disturbances, such as easterly waves, tropical storms, and hurricanes. Rainfall from these storms is very important to the trade wind zone which otherwise would be relatively dry. Fronts, such as one encounters in middle latitude storm, are absent in these disturbances because they occur in warm air masses and no cold air mass is present; as a result, there are no sharp contrasts in temperature and density to establish a front.

Global Circulation of the Atmosphere and Oceans

Although the sun is the ultimate energy source that powers the atmosphere-ocean circulation system, the immediate cause of weather phenomena is air movement. Air moves because of differences in atmospheric pressure, and weather is the product of this movement. The prevailing wind patterns maintain surface currents in the oceans, driving the currents by friction between the winds and the water surface. The prevailing circulation pattern of the atmosphere and the oceans plays a major role in determining climate. This pattern is referred to as the *general* or *primary circulation.*

Atmospheric Pressure and Wind

Variations in atmospheric pressure from place to place, and from time to time, determine the direction and velocity of air movement. Barometric readings taken at weather stations throughout the world indicate these pressure differences. The standard instrument is the *mercurial barometer*, invented in 1643 by Evangelista Torricelli, a student of Galileo. Atmospheric pressure is measured in *inches of mercury* or in *millibars* (mb). Average sea-level pressure is about 29.92 in (760 mm) or 1013.2 mb. Because the horizontal pressure gradient is of primary concern, it is necessary to reduce the station pressures taken at various altitudes to sea level. It is this horizontal difference in sea-level pressure which determines the direction and velocity of air movement.

Pressure patterns are indicated on a weather map by lines called *isobars* that connect points of equal barometric pressure. Isobars are much like contour lines on a topographic map; a high pressure area is like a hill, and may be termed a ridge or dome of high pressure; likewise, a low is analogous to a valley and lies in a trough of low pressure. As water flows down the slope of

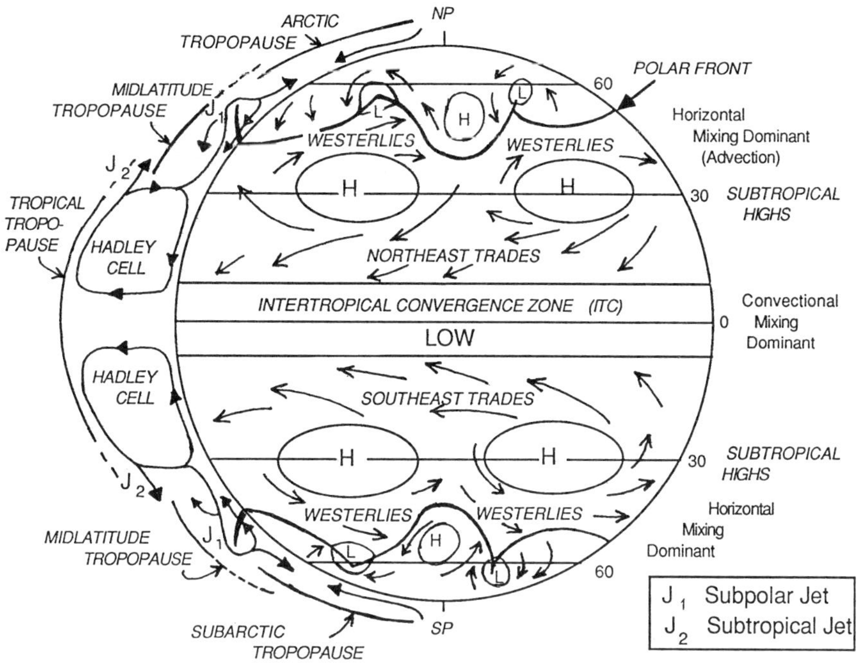

Fig. 1–5. Wind and pressure belts as they might appear on a hypothetical model of the Earth's atmosphere.

a hill, air flows down the slope of a high pressure ridge. This pressure slope is called the *pressure gradient.* The rate of change with distance—the "slope" of the pressure gradient—determines the velocity of the wind. On a weather map, areas with steep pressure gradients and strong winds are indicated by closely spaced isobars, just as a steep hill is indicated by closely spaced contour lines on a topographic map.

If only pressure were involved, wind direction would be parallel to the pressure gradient; this is not the case, however, because the rotation of the earth introduces the *Coriolis effect,* which deflects the wind to the right of the pressure gradient in the Northern Hemisphere, and to the left in the Southern Hemisphere. As shown in figure 1-5, air circulates clockwise around a high pressure center in the Northern Hemisphere and counterclockwise in the Southern Hemisphere; around a low pressure center, the circulation is reversed and runs counterclockwise north of the equator and clockwise to the south. A low pressure center is referred to as a *cyclone,* and atmospheric circulation around it is called *cyclonic*; a high pressure center is an *anticyclone,* and circulation associated with it is *anticyclonic.*

Hypothetical Model of Atmospheric Circulation

On a nonrotating earth, the maximum solar energy would be received at the equator, the surface air would be heated, expand, and rise to the tropopause, and it would then flow northward and southward toward the poles. Cooling would take place as the air moved poleward, so it would settle at the poles and flow back at the surface toward the equator; thus, a north-south flow of air would occur in a simple convective cell.

On the rotating earth, the Coriolis effect comes into play and prevents such a simple system from developing. Near the equator, the simple convective cell model is approximated in the *Hadley Cell*, illustrated in figure 1-5, but the constant Coriolis deflection of the diverging upper air prevents it from reaching the poles. In the real atmosphere, the air in the high level, northward and southward components of the Hadley Cell circulation loses heat and moisture, becomes more dense, and settles back to the surface at about 30°N and 30°S where it feeds the *subtropical highs*. The subtropical high pressure cells are best developed over the oceans in summer and are among the most persistent features of the general circulation. Unlike the shallow cold-induced (*thermal*) continental highs of winter, these *dynamic highs* are produced by movement of air in the atmosphere.

From the subtropical highs, surface air moves poleward as the *westerlies*, and equatorward as the *trade winds*. Additional air is added to the subtropical highs from the upper westerlies to compensate for the divergence of the low level westerlies; thus, the system does not run down. A major factor in heat transfer in the westerlies is the great moving low pressure systems—the *middle latitude cyclonic storms*. Consequently, atmospheric mixing and heat transfer are primarily convective in the tropics and subtropics, but horizontal mixing dominates in higher latitudes.

Older models of atmospheric circulation assume a high pressure cell above the polar regions because the air is cold and dense. In those models, cold, sinking air moves away from the poles and is deflected into the polar easterlies. Actual observations in the Arctic indicate that a weak surface high can be detected only in spring and summer. In the winter the Arctic Basin is dominated by low pressure, and shallow cold continental highs are located farther south in Siberia and western Canada. It appears that the *polar easterlies*, shown on many models of global circulation, are in large measure easterly winds associated with the poleward sides of moving cyclonic storms (fig. 1-5). In the Northern Hemisphere, it is doubtful that they can be considered a prevailing wind belt comparable to the westerlies or the trades. A better case can be made for a polar high and polar easterlies in the Southern Hemisphere. Reports from some coastal stations in Antarctica indicate strong and persistent easterly winds with frequencies as high as 70 percent. These winds probably originate from the downslope movement of cold, heavy air from the ice-covered Antarctic plateaus where shallow anticyclones often form.

Compared to the deep warm subtropical highs, the cold high pressure areas in the polar and subpolar regions are shallow phenomena, and the polar easterlies, where observed, generally extend to less than 5 km (16,000 ft) above the surface. Almost everywhere the weak polar easterly circulation is overlain by westerly flow, and a weak low pressure area is located over the North Pole at elevations between 6 and 12 km (20,000 and 40,000 ft).

In winter, as seen in figure 1-6, two deep lows dominate the North Atlantic and North Pacific Oceans, the *Icelandic Low* and the *Aleutian Low*, respectively. What separates them are the great thermal high pressure areas in northeastern Asia and northern North America. Rather than being semipermanent low pressure cells, they are areas of average low pressure reflecting frequent passages of cyclonic storms, which tend to converge on the North Pacific and North Atlantic as they move along the polar front from the continents to the west. Since there are no intervening land masses at that latitude, the subpolar low in the Southern Hemisphere forms a continuous circumpolar trough.

The subpolar low pressure troughs mark zones of convergence where air of tropical origin meets air from the polar regions. Because of temperature contrasts between the polar and tropical air masses, a sharp temperature boundary is formed—the *polar front*. The great cyclonic storms of middle and high latitudes form along this polar front. Generally, these cyclones move in an easterly direction, downstream in the westerlies, and they often alternate with high pressure anticyclones.

The westerlies, and the attendant cyclonic activity, are much more vigorous in winter, being driven by the great contrast in temperature between the tropics and the polar regions. Middle latitude cyclones are dynamic systems produced by atmospheric movement, and they are somewhat analogous to whirlpools in a stream, moving downstream in the westerlies.

The Upper Westerlies and the Jet Streams

The surface westerlies are much less constant than the trade winds and can show great variations in a matter of hours or days both in velocity and in direction of flow. This surface variability principally is the result of the moving cyclones and anticyclones already discussed. In the upper troposphere, the westerly flow follows a much more persistent and simple pattern. An important element in this upper westerly flow is the *jet stream*.

A jet stream, as defined by the World Meteorological Organization, is a well-defined flow in the atmosphere that exceeds a velocity of 30 m per second (67 mph or 108 km per hour). Jets are well developed in the upper westerlies, particularly in winter, and speeds of over 134 m/sec (300 mph)have

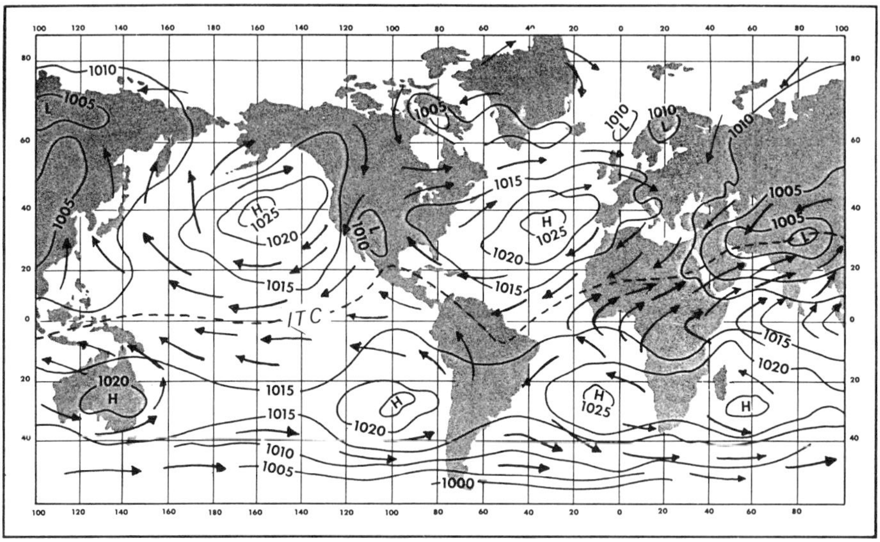

Prevailing Pressure Systems of the World, July

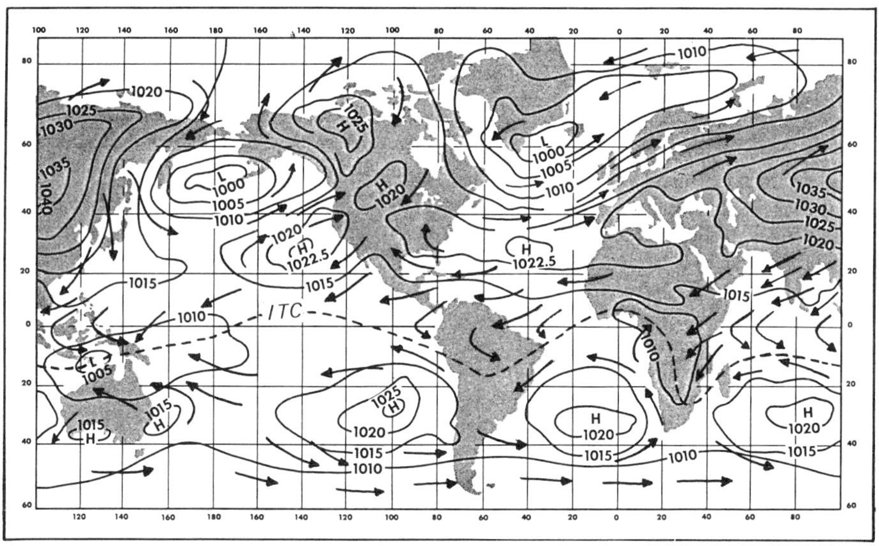

Prevailing Pressure Systems of the World, January

Fig. 1–6. Mean sea-level pressure and winds for July and January. The variability of the pattern is much more apparent in the Northern Hemisphere because of the marked seasonal variation in pressure on the large land masses at middle and high latitudes. (Modified from U.S. Air Force Manual 105-5, *Weather for Aircrew Trainees,* 1956.)

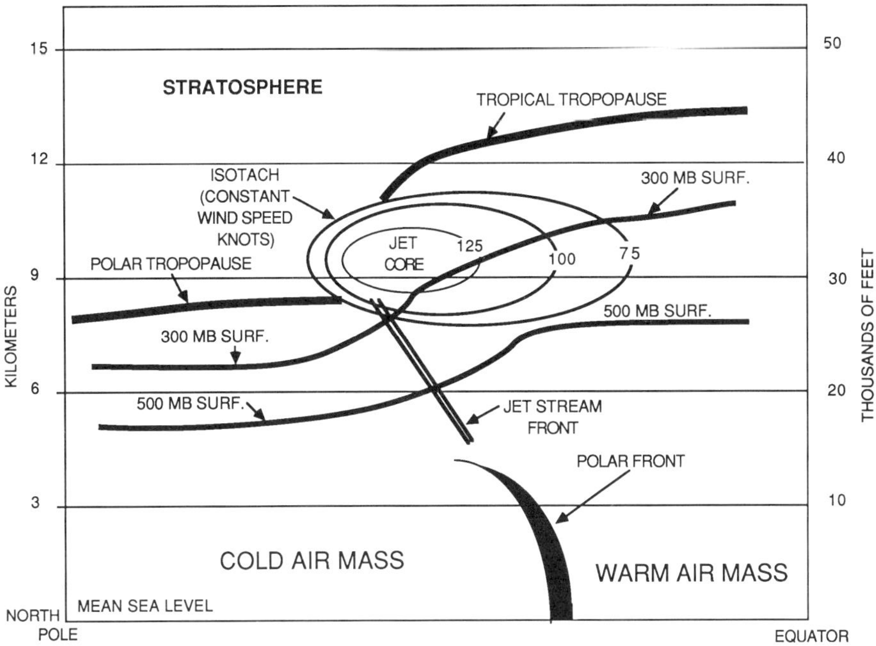

Fig. 1–7. Schematic north-south cross-section of the jet stream looking eastward or downstream. (Modified from U.S. Air Force Manual 105-5, *Weather for Aircrew Trainees,* 1956.)

been recorded. The strongest occur in the Western Pacific Ocean near southern Japan during the Northern Hemisphere winter.

The jet streams constitute the high velocity core of the westerly flow at heights between 9 and 12 km (30,000–40,000 ft). In the Northern Hemisphere, two jets are usually present in the winter, the *polar front jet* and the *subtropical jet,* located where subpolar and subtropical breaks occur in the tropopause (fig. 1-7). Similar jets are present in the Southern Hemisphere westerlies. The configuration of these jets, and the paths they follow around the earth, are important in determining weather patterns in the middle latitudes.

The jets often follow meandering paths from west to east, and frequently they shift their latitudinal positions. These great meanders are called *Rossby waves,* after the Swedish meteorologist who first described them, and they separate cold air of polar origin from warm air of tropical origin; thus, the name *polar front jet* is often applied to the one occupying the position farthest poleward. When the amplitude of the Rossby waves is small, *zonal flow* prevails; that is, air flow is mostly west to east, the warm air remains south of

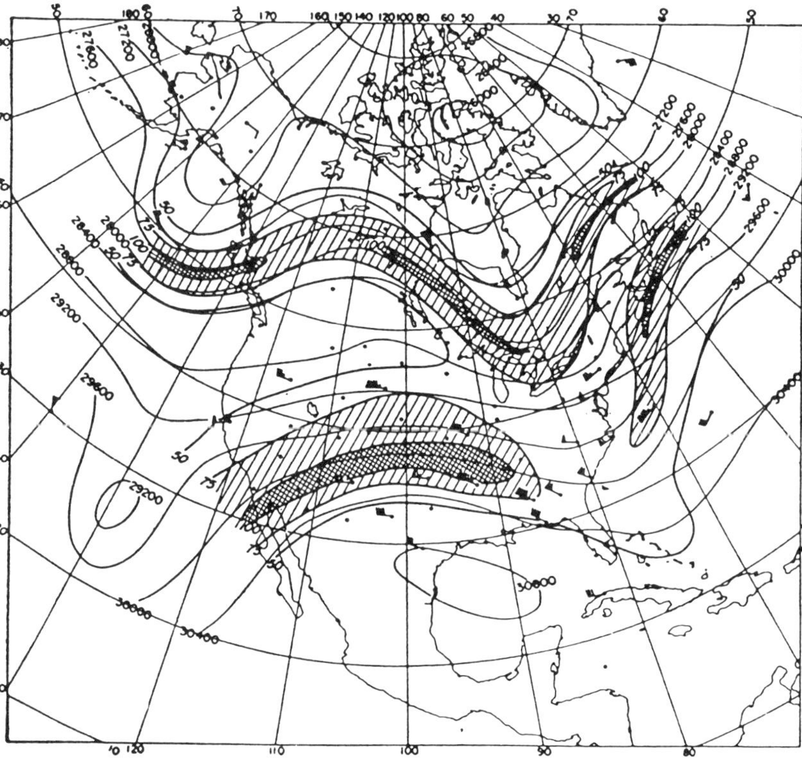

Fig. 1–8. Pressure and winds at the 300 mb (8.86 in) level between 8,230 m and 9,266 m (27,000–30,400 ft) on October 25, 1950, 0300 GMT. The subpolar jet lies to the north of the Canadian border, and the subtropical jet is located to the south. Hatched areas denote regions with wind speed exceeding 139 km per hour (75 knots); crosshatched areas have wind speeds exceeding 185 km per hour (100 knots). (U.S. Air Force Manual 105-5, *Weather for Aircrew Trainees,* 1956.)

the cold air, and weather patterns are stable. When large amplitude waves exist, cold air flows equatorward and warm air flows poleward in what is termed *meridional flow.* Strong meridional flow brings frequent and often severe weather changes in the middle latitudes. These waves in the westerlies are an important factor in balancing the heat deficit in polar regions and the heat surplus in the lower latitudes through horizontal mixing (fig. 1-8).

Middle Latitude Cyclones and Associated Weather

Because of their importance in middle latitude weather, the cyclones of the westerlies bear further consideration. These depressions seem to form because of divergence in the westerly flow of the upper troposphere. When air

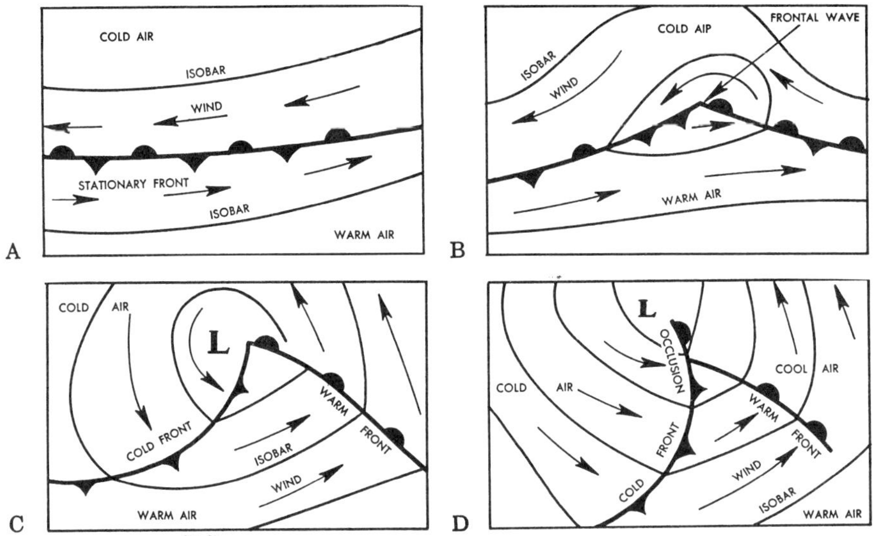

Fig. 1–9. Life cycle of a typical middle latitude cyclonic storm. (A) Cold polar air is separated from warm tropical air by a stationary front located in a trough of low pressure. (B) Low pressure intensifies, forming a low-pressure area with counter-clockwise circulation around the center. This pressure pattern causes cold air to push southward to the west of the center, producing a cold front, and warm air to push northward to the east, producing a warm front. (C) The process continues until a fully developed wave cyclone is formed, with a warm sector south of the fronts and a cold sector to the north. (D) The cold front advances faster than the warm front, pinching out the warm sector and eventually overtaking the warm front to form an occlusion. During this process the storm center (low) moves east to west with an average speed of 48 kph (30 mph). (Air Force Manual 105-5, *Weather for Aircrew Trainees,* 1956.)

diverges aloft, low pressure and convergence are induced in the lower level westerly flow. This convergence causes air of polar origin to move equatorward on the western side of the cyclone, and air of tropical origin to move poleward on the eastern side. Because of the temperature contrast between these two air masses, a front develops between them—the familiar warm and cold fronts on the weather map (fig. 1-9). These storms tend to follow tracks parallel to the orientation of the upper level jet stream (fig. 1-10).

The origin of these storms is dynamic, and they are more numerous in winter when the westerlies are stronger and winter storm tracks favor a more equatorward position. If air mass and air flow conditions are favorable, violent weather, including severe thunderstorms and tornadoes, may

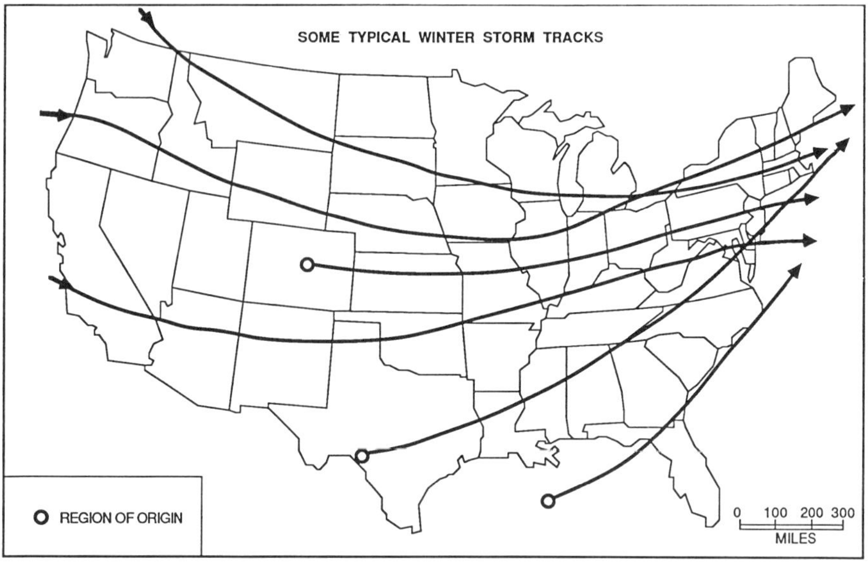

Fig. 1–10. Typical winter storm tracks in the United States. The tendency of storm tracks to converge on northeastern states assures relatively rainy winters there, whereas wider spacing of storm tracks farther west assures drier winters in the western corn belt and the Great Plains.

develop in association with cyclonic storms. These conditions frequently occur in spring and early summer when the temperature gradient from the equator to the pole remains steep. In late summer and fall, the zonal temperature contrasts are much smaller, most storm tracks are farther poleward, storms are less frequent, and violent weather is much less likely to occur.

Thunderstorms

A *thunderstorm* is any storm containing lightning and thunder; the U.S. Weather Service defines a *thunderstorm day* as "a day on which thunder is heard." An estimated forty thousand thunderstorms occur on Earth each day, most of them in the humid tropics where they account for a major share of the precipitation. It is in the central and southeastern United States, however, where the most violent thunderstorms and tornadoes develop, most frequently in association with larger cyclonic disturbances (fig. 1-11). Thunderstorms may occur in any conditionally or convectively unstable air mass; that is, lifting of the air mass will cause it to become unstable and to rise at an accelerated rate. Instability may be induced if the air mass is warmed from below by the warm ground; consequently, thunderstorms are more

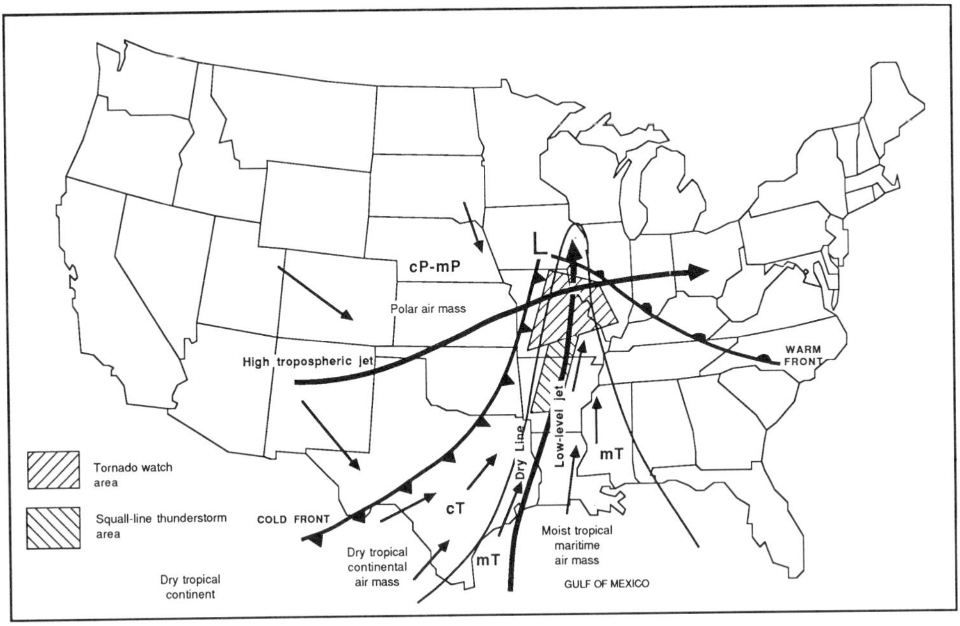

Fig. 1–11. Meteorological conditions associated with the formation of severe thunderstorms and tornadoes associated with a middle latitude cyclonic storm. Air masses are indicated as: cP—polar continental; mP—polar maritime; mT—tropical maritime; and cT—tropical continental.

frequent in the spring and summer in middle latitudes. As unstable air rises, adiabatic cooling causes rapid condensation and releases the latent heat of vaporization. This release of energy is essential for producing thunderstorms and associated weather phenomena, including tornadoes and hail. Hail is of special concern to farmers because it may cause extensive crop damage.

Thunderstorms can be grouped into several broad categories: (1) *air-mass thunderstorms*, (2) *frontal thunderstorms*, (3) squall-line thunderstorms, and (4) *mesoscale convective complexes (MCCs)*. Air-mass thunderstorms develop in warm, tropical maritime air masses and are not associated with fronts or squall lines. Their life span is short, and they seldom produce severe weather or damaging hail. Frontal thunderstorms form along fronts as conditionally unstable air is lifted; they are similar in intensity to air-mass storms and generally do not produce severe weather. Squall-line thunderstorms and MCCs account for most of the severe weather associated with thunderstorms, including most tornadoes and damaging hail. These spectacular and dangerous storms merit special attention (fig. 1-12).

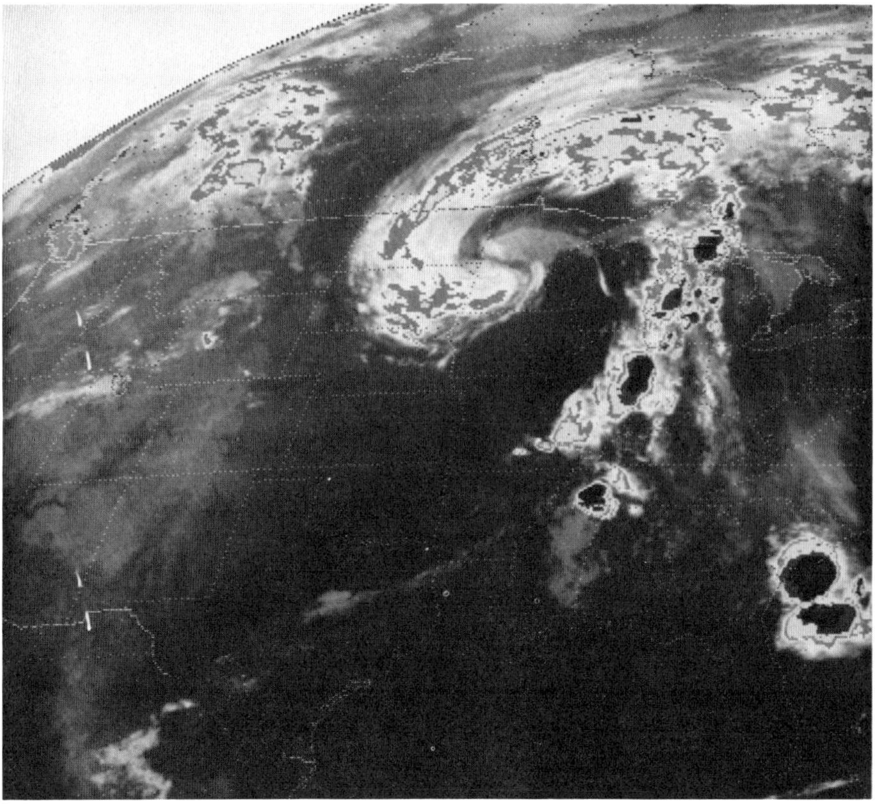

Fig. 1–12. An IR satellite image of a cyclonic disturbance containing mesoscale thunderstorm cells centered over the upper Midwest on 20 June 1979 at 1230 GMT. The severe thunderstorms are indicated by the bright areas with dark centers, indicating extremely cold tops of the very thick cumulonimbus clouds, which penetrate into the lower stratosphere. (U.S. Weather Service.)

Severe Thunderstorms and Tornadoes

A *squall line* consists of a line of thunderstorms that forms in the warm sector of a cyclonic storm some 50–300 km (30–180 mi) ahead of the cold front. It is along these squall lines that some of the most violent weather occurs, particularly when a number of individual thunderstorm cells grow in size and organize into a large convective system, a mesoscale convective complex. Such super-thunderstorms are formed by the coalescence of many individual thunderstorm cells, and they often cover an area of more than 100,000 km^2 (3,861 mi^2), a thousand times larger than the average air-mass thunderstorm (fig. 1-12).

Frequently, conditions favorable for the formation of these mesoscale storms and their attendant tornadoes occur in the area of the United States lying north of the western sector of the Gulf of Mexico. Here, all the ingredients often coalesce—a tongue of warm moist tropical maritime air (mT) extending northward from the Gulf of Mexico with a low level jet, warm dry air from the arid southwest (cT), and cool dry polar air from the northwest and west (cP or mP). These elements may combine to produce severe squall-line and mesoscale thunderstorms; with an additional component, a high level jet stream, severe tornado outbreaks may occur.

Figure 1-11 illustrates how all these components set the stage for severe thunderstorm and tornado development. Let us visualize an open wave cyclone over the central Great Plains, with a cold front extending southward, and a warm front toward the southeast. A relatively shallow tongue of warm, moist tropical maritime (mT) air, with a low level jet at its core, is pushing northward into the warm sector of the storm. Warm dry tropical continental (cT) air from the arid southwest is flowing into the warm sector between the cold front and the mT air. The boundary between these two air flows is the *dry line*. West of the line, the dry air is at the surface; ahead of it, the dry subsiding air overlies the moist surface air. A strong upper-troposphere jet is blowing from the southwest, and an upper level trough of low pressure lies to the west of the storm system.

A temperature inversion, produced by subsidence in the dry air aloft, overlies the moist, conditionally unstable surface air. This inversion acts like the lid of a pressure cooker, preventing convection from developing into benign local thunderstorms and causing a buildup of potential convective energy in the moist air below. The initiating factor for severe thunderstorms is a lifting mechanism for forcing the warm, moist air upward through the inversion and into the cold, unstable air aloft. The advancing denser, dry air behind the dry line furnishes such a mechanism.

The moist surface air, forced upward, cools at the dry adiabatic lapse rate until it reaches the condensation level. If uplift continues, cumulus clouds will form, and the moist air will cool more slowly than the surrounding dry air because of the release of latent heat. If these conditions force the warm air through the inversion layer, and if it penetrates far enough into the cold, unstable air to reach the *level of free convection*—the level above which it will become warmer than its surroundings—it will rise convectively, and towering cumulus clouds will build up to the level of the high tropospheric jet. The function of the jet is to remove air from the top of the convective system; thus, this ventilating action of the jet causes moist air from below to enter the system at a more rapid rate, assuring a more rapid release of latent heat and intensifying the storm. At their maximum development, the tops of these storm cells push well into the lower stratosphere and may reach heights of more than 18 km (60,000 ft).

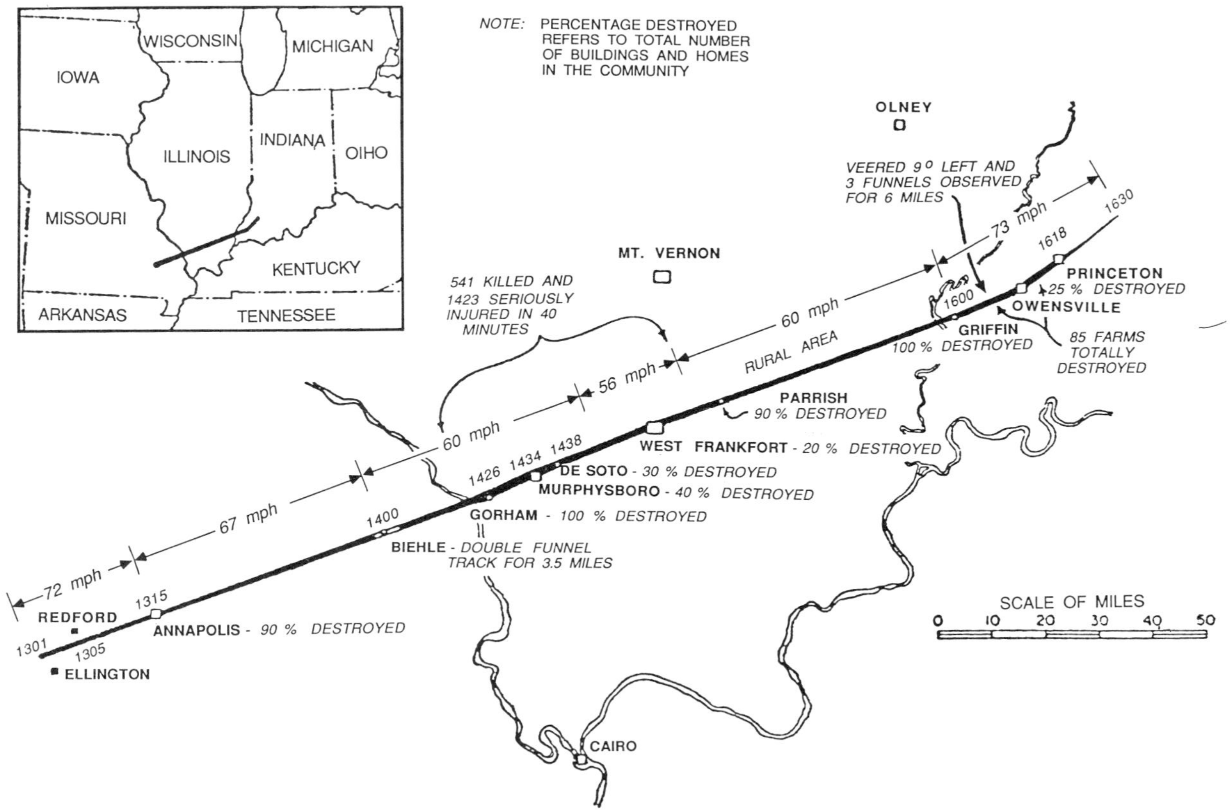

Fig. 1–13. Path of the Tri-State Tornado of 25 March 1923. (From J.W. Wilson, and S. A. Changnon, Jr., *Illinois Tornadoes,* Illinois State Water Survey Circular 103, 1971, 32.)

Three Days That Made Tornado History

About a hundred thousand thunderstorms occur each year in the United States; of these, approximately seven hundred to twelve hundred produce tornadoes. Most are small, last only a few minutes, and cause only minor damage. Fortunately, few tornadoes are large storms, with paths exceeding 2 km (>1 mi) in width and hundreds of kilometers in length, and life spans exceeding three hours. There have been two outstanding outbreaks of tornado activity in the United States: those of 18 March 1925 and 3–4 April 1974.

The Great Tri-State Tornado. The world's most deadly tornado occurred on 18 March 1925, when the great Tri-State Tornado roared through Missouri, Illinois, and Indiana, spreading vast destruction and causing terrible loss of life. First sighted at about 1 P.M. near Ellington in the sparsely settled hill country of southeastern Missouri (fig. 1-13), for the next three and one-half hours it followed a remarkable straight course, traveling toward the northeast for 552 km (219 mi). Along this path, the funnel did not leave the ground until finally it dissipated in southwestern Indiana, making this the longest continuous tornado path on record. During its brief lifetime it killed 695 people, injured 2,027 more, and caused an estimated $140 million in property damage (in 1988 dollars). The death toll was the greatest ever recorded for a single tornado, and the property damage places it among the three or four most destructive tornadoes. Had the storm encountered a large city, or had it passed over more densely settled suburban and rural areas, both death toll and damages would have been enormous.[2]

The Tornado Outbreak of 3–4 April 1974. The sixteen-hour outbreak of tornado activity of 3–4 April 1974 was the most severe ever recorded, in terms of the number of tornadoes reported, although the death toll of 335 was less than half that of the great Tri-State Tornado. Of the 147 tornadoes reported on those two days, most caused relatively little damage, with less than 5 percent of the storms causing more than half of the deaths. The worst single tornado occurred at Xenia, Ohio, where 35 persons were killed and 1,150 injured. Figure 1-14 shows the paths of the 148 tornadoes reported during the two-day outbreak. (One of these storms was later reclassified as not a tornado but only a severe thunderstorm.)[3]

[2]I have a personal interest in the Tri-State Tornado, having survived the storm, which destroyed my hometown of Murphysboro, Ill. The town suffered the greatest loss of life among the many communities along the storm's path. I grew up hearing tales of the storm, in a community where almost every family, tragically, had been affected, and where events were apt to be dated "before the Storm" or "after the Storm."

[3]For a detailed account of this tornado outbreak, see Robert F. Abbey and T. Theodore Fujita, "Tornadoes: The Tornado Outbreak of 3–4 April 1974," in Edwin Kessler (ed.), *The Thunderstorm in Human Affairs* (Norman: University of Oklahoma Press, 1983).

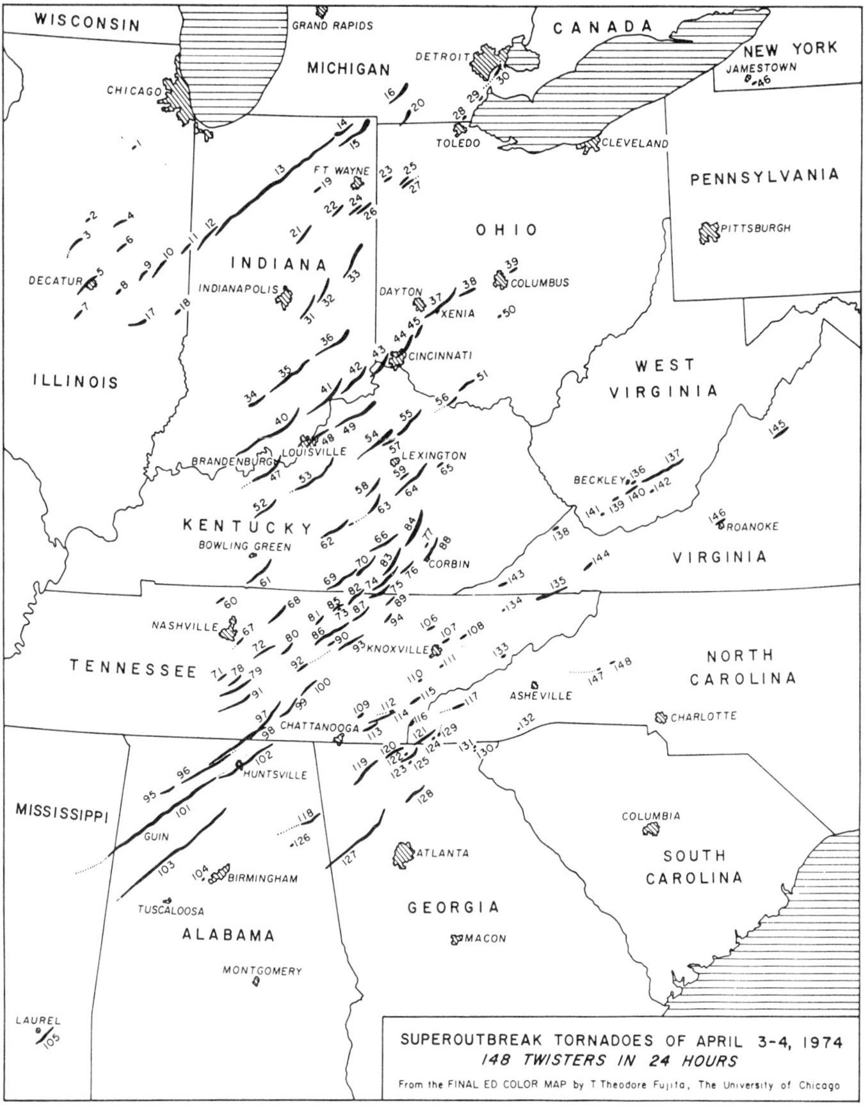

Fig. 1–14. The record outbreak of tornadoes of 3–4 April 1974, during which 148 tornadoes were reported officially. (Based on a map of T.T. Fujita and reproduced from Kessler 1983, 41.)

Subtropical Highs—The Desert Makers

The great subtropical high pressure cells of the ocean basins play a particular role both in the formation of the great low latitude, west coast deserts, and in the monsoonal circulation characteristic of continental east coasts. These cells represent an intensification of the two ridges of average high pressure that circle the globe at about 30° latitude in each hemisphere. The cellular pattern is well defined especially in summer because of relatively low continental pressure brought about by surface heating. Five of these cells can be recognized on a world pressure map (fig. 1-6): (1) the Azores High in the North Atlantic, (2) the South Atlantic High, (3) the Hawaiian High in the North Pacific, (4) the South Pacific High, and (5) the Indian Ocean High.

The average positions of the subtropical highs and other wind and pressure belts of the oceans play a role in the general nature of climates on adjacent continents. To the east of each of these semipermanent anticyclones lie some of the world's great deserts:

Cell	*Desert*
Azores High	Sahara Desert
Hawaiian High	Sonoran Desert
South Atlantic High	Namib and Kalahari Deserts
Indian Ocean High	Great Australian Desert
South Pacific High	Peruvian-Chilean (Atacama) Desert

These deserts are caused by intense subsidence in the atmosphere and offshore winds induced by the configuration of the subtropical anticyclones. The onshore flow around the west side of the highs produces humid climates on continental east coasts—the monsoonal circulation discussed below.

Climates that are more humid, showing well-marked wet and dry seasons, border the low latitude deserts on their poleward and equatorward sides. On their poleward side, the *Mediterranean climate*—also called *dry-summer subtropical*—has a moist winter and a dry summer; the reason is the poleward shift of the high in summer, and an equatorward shift in winter. In summer, areas poleward of the deserts experience the subsidence and offshore winds of the subtropical high; in winter, they come under the influence of the westerlies and rain-producing cyclonic storms. These dry summer climates are found in the Mediterranean Basin, California, Middle Chile, southern and southwestern Australia, and the Cape Town area of South Africa. Equatorward of the deserts, the climate is *wet-dry tropical* with a moist summer, produced by the poleward shift of the intertropical convergence zone (ITC), and a dry winter, brought about by the equatorward shift of the subtropical high. We will deal with these climates more fully in chapter 2.

The Trade Winds and the Intertropical Convergence Zone (ITC)

The most persistent of the planetary wind systems, the trades, emanate from the equatorward side of the subtropical highs. They occupy most of the area between about 30°N to 30°S latitude, and they shift widely with the seasons, so they influence almost half of the Earth's surface at least sometime during the year. The name *trade winds* derives from the navigators of sailing ships who depended on these winds to cross the oceans from east to west. In the Atlantic, outbound ships for the New World would follow the northeast trades from the African coast to the West Indies, and return to Europe on the westerlies. This navigation pattern gave rise to the infamous "Triangular Trade"—from England to West Africa with trade goods; from Africa to the West Indies with slaves; and back to Europe with sugar, rum, and cotton. A similar trade pattern was followed by the Spanish between Acapulco, Mexico, and the Philippines: the famous Manila galleons sailed westward on the trade winds, and returned on the westerlies to California, then traveled south to Acapulco.

Although dependable, the trade winds are relatively shallow toward the eastern margins of the ocean basins, where intense subsidence in the subtropical highs produces an inversion that may be only a few hundred meters above the surface. Here the marine surface air of the trades is too shallow to produce much precipitation. Where cold ocean currents exist, they intensify the aridity and cause foggy conditions.

As the trades blow westward across the warm tropical oceans, they absorb heat and evaporate moisture, the inversion level is pushed upward by the more buoyant air, and cumulus clouds build up to the inversion level. These "trade wind cumulus" are the trademark of the wind belt. Toward the west sides of the tropical oceans, the trade wind inversion increases in height and may disappear, the tropical maritime air becomes conditionally unstable, and climates become more humid.

Easterly Waves and Hurricanes

Where the trade winds blow over warm water and become moist and conditionally unstable, wave-like disturbances develop and bring periods of rainfall. These are called *easterly waves* because they move from east to west at speeds averaging about 18 km per hour. Ahead of an easterly wave, low level divergence and subsidence from above reinforces the trade wind inversion and brings clear dry weather (fig. 1-15). On the eastern side of the wave, convergence induces uplift and breaks the inversion; this produces thunderstorms and precipitation. The passage of an easterly wave may take several days, with two or three days of almost cloudless weather, followed by two or three days of thunderstorms and rain. Easterly waves are very important in the water budget of tropical lands within the trade wind belt,

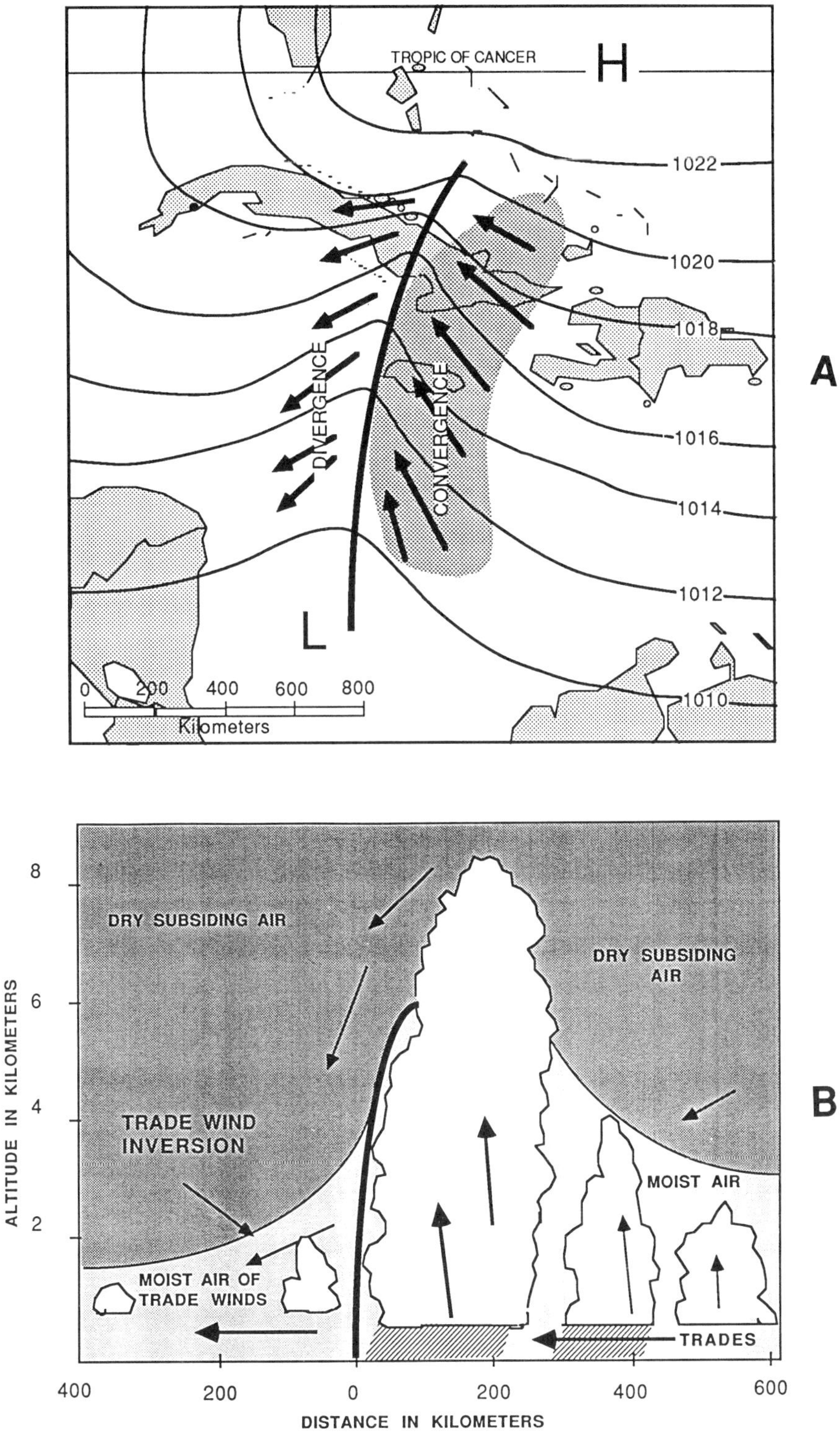

Fig. 1–15. (a) Easterly wave in the Caribbean; (b) cross-section of an easterly wave. Horizontal scales are not the same; vertical scale exaggerated in (b).

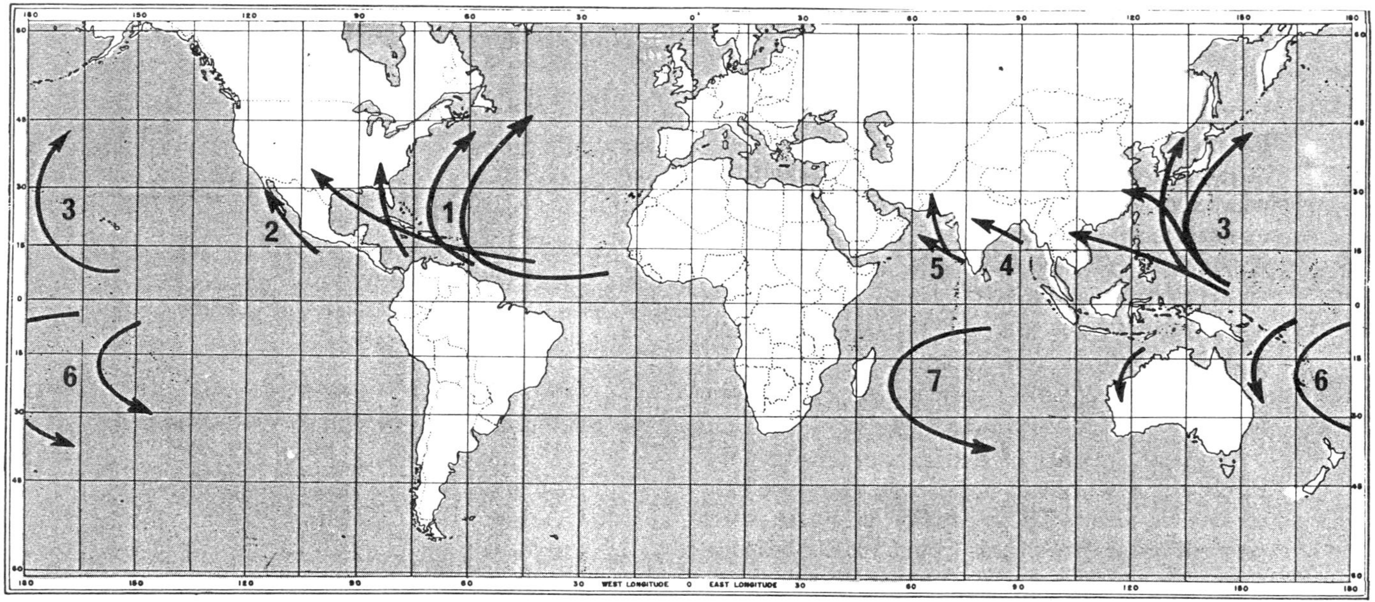

Fig. 1–16. The seven regions of the world where tropical cyclones form: (1) the western Atlantic and Caribbean where the storms are called *hurricanes*; (2) west coast of Mexico; (3) Western Pacific Ocean north of equator, where they are called *typhoons*; (4) and (5) Bay of Bengal and Arabian Sea (*cyclones*); (6) Southwestern Pacific; and (7) Western Indian Ocean near Madagascar (Malagasy). Arrows indicate typical storm tracks in each region, although individual storm paths may vary widely from those indicated. (Modified from U.S. Air Force Manual 105-5, *Weather for Aircrew Trainees,* 1956.)

particularly the low-lying areas where there is no topographic uplift to induce precipitation from the winds.

A few of these easterly waves intensify and develop closed isobaric patterns to become true cyclones. They are *tropical depressions* if winds increase to 37–63 km per hour (20–34 knots), and *tropical storms* when winds are between 64 and 119 km per hour (35 and 64 knots). If winds reach sustained velocities exceeding 119 km per hour (64 knots), they become *hurricanes* in the North Atlantic and Eastern Pacific oceans, and *typhoons* in the Western Pacific. In other areas they are known as *cyclones* (India and Southwestern Pacific Ocean) and locally by still other names (fig. 1-16).

In a hurricane, not only are winds stronger than in a tropical storm, but also the structure of the storm is different. A distinguishing feature is the *eye* of the storm, an area averaging from 20 to 50 km wide (12–30 mi), with broken clouds or clear skies and light winds. Barometric pressure in the eye is extremely low, sometimes below 900 mb (26.58 in). The world's lowest recorded sea-level pressure is 870 mb (25.69 in), recorded on 12 October 1979 in the eye of typhoon Tip in the Western Pacific Ocean south of Japan. The nature of the eye is still not completely understood, but it is an area of subsiding air that produces relatively clear weather and high air temperatures—the so-called *warm core* of the storm. Surrounding the eye is a wall of clouds, marking the area of strongest winds, greatest uplift, and heaviest precipitation. The presence of the eye seems to be essential for maintaining high wind velocities; at the same time, the strong winds appear to be critical for maintaining the eye. When these storms move over land, friction slows the winds, the eye disappears, and the hurricane becomes just a tropical depression.

Most hurricanes form in the trade wind belt between 5° and 10° N and S of the equator, and they move westward with the trade winds; eventually, they tend poleward around the western margins of the subtropical high (fig. 1-16). Some eventually may reach the westerlies and curve toward the east where, for the most part, they lose energy over cooler water or land and diminish in intensity. Water temperature seems to be critical for the formation of these storms, and most of them originate over ocean areas where water is at least 27° C (fig. 1-17). Hurricanes do not occur in the South Atlantic because in areas where hurricanes would normally form, water temperatures are below this critical value.

In a number of ways hurricanes and other tropical disturbances are different from middle latitude cyclones. First, the tropical storm derives its energy from warm water and from the release of the latent heat of vaporization, whereas horizontal temperature contrasts drive the middle latitude storm. Second, fronts are not present since the tropical disturbance involves only a tropical maritime air mass, and there is no source of cold air nearby. Third, the isobar pattern in the hurricane is circular, and the storm is smaller than a middle latitude cyclone (fig. 1-18).

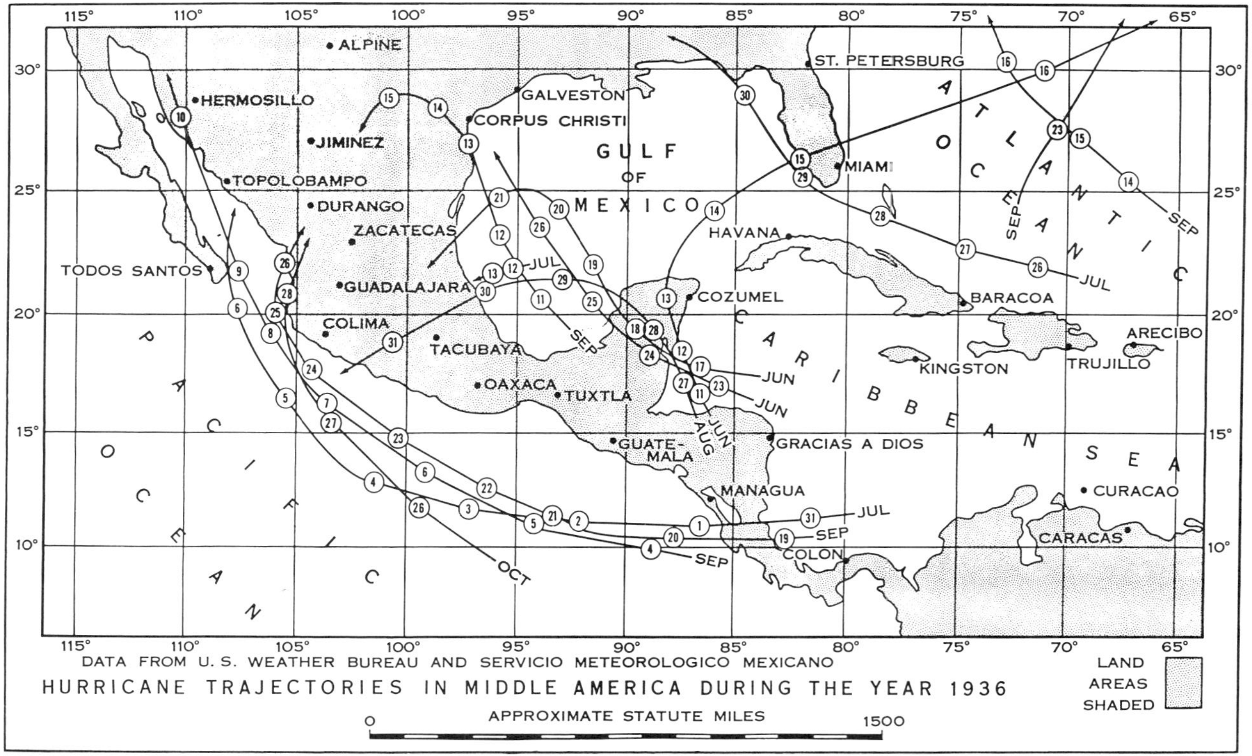

Fig. 1–17. Hurricanes in Caribbean and Mexican west coast waters during 1936. The fact that many storms originate in the Caribbean and cross to the west coast indicates the close meteorological connection between these two areas. (From W. E. Akin, "The Geography of the Río Tepalcatepec Basin, Mexico," unpublished master's thesis, Indiana University, 1949.)

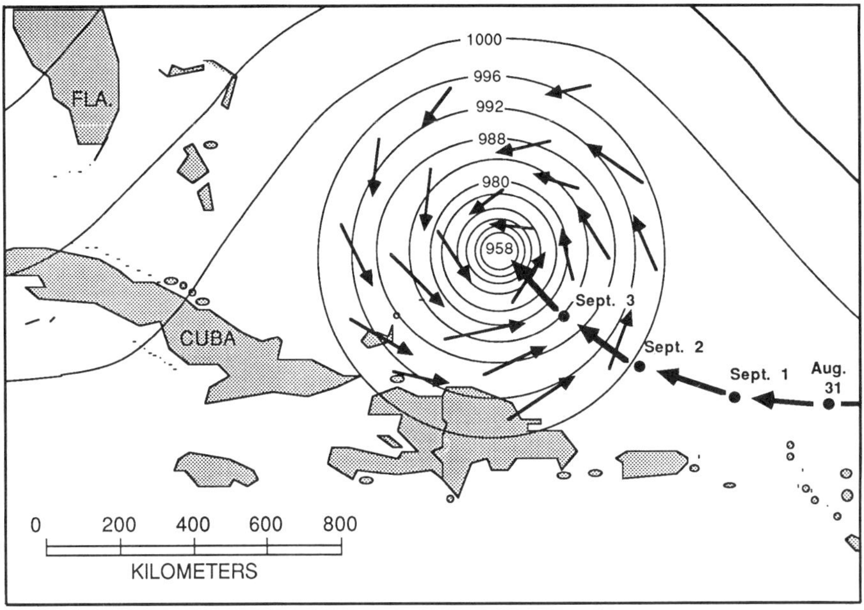

Fig. 1–18. Fully developed tropical hurricane in the Western Atlantic Ocean, showing barometric pressure in millibars, wind patterns, and storm path between 31 August and 3 September.

There is no doubt that hurricanes are the most destructive of weather phenomena. Although their winds may not reach the velocities of a severe tornado, the potential destruction extends over a much larger area—tens of thousands of square kilometers in the tropical storm, compared to an average of several square kilometers in a tornado. A large share of the hurricane damage and most of the deaths come not from the wind but from the piling up of water along coastal areas by what is termed a *storm surge.*

Storm surges may raise sea level as much as 12 m (39 ft) and with the accompanying high waves, they can spread destruction miles inland from low-lying coastal areas. The greatest natural disaster in the United States occurred in 1900, when a hurricane and storm surge demolished the city of Galveston, Tex., and drowned five thousand people. In November 1970, a cyclone (hurricane) struck the coast of Bangladesh, at the head of the Bay of Bengal, inundating the coastal lowlands and killing more than five hundred thousand people, a death toll from a natural event equaled only by some earthquakes in densely settled areas of China and India.

Monsoons

The word *monsoon* is derived from Arabic *mausim,* season, a term used for centuries by Arab sailors to describe the seasonal change in wind direction in the Arabian Sea and the Western Indian Ocean. In that area winds are northeasterly in January and southwesterly in July (fig. 1-6), and they are used to this day by Arab sailors who sail out along the east coast of Africa to Zanzibar on the winter monsoon and return on the summer monsoon.

Although monsoons are developed best along the southern and eastern margins of Asia, and in East and West Africa, monsoon climatic patterns also are encountered in parts of South America and Australia. Monsoonal tendencies also are present in the eastern United States, but they do not dominate the wind and rainfall patterns as they do in the monsoon lands of Asia and Africa.

This annual cycle of prevailing surface winds can be explained in part by the contrast in seasonal temperatures between the continents and the adjacent oceans. During summer, continents warm rapidly, and intense low pressure develops over them; at the same time, the subtropical high pressure cells over the oceans show a slight increase in pressure (fig. 1-6); thus, a steep sea-to-land pressure gradient is established, and moist tropical maritime air will flow inland, bringing with it thunderstorms and general rains. In India, the onset of the monsoon rains often is so sudden that it is called the "burst of the monsoon."

The winter monsoon (sometimes called the *antimonsoon*) is in part a response to the reversal of the pressure situation between the continents and the oceans. In January, the cold high pressure cells now dominate the continents in the Northern Hemisphere, the subtropical highs of the Northern Pacific and Atlantic Oceans have diminished in size, and the subpolar lows dominate the northern oceans. In response to the land-to-sea pressure gradient, there is an outward movement of cold heavy air, especially from central Asia, so winters in monsoon lands are relatively dry. In some areas, however, such as parts of Vietnam, the eastern side of the Indian Peninsula, and in western Japan, the winter monsoon winds pass over relatively warm waters and bring winter precipitation. In 1972, the Winter Olympic Games were held at Sapporo in northwestern Japan, taking advantage of the dependable winter snowfall produced by winds crossing the Sea of Japan from mainland Asia.

Monsoons are an integral part of the general circulation of the earth's atmosphere. Sometimes they are likened to the daily wind shifts from ocean to land in the afternoon and from land to ocean at night—the well-known sea breeze and land breeze phenomena—but this model is an oversimplification. For example, figure 1-6 shows the ITC over northern India in July, with an intense low pressure area over the Indus River valley. In response to this low, the Southern Hemisphere subtropical high drives the southeast trade winds of the Indian Ocean across the equator, where Coriolis effect deflects them into the southwestern monsoon. A similar seasonal pattern develops in West Africa.

The major impact of the monsoonal circulation is to produce climates with very distinct wet and dry seasons. Extremely high rainfall may occur where monsoon winds encounter steep mountainous terrain. Northeastern India furnishes an excellent example of very high monsoon rainfall. Here, the winds of the summer monsoon rise against the Khasi Hills of Assam and the Arakan ranges of Burma and cool adiabatically as they ascend the steep slopes. At Cherrapunji, a station on the southern flank of the Khasi Hills at 1,300 m (4,300 ft), the average annual rainfall is 11,430 mm (450 in), and as much as 26,470 mm (1,042 in) has fallen in a single year, mostly in the six-month summer monsoon season (fig. 1-19).

Modern analysis of the Asiatic-African monsoon indicates a rather complex relationship among the general atmospheric circulation, the regional pattern, and surface and upper air conditions. Over less mountainous terrain, such as the plains of northern India, the monsoon does not produce a simple weather pattern as we might expect. The fact is that the monsoon rains are highly variable from year to year, and drought often interrupts the rainy season although monsoon air flow continues. This variation in monsoon rainfall, independent of the prevailing surface air flow, suggests factors involved that are not explained by the simplified seasonal circulation model.

The traditional "sea breeze" model depicts only the flow of warm moist air in the lower troposphere, a layer only about 3 km thick. In spite of the presence of the summer thermal low pressure area, the northwestern part of the Indian subcontinent is dry. A thin wedge of monsoon air penetrates the area, but it is overlain by subsiding continental air and a temperature inversion that prevents convection, and there is little summer rain.

Over less mountainous terrain, cyclonic disturbances in the monsoon flow, *monsoon depressions*, play a major role in producing precipitation. These weak cyclones develop over the northern Bay of Bengal, and move westward along the monsoon trough parallel to the Himalayan front and the Ganges River valley. Monsoon depressions often persist for a week or more, and they bring widespread and locally heavy rain. Without these depressions, the Indian subcontinent would be much drier, in spite of the monsoonal flow.

Africa also is included in the great monsoon belt adjacent to the Indian Ocean. In Ethiopia, monsoon winds contribute to the heavy summer rains, the source of the annual flooding of the Nile River in Egypt. Although Egyptians have recorded Nile floods for thousands of years, it was not until the late nineteenth century that explorers discovered the actual source of the flood waters from the Blue Nile—the Ethiopian highlands. Other well-developed monsoon rainfall regimes occur in parts of the East African Highlands and in West Africa north of the Gulf of Guinea.

Outside the tropics, we find monsoon regimes on the eastern margins of the continents, particularly in Asia. Monsoon patterns are present also on the southeastern coast of the United States and north and west of the Gulf of Mexico. In summer, circulation around the western side of the great sub-

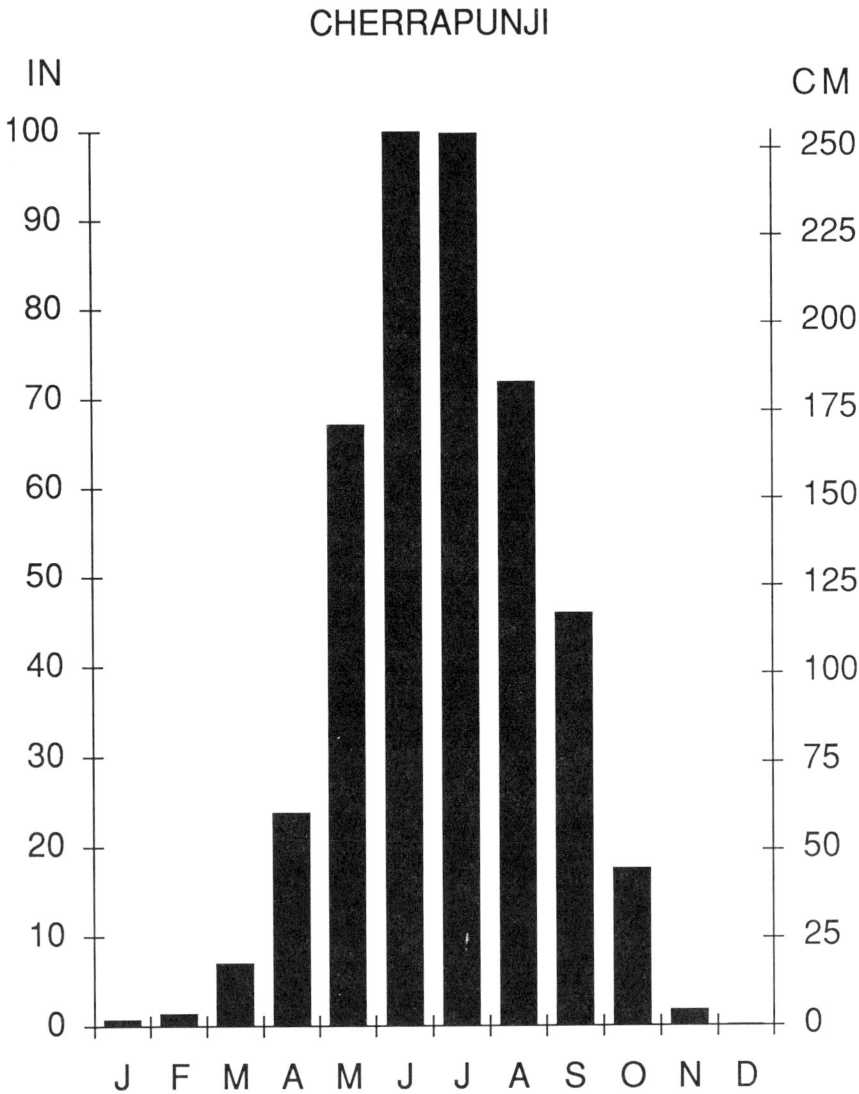

Fig. 1–19. Monsoonal seasonal rainfall pattern at Cherrapunji, India, one of the rainiest stations on earth, with an average annual rainfall of 1,143 cm (450 in). Monsoon winds in the warm half of the year move across the Bay of Bengal and are forced up by constricting mountain ranges and increasing elevation. This station holds the record for the world's greatest twelve-month rainfall of 2,647 cm (1,042 in), and one-month rainfall of 930 cm (366 in).

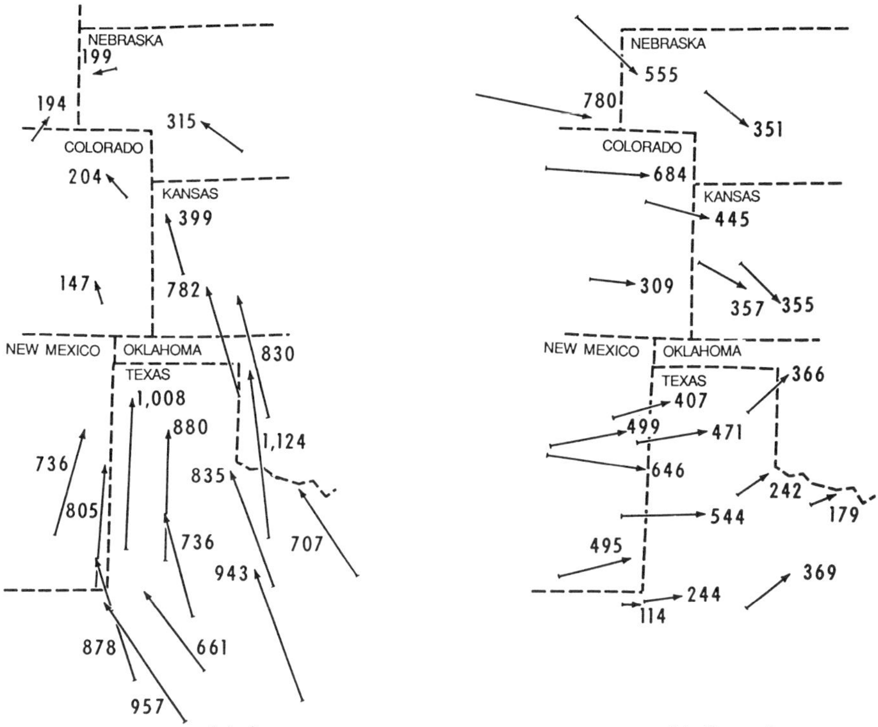

(a) June (b) December

Fig. 1–20. Seasonal resultant wind patterns on the southern Great Plains for (a) June and (b) December indicate a monsoonal tendency to the wind patterns along the western margins of the Gulf of Mexico. Arrows indicate direction of wind movement and numbers indicate miles of movement in 100 hours. Tip of arrow indicates location of station. (From USDA, *Wind in the Southwestern Great Plains*, Conservation Research Report No. 6, 1965.)

tropical highs favors an onshore flow of moist tropical maritime air. In May and June on the southern Great Plains, the southerly and southeasterly flow of air corresponds with the rainiest part of the year (fig. 1-20). In winter, northwesterly and westerly flows predominate. Rainfall patterns reflect the monsoonal circulation in eastern China and in part of the United States east of the Rocky Mountains. The western corn belt receives over 70 percent of its precipitation in the warm half of the year, and parts of north China receive more than 90 percent in the warm season. Stations in the eastern United States show less concentration of rainfall in summer because of winter rainfall produced by frequent cyclonic disturbances, which tend to follow a more southerly path in winter.

'Climate' is a composite idea, a generalization of the manifold weather conditions from day to day throughout the year. . . . Certainly no picture of it is at all real unless it is painted in all the colours of the manifold variations of weather and the rounds of the seasons which are the really prominent features; it is quite inadequate to give merely the mean state of any element.

W. G. KENDREW, *Climatology*

Chapter 2

GLOBAL PATTERNS OF CLIMATE

Climate is more than just the average state of the atmosphere; rather, it is the long-range pattern of weather. It takes into account seasonal changes, prevailing weather patterns, and probabilities of climatic events calculated from past climatic records. Seasonal uniformity of temperatures and rainfall marks some climates, whereas others are characterized by extremes.

Modern climatologists depend for their analyses on accumulations of global weather records, which are made possible by the development of a wide range of meteorological instruments, from simple thermometers to sophisticated weather satellites. It appears that Galileo constructed the first thermometer in 1593, and we know that his pupil, Torricelli, invented the mercurial barometer in 1643. The earliest systematic weather records in the United States and Europe were made in the first half of the eighteenth century, but it was not until the middle of the nineteenth century that the modern period of *instrumental climatology* really began. Even today, large areas in the tropics, deserts, and polar regions are without adequate surface observations, and where stations do exist, the network may be sparse and records often short and discontinuous. Rainfall and temperature data obtained from satellites in space are filling some of these gaps.

Climatological research encompasses not only the modern period of instrumental climatology, but a major part of Earth history. For climatologists to piece together the intriguing climatic history of the Earth before systematic instrumental records became available, they must depend on noninstrumental and largely indirect evidence of past weather patterns. Fossils and other bits of geological evidence, such as ancient soils and glacial deposits, indicate climates that prevailed during the vast span of geologic time. The study of ancient climates is called *paleoclimatology,* the prefix *paleo-* meaning old or ancient. It overlaps the fields of *paleontology* (ancient life), *paleobotany* (ancient vegetation), and also anthropology and archaeology.

Historical records predating systematic weather observations often record such climate-related events as freezing dates of rivers, presence or absence of sea ice, height of floods, crop failures from drought or early frosts, unusually cold winters or hot summers, and the nature of the original vegetation cover. This study and analysis is called *historical climatology;* we will explore it more thoroughly in chapter 4.

Climatic Elements

The weather elements described in chapter 1 are the raw materials of climatology. They include (1) temperature, (2) atmospheric moisture (humidity, condensation, clouds, and precipitation), (3) atmospheric pressure, and (4) winds. The role of the climatologist is to collect, analyze, classify, and explain the patterns of these elements in space and time, and to evaluate the dynamics of the *climatic controls* that create these patterns.

Climatic Controls

The most obvious climatic control, latitude, is related closely to global variations in the amount of solar energy reaching the surface. The ancient Greeks developed the earliest climatic classification, a simple latitudinal division of climatic zones designated as frigid, temperate, and torrid. The term *climate* is derived from the Greek *klima,* meaning slope or curvature of the earth's surface from north to south. The unequal distribution of land and water on the earth distorts this latitudinal control.

There are other major climatic controls: (1) distribution of land and water; (2) geographic position relative to ocean and continental influences; (3) atmospheric circulation, involving semipermanent wind and pressure belts and the movement of air masses; (4) ocean currents (cold and warm); (5) elevation above sea level and its effect on temperature; and (6) topographic barriers, especially high mountains. An interrelationship, sometimes complex, exists among these climatic controls, and their roles vary from climate to climate.

Describing and Classifying Climates

Temperature and precipitation records serve as the most available and obvious choices for describing and classifying climates. The effectiveness of precipitation—that is, how much of it is taken in by soil and used by plants—is closely related to temperature. Effectiveness is determined by the rate of evaporation from free-water surfaces and from transpiration of soil moisture by plants. A combined term, *evapotranspiration*, applies to the total water loss through these two processes.

As a general rule, the higher the temperature, the higher the rate of evapotranspiration, and the less effective is the precipitation. For example, an annual rainfall of 630 mm (25 in) produces a humid climate in northwestern Europe but results in semiarid conditions in the lowland tropics.

Over a span of years, weather records give us a composite picture of weather variations. Statistical analysis of these records serves as a guide for many weather-related human activities. Variations from "average" weather conditions may be just as significant as the means. For example, an engineer must take into account the probability of receiving rare, but high intensity, rainfall when designing a drainage system for carrying storm runoff from an urban development. The National Weather Service and other government agencies assist in such design problems by publishing computed frequencies of rainfalls of various intensities in different locations.

Rainfall frequencies usually are given as *recurrence intervals*, measured in years—a rainfall of (25.4 mm) (1 in) in one hour might have a statistical probability of occurring at a particular station once in two years, so it would be called a "two-year, one-hour rainfall"; a rainfall of 50.8 mm (2 in) in one hour might have a frequency of one year in fifty and would be designated as a "fifty-year rainfall." Keep in mind that this is only the statistical probability, and a fifty-year rainfall might occur two years in succession, or even more than once in a single year. An engineer, by knowing the frequency for high intensity rainfalls of an area, can design a drainage system for any recurrence interval, constrained only by practicality and cost.

Descriptive climatic data have many other uses, such as planning for flood control, determining crop insurance rates, budgeting for snow removal, and planning resort areas. Also, climatic classifications are based on descriptive data, such as average monthly temperature and precipitation, evapotranspiration and precipitation ratios, and air mass frequencies.

Classifying Climates

What is the value of climatic classification? Is one system superior to another? What is the best system? The principles of any scientific classification are stated concisely in the U.S. Department of Agriculture publication *Soil Classification: A Comprehensive System* (1960):

> Classifications are contrivances made by [scientists] to suit their purposes. They are not themselves truths that can be discovered. Therefore, there is no true classification; a perfect one would have no drawbacks when used for the purpose intended; the best classification is that which best serves the purpose or purposes for which it is to be used. [P.6]

It is obvious there is no "best" climatic classification for all purposes. This chapter describes two classifications that present the panorama of climatic patterns on the earth.

Climatic statistics determine boundaries of climatic maps. These boundaries are, therefore, arbitrary statistical lines not easily discernible in the

Fig. 2–1. Microclimates can vary considerably within short distances, as demonstrated by the vegetation contrasts between the north and south sides of this tributary valley to Swift Current Creek in southwestern Saskatchewan, Canada. More direct sunlight on the south-facing north slope (*right*) causes higher soil temperatures, increased evaporation, more rapid snow melt in spring, and generally drier conditions than on the more shaded north-facing slope (*left*). These drier conditions account for the sparse vegetation and more vigorous erosion on the south-facing slope. (Photo by W. E. Akin.)

natural landscape. For the most part, one climate grades into another across transitional zones, and relatively sharp boundaries occur only where sudden changes take place in terrain or elevation.

Finally, we should make a distinction between *macroclimate* and *microclimate*. Climatic data used in most general classifications are from standard weather instrument shelters situated about 1.5 m (5 ft) or more above the ground. Climates based on data from standard weather stations are called macroclimates, whereas microclimate is the climate nearer the ground. A variant in the use of the term *microclimate* involves the application of data derived from special situations; for example, the climate at different levels above the ground within a forest, or the climate of a city as opposed to that of the surrounding countryside.

Figure 2-1 illustrates the impact of microclimate on local conditions. This photograph shows the contrast in vegetation cover between a southward-facing slope and a northward-facing one, reflecting the difference in sunlight and moisture conditions across the valley. Microclimates can be altered radically when forests are removed, land is cultivated, or cities are built, whereas the effect of such activities on the macroclimate of a region may be small if the affected areas are limited in size. Chapters 5 and 6 touch on the role of certain microclimates in the distribution of natural vegetation.

THE KÖPPEN CLIMATIC CLASSIFICATION SYSTEM

In 1918 a botanist, Waldimir Köppen, of the University of Graz, Austria, published the first version of his climatic classification. He recognized climate as a major factor in the distribution of vegetation, and he used the relationship as an indicator of major climatic gradations. The Köppen classification, therefore, was strictly an empirical system, which matched known vegetation patterns with observed temperature, rainfall averages, and seasonality.

Although most climatic classification systems depend on vegetation as the most important indicator of climatic conditions, particularly where weather records are inadequate, it is obvious there is not a perfect relationship between the two phenomena. For example, Köppen's climatic boundaries (fig. 2-2) rather closely follow vegetation boundaries in Africa, (see chapter 6, fig. 6-1), but in South America, the fit is not so good. A probable explanation for this variation between continents is that Köppen used Africa as his model for drawing climatic boundaries on the basis of vegetation. This lack of perfect correlation between vegetation and climate should not be surprising, because climate is only one of the environmental constraints on the type of plants that will grow in any given area. Humans also have had a long and profound impact on the vegetation in all parts of the inhabited world, an impact that has altered dramatically the character of the "natural" vegetation.

Failure of climatic boundaries to match perfectly with broad vegetation formations does not invalidate the system. Correlation is much better when core areas of both climatic zones and vegetation formations are compared, rather than their boundaries. Köppen named his climates, in part, according to vegetation formations; for example, the *tropical rainforest climate* and *tropical savanna climate*. Preferred climatic terms for these two climates today are, respectively, *wet tropical* and *wet-dry tropical*.

A major advantage of the Köppen system is its employment of symbols that eliminate cumbersome descriptive terms by the use of simple combinations of letters, each letter having a precise meaning. For example, in the modified system used in this chapter, the symbol *Daf* can be interpreted as follows: *D*—cold winter, with at least one month averaging below 0°C (32°F), and at least one month averaging above 10°C (50°F); *a*—hot summer, with the warmest month of summer averaging above 22°C (71.6°F) and *f*—no dry season. The *Daf* climate is called *humid continental—warm*

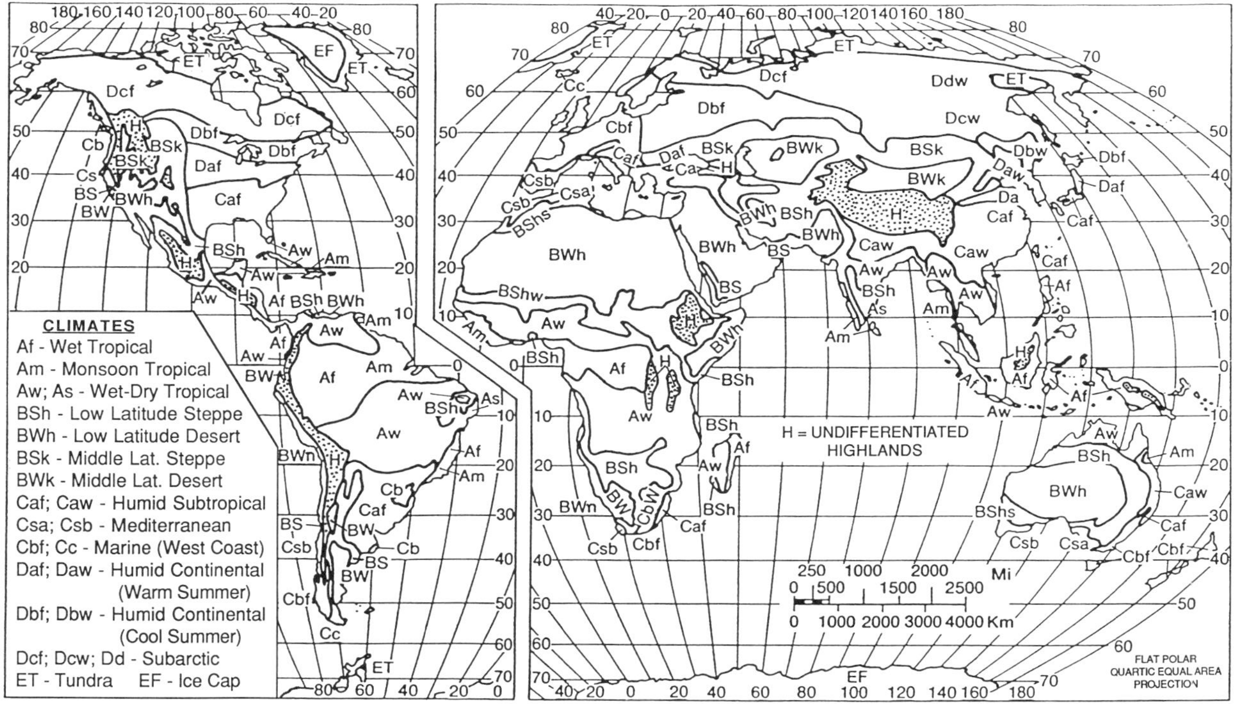

Fig. 2–2. Major world climates according to Trewartha's modification of the Köppen classification system. Except for *tundra climate*, vegetation-based terms for climates have been changed to climatic terms: *i.e.*, *wet tropical* for *tropical rainforest climate*; *wet-dry tropical* for *tropical savanna climate*.

summer; in North America its location corresponds to a major part of the corn belt of the United States.

Trewartha's Modification of the Köppen System

In 1954, the American geographer Glenn Trewartha modified the Köppen system in response to criticism of the classification, particularly as applied to the United States. Why was modification desirable? Trewartha noted that the Köppen system classified the western areas of Oregon and Washington, with their splendid forests of Douglas fir, in the same climate (*Cs*) as southern California, with its scrub vegetation. Also, it placed southern New England and a part of the American corn belt in the humid subtropics (*Cfa*), along with the Gulf Coast. Trewartha's modification classifies the climate of the northwest coast of North America as west coast marine (*Cb*), and that of all of New England and the corn belt as humid continental (*Daf* and *Dbf*) (fig. 2-3).

Trewartha also changed the order of the second and third symbols to emphasize the climate's most distinctive characteristic; thus, in *Daf, a* emphasizes the warm summer, over the *f* which pertains to seasonality of precipitation. Likewise, the indication for the *dry-summer subtropical* climate is *Cas* in the original system, but changes to *Csa* in the modification, the emphasis being placed on *s,* indicating the dry summer, the outstanding characteristic of this climate.

Climatologists have debated at length about the relative merits of emphasis on climatic boundaries rather than on the core of climatic areas. As John E. Oliver comments; "Much of the literature dealing with climatic clas-

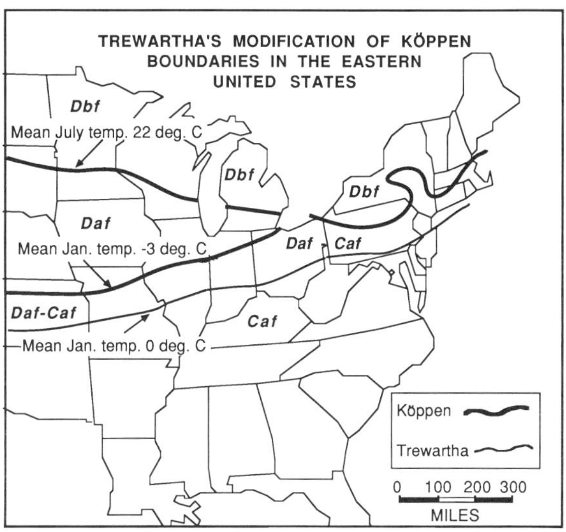

Fig. 2–3. Climatic boundaries in the eastern United States according to Köppen, compared with Trewartha's modification of the boundary values between *C* and *D* climates.

sification appears more concerned with the limits of a climate group than with the climate itself. . . . Whether such an emphasis is valid is debatable, for some might feel the desert itself is more important than the desert boundary" (1973, 169). In the remainder of this chapter, we will consider the criteria for drawing climatic boundaries but the main emphasis will be on characteristics of typical or "core" climate, and on zones of transition to neighboring climates.

Tables 2-1 and 2-2 summarize the Köppen symbols, as modified by

TABLE 2-1 Climatic Types Based on Trewartha's Modification of Köppen's System of Climatic Classification

A Humid Tropical Climates
- Af Wet Tropical
- Am Monsoon Tropical
- Aw Wet-Dry Tropical (dry winter)
- As Wet-Dry Tropical (dry summer)

B Dry Climates
- BS Steppe or Semiarid
  - BSh Tropical and Subtropical phase (Low Latitude Steppes)
  - BSk Middle Latitude Steppe
  - BSn Low Latitude Coastal Deserts—frequent fog; temperatures modified by cold ocean currents
- BW Desert or Arid
  - BWh Tropical and Subtropical Desert (Low Latitude Desert)
  - BWk Middle Latitude Desert
  - BWn Low Latitude Coastal Deserts—frequent fog and cool temperatures because of cold ocean currents

C Humid Mesothermal Climates
- Cs Mediterranean or Dry Summer Subtropical
  - Csa Hot Summer Mediterranean
  - Csb Cool Summer Mediterranean
- Ca Humid Subtropical
  - Caf no dry season
  - Caw dry winter
- Cb Marine or West Coast Marine—wet winters and cool, moist or relatively dry summers
- Cc Marine with mild moist winters and extremely short, cool summers

D Humid Microthermal Climates
- Da Humid Continental Warm Summer
  - Daf no dry season
  - Daw dry winter
  - Das dry summer (rare)
- Db Humid Continental Cool Summer (*f, w,* and *s* same as in *Da*)
- Dc Dd Subarctic (*f, w,* and *s* same as in *Da*)

E Polar Climates
- ET Tundra
- EF Ice Cap

H Undifferentiated Highlands

Table 2-2 Definitions of Symbols Used in the Köppen System of Climatic Classification as Modified by Trewartha

Symbol	Definition
Symbols for the Major Climatic Divisions	
A	Mean temperature for the coolest month is above 18°C (64.4°F). Used only if climate is humid (see fig. Z-8).
B	Potential loss of moisture through evaporation and transpiration exceeds precipitation. Use figure 2-8, or formulas from text to determine boundary. BS: Semiarid (Steppe); dry/humid boundary precipitation value varies with mean temperature (See fig. 2-8). BW: Arid (Desert); S/W boundary is 1/2 of BS/humid boundary value.
C	Mean temperature for warmest month is above 10°C (50°F); mean temperature for coolest month is above 0°C (32°F), but below 18°C. [Köppen value for coolest month above −3°C (26.6°F).]
D	Mean temperature for warmest month is above 10°C (50°F); mean temperature for coolest month is below 0°C (32°F). [Köppen value for coolest month: below −3°C (26.6°F).]
ET	Mean temperature for warmest month below 10°C (50°F), but above 0°C (32°F).
EF	Mean temperature for warmest month is below 0°C (32°F).
Temperature Symbols used to Subdivide *C* and *D* Climates	
a	Mean temperature of warmest month above 22°C (71.6°F).
b	Mean temperature of warmest month below 22°C (71.6°F); at least four months above 10°C (50°F).
c	Mean temperature of warmest month below 22°C (71.6°F); 1 to 3 months above 10°C (50°F).
d	Same as *c* for warm temperatures; mean temperature of coldest month below −38°C (−36.4°F).
i	[Köppen only: Isothermal; less than 5°C (9°F) range between temperature of coolest and warmest month.]
Temperature Symbols used to Subdivide *B* Climates	
h	Mean temperature for coldest month above 0°C (32°F). [Köppen value: Mean annual temperature above 18°C (64.4°F).]
k	Mean temperature for coldest month below 0°C (32°F). [Köppen value: Mean annual temperature below 18°C (64.4°F).]
k′	[Köppen only: Temperature of warmest month is below 18°C (64.4°F).]
n	Cool coastal deserts in low latitudes; frequent fog. No temperature criteria, but generally no month is hot (above 22°C or 71.6°F).
Symbols Indicating Seasonality of Precipitation	
f	No dry season. In *A* climates, mean precipitation for driest month exceeds 60 mm (2.4 in). Used in *C* or *D* climates where *w* and *s* not applicable and the driest month of summer receives more than 30 mm (1.2 in).
s	Summer Dry. Mean precipitation for wettest month of winter is at least three times that of driest month of summer. Precipitation for driest month must average less than 40 mm (1.6 in).
w	Winter Dry. Mean precipitation for wettest month of summer is 10 times that of driest month of winter.
m	Monsoon: Short dry season but precipitation amounts are comparable to *Af*. Boundary between *Am* and *Aw* is determined by graph or formula (fig. 2-6). Symbol is used only with *A*.

Note: Original Köppen values are given where they differ from the modification.

Trewartha; differences between the modified system and the original system are indicated. The major groupings of climates are alphabetical; *A, humid tropical* or *humid megathermal* (no winter); *B,* dry; *C, humid mesothermal* (mild winter); *D, humid microthermal* (severe winter); and *E,* polar (no summer).

THE HUMID TROPICS (*Af, Am, Aw*)

The humid tropics occupy more than one-third of the Earth's total surface (about 20 percent of the land area, and just over 40 percent of the oceans), making them more extensive than any other group of climates. They have no true winter, in the sense of a cool or cold season, and the difference in the mean temperatures between the warmest and coolest month is considerably less than the difference in temperature between night and day. Local inhabitants recognize seasons of the year in terms of "rainy season" and "dry season." Köppen chose the cool-month isotherm of 18°C (64.4°F) for the poleward limit of the humid tropics, on the assumption that it approximated the range of certain tropical plants, called *megatherms,* which require constant high temperatures for survival.

Humid tropical climates occupy that part of the earth dominated by air movement toward the great intertropical convergence zones (ITCs), or by summer monsoons along the tropical margins of the great continental land masses.[1] Along the eastern littorals of continents, particularly in southeast Asia and North and South America, the humid tropical climates extend to the margins of the astronomical tropics, and in limited areas they spread poleward of the tropics of Cancer and Capricorn; here, they border the *humid subtropical* (*Ca*) climates. Along west coasts, and in some continental interiors, rainfall rather than cooler temperatures restricts their latitudinal extent. In western South America and Southwest Africa, where cold ocean currents help push the dry (*B*) climates equatorward, the humid tropical climates extend only a short distance south of the equator.

The humid tropics fall into three climates, divided on the basis of rainfall amounts and seasonal patterns: the *wet tropical* (*Af*); *monsoon tropical* (*Am*); and *tropical wet and dry* (*Aw*). All months have a mean temperature above 18°C (64.4°F), as indicated by the symbol *A,* except in tropical highlands where altitude may depress average temperatures below this figure. Otherwise, they have similar climatic regimes and many of the environmental characteristics of the adjacent lowland tropics. The original Köppen system included these highlands in the *C* climates—usually *Cwi, i* indicating an annual range of mean monthly temperatures of less than 5°C (9°F).

[1]From the standpoint of world climates, it is best to define *continents* in terms of hemispheric land masses rather than in a traditional sense. The Western Hemispheric land mass includes both North America and South America, and the Eastern Hemispheric land mass comprises Africa, Europe, and Asia.

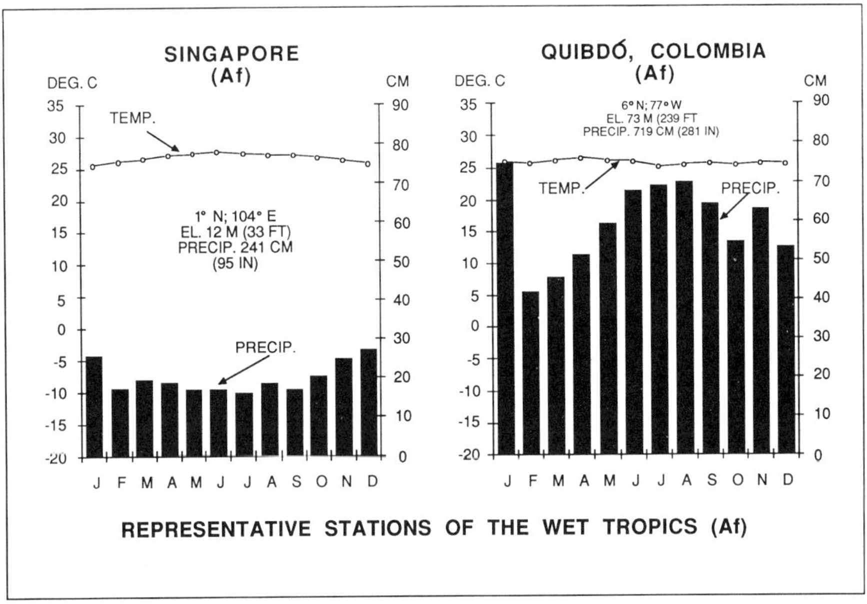

Fig. 2–4. Singapore and Quibdó are typical of *Af* stations in equatorial locations where the Equatorial Trough (an ITC) is always in control of the weather. In these areas there is a very small range in the monthly mean temperatures, and the precipitation is relatively evenly distributed throughout the year. Quibdó is located along the Pacific coast of Colombia, the rainiest area in South America, where moist winds blow from the ocean and are uplifted against the mountains behind the coast.

Wet Tropical Climates (*Af*)

Conditions within the wet tropical climate are always warm and moist. At lower elevations, monthly mean temperatures lie between 27°C and 30°C (80°F and 86°F), with an annual range of less than 3°C (5°F) between the coolest and warmest months. It is truly a climate where winter never comes. At most stations, mean annual rainfall will exceed 1,500 mm (59 in), with no month receiving an average of less than 60 mm (2.4 in). Although by this definition there is no dry season, almost always there is a rainy and less rainy season (fig. 2-4). Tropical rainforests occupy large areas of land within the *Af* climate; hence its designation as "tropical rainforest climate," by Köppen and many of his followers.

A major part of the area occupied by *Af* climates lies within 10° to 15° of the equator, where the equatorial low pressure zone of the ITC dominates the weather patterns. Paralleling the other humid tropical climates, its widest latitudinal span occurs on the east sides of the great continental land masses. A few outliers of *Af* climate lie at the very edge of the tropics in southeastern

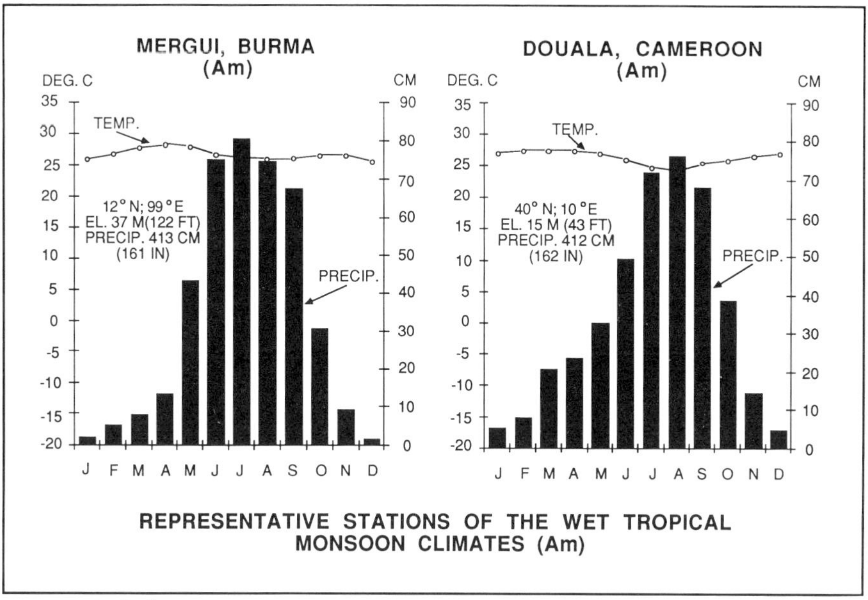

Fig. 2–5. Mergui, on the west coast of the Malay Peninsula, and Douala, on the west coast of Equatorial Africa, are typical tropical monsoon stations (*Am*). Their seasonal rainfall pattern is associated with onshore monsoonal winds of summer, which are forced against mountain ranges paralleling the coast, and offshore winds in winter.

Brazil and Madagascar, but the climate is most widespread in three areas: (1) the Amazon Basin and adjacent parts of South America; (2) central and western Africa, particularly in parts of the Zaire (Congo) Basin; and (3) the numerous islands and peninsulas of Southeast Asia.

Monsoon Tropical Climates (*Am*)

Poleward of the core of the wet tropical (*Af*) climatic areas, particularly on the eastern margins of the continental masses, are climates with average annual precipitation equal to or exceeding that of the *Af* type (fig. 2-5). Unlike the *Af* climates, however, these areas have a well-defined dry season that gives them a rainfall regime like that of the *wet-dry tropical* (*Aw*) climates, considered in the next section. Because of the heavy seasonal rainfall and the relatively short dry season, the soil remains moist enough throughout the year to support the growth of an evergreen rainforest.

In the Köppen scheme, the criterion for separating *Am* climates from *Af* and *Aw* is the relationship between the total amount of precipitation and the amount occurring in the driest month. In both *Aw* and *Am* climates, the average precipitation for the driest month is less than 60 mm (2.4 in);

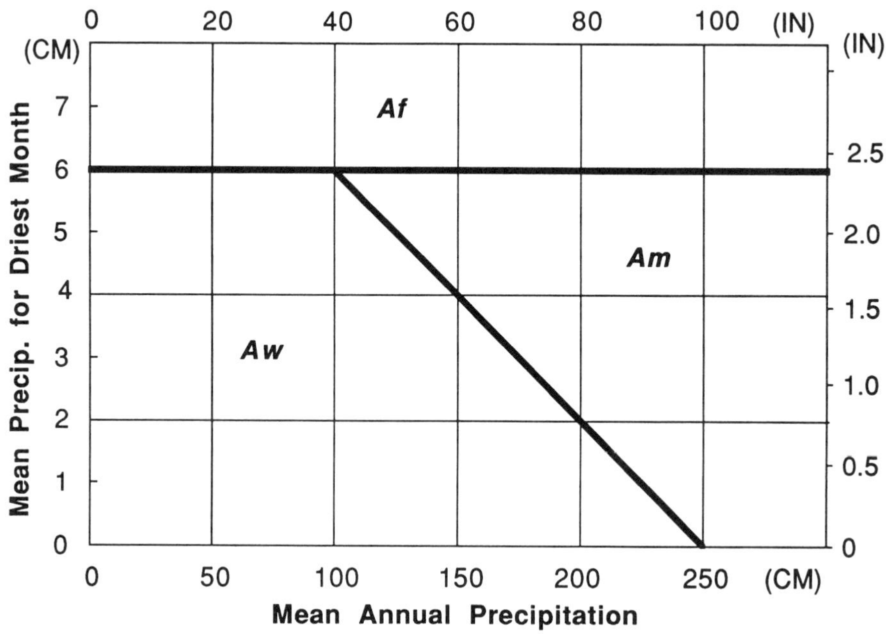

Fig. 2–6. Graph for determining boundaries between *A* climates. Note that if no month averages below 60 mm (2.3 in), the station is *Af*; if a station has a mean annual rainfall of 1,500 mm (60 in), it will be *Am* if the driest month averages more than 40 mm (1.6 in), or *Aw* if the average is less. All stations with a mean annual total of 2,500 mm (97.5 in) will be classified as *Am* if at least 1 month averages less than 60 cm (2.3 in).

the two are then separated by means of the graph in figure 2-6. Note from the graph that all stations with a dry month mean precipitation of more than 60 mm are classified as *Af*, and those with a dry month mean of less than 60 mm but a mean annual rainfall greater than 100 mm (40 in), are *Am*. All stations with less than 60 mm for the driest month, but with an annual mean of more than 2,500 mm (98 in) are *Am*, regardless of the severity of the dry season. Because the distinction between the *Af* and *Am* subtypes is rather arbitrary, and because there is no marked change in vegetation cover, the *Af*/*Am* boundaries are not shown in figure 2-2.

Monsoon tropical areas exist on the coastal margins of northeastern South America, West Africa, and Southeast Asia; however, large areas of monsoon Asia, Africa, and South America have climates that fall into the *Aw*, *Cw*, or highland (*H*) classifications.

Tropical Wet and Dry Climates (*Aw*)

Lying poleward of the core of the wet tropical climates is a broad area in which annual precipitation averages between about 1,500 mm (59 in) and

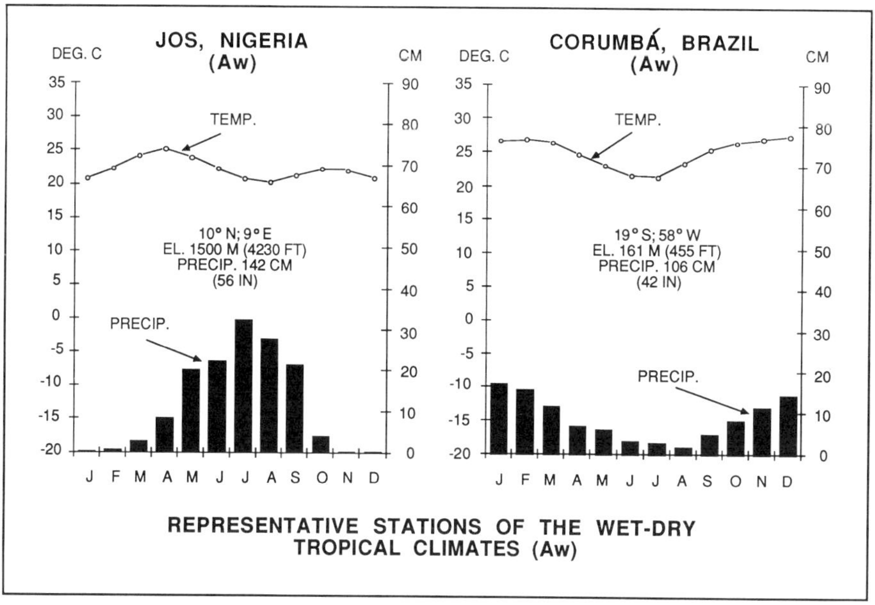

Fig. 2–7. Jos is in the Northern Hemisphere, and Corumbá lies south of the equator; both are typical *Aw* climates. The rainy season occurs in the high sun period (summer) at both stations.

760 mm (30 in), and where there is a severe dry season. Rainfall is concentrated in the high sun (summer) season, and the dry season coincides with the low sun (winter) season (fig. 2-7). The letter *w* in the *Aw* code designates a "winter" dry period. At least one month must average below 600 mm (2.4 in), although at many stations the dry season is almost absolute. A few stations have *As* climates, where winter rain comes on the winter monsoon.

The wet and dry tropics constitute a broad transition zone between wet equatorial areas and the great low latitude steppes and deserts—the *Bh* climates of the Köppen system. High sun rainfall brings a rainy season during which conditions are much like those of the *Af* regions, whereas the drought of the low sun season is reminiscent of the deserts and steppes. The annual pattern of rainfall is the result of shifts in the semipermanent wind and pressure belts discussed in chapter 1. Essentially, the *Aw* areas come under the dominance of the ITC of the doldrums during the rainy season, and under that of the great subtropical highs and the dry margins of the trade winds during the dry season.

Major areas of *Aw* climate exist in Africa, South America, Southeast Asia, and Australia, where tropical savannas are important components of natural vegetation. Savannas are mixtures of grass and trees, the basis for the designation "tropical savanna climate" in older terminology. The correlation between climate and vegetation, however, is even less satisfactory here than in areas of *Af* climate. Large areas of deciduous, semideciduous, and evergreen forests occur in other areas of *Aw* climate, lending credence to the theory that periodic burning, rather than climate, produces tropical savannas. The dry season, of course, plays a major but indirect role by drying out the vegetation and thus setting the stage for fire to become an important ecological factor. In chapters 5 and 6, we will explore further the role of fire in tropical savannas and other plant communities.

The Dry Climates (*B*)

A climate is dry if potential loss of water through *evapotranspiration* exceeds the average annual precipitation.[2] Moisture deficiency occurs at some time during the year in almost all climates, but when total values are considered for an average year, humid climates will have a water surplus. For determining *humid/dry boundaries* for classification purposes, this criterion seems logical, but it is difficult to measure. Later we will consider a method for computing the value of evapotranspiration.

The Köppen formulas and graphs for determining the *humid/dry* boundaries are shown in figure 2-8. Some calculated boundary values for selected mean annual temperatures are given in table 2-3. These values prove that moisture requirements increase as temperature increases, and that precipitation is more efficient during the cool season. The amount of precipitation determining the boundary between humid and dry climates is computed for several stations having diverse temperature and rainfall regimes; table 2-4 summarizes these data.

Köppen divides the *B* climates into semiarid steppe (*BS*) and arid desert (*BW*). (*S* stands for steppe, and *W* for *wuste,* the German word for desert.) The *steppe/desert* boundary is arbitrarily set at one-half the mean annual precipitation of the *steppe/humid* boundary. A further division is between the low latitude steppes and deserts (warm), and those of the middle latitudes (cool). Low latitude dry climates (fig. 2-9) have no month with an average temperature below 0°C (32°F), a characteristic noted by the symbol *h* (German *heiss,* hot). The low latitude steppe is *BSh,* and the low latitude desert, *BWh.* Middle latitude steppes and deserts are *BSk* and *BWk*, respectively; the *k* (German *kalt,* cold) indicates that at least one month has a mean temperature below freezing. In the original Köppen system a mean annual temperature of 18°C (64.4°F) was the boundary between the warm and cool dry climates, a value that gives slightly different results.

Although low and middle latitude steppes and deserts have the common characteristic of dryness, their origins are quite different. Low latitude des-

[2] *Transpiration* is the loss of water that moves from soil moisture through the roots and tissues of plants, and evaporates from openings, or *stomata,* in plant leaves. In most areas transpiration is by far the largest component of total evapotranspiration.

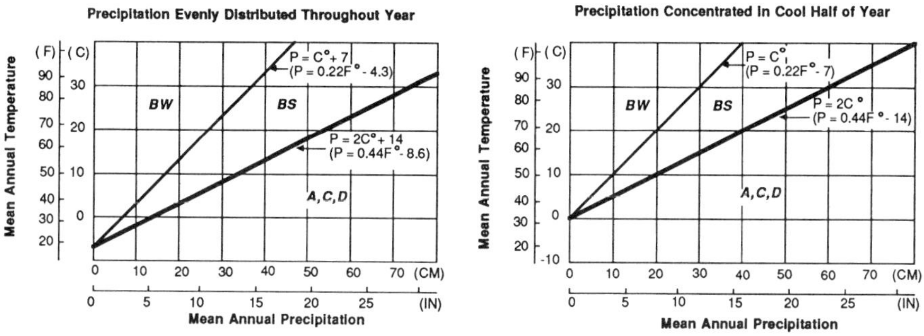

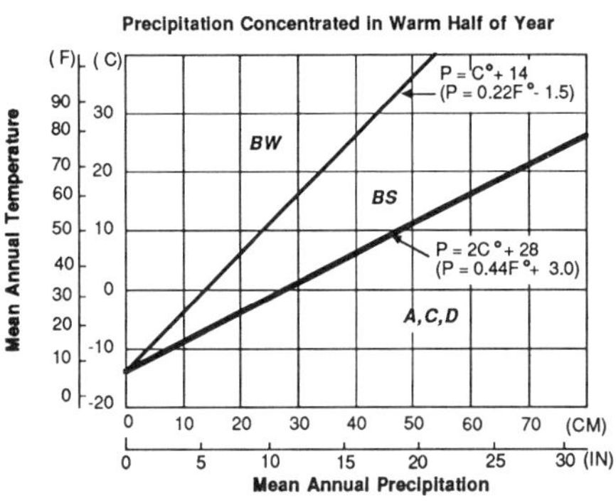

Fig. 2–8. Graphs for determining boundaries between dry and humid climates according to the Köppen formulas. *P* is mean annual precipitation required to determine the *ACD/BS* and the *BS/BW* boundary values at temperature *T*. Formulas using English units are given in parentheses (see table 2–3).

TABLE 2-3 Humid-Dry Boundary Values in Millimeters (Inches) of Precipitation According to the Köppen Formulas

	Mean Annual Temperature				
Seasonality of Precipitation	10°C (50°F)	15°C (59°F)	20°C (68°F)	25°C (77°F)	30°C (86°F)
Winter	200 mm (7.8 in)	300 mm (11.8 in)	400 mm (15.7 in)	500 mm (19.7 in)	600 mm (23.6 in)
No Seasonality	340 (13.3)	440 (17.3)	540 (21.2)	640 (25.2)	740 (29.1)
Summer	380 (15.9)	580 (22.8)	680 (26.8)	780 (30.7)	880 (34.7)

TABLE 2-4 Humid-Dry Climatic Boundary Values Computed for Selected Stations Using the Köppen Formulas

Station	Annual Temp. (°C)	Annual Precip. (mm)	Summer Concentration (%)	B/Humid Boundary Value (mm)	Köppen Class.
Omaha, Neb.	10.6	712	75	492	Da
Los Angeles, Cal.	17.4	377	12	354	Cs
Phoenix, Ariz.	21.5	193	46	570	BWh
Manta, Ecuador	24.7	213	85	774	BWh

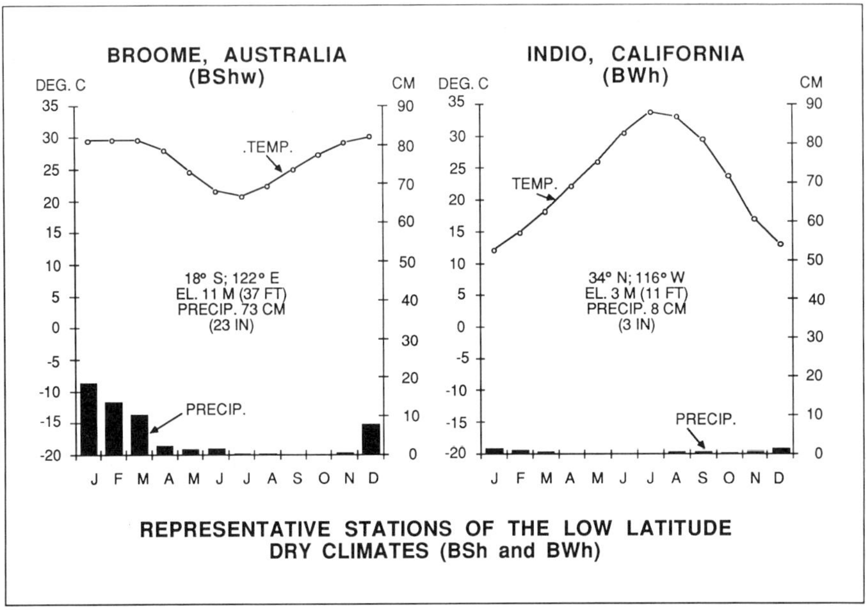

Fig. 2–9. Broome, located near the *BS*/*Aw* boundary in northwestern Australia, has a rainfall pattern that reflects the seasonal distribution in the *Aw* climates; thus, the station designation is *BShw* (Low Latitude Steppe), indicating a high sun concentration of precipitation. Indio, in California's Imperial Valley, has a *BWh* (Low Latitude Desert) climate with a winter concentration of precipitation similar to that of the *Cs* climate of coastal California.

erts and steppes occupy both continental interiors and west coasts between about 10° and 35° latitude. They are a product of the subsiding and diverging air of the subtropical highs and the drier trade winds. Middle latitude dry areas (fig. 2-10) lie in the rainshadow of mountains or occur at great distances from oceanic sources of moisture. Located in the westerlies, they extend eastward from continental interiors in middle latitudes and to some east coast locations in China and Argentina.

Cold ocean currents border many coastal deserts in low latitudes, causing cooler temperatures, increased humidity, frequent fogs, and low stratus clouds. Köppen recognizes these exceptions to the general conditions of low humidity, abundant sunshine, and high summer temperatures, which are usually associated with steppes and deserts; he designates the cool littorals *BWn* and *BSn*. The *n* indicates foggy deserts and derives from the German *nebel,* fog. Areas inland from these cold water coasts generally are not foggy, but they have relatively cool summers and mild winters with cool month means above 0°C (32°F) and warm month means below 22°C (71.6°F). The

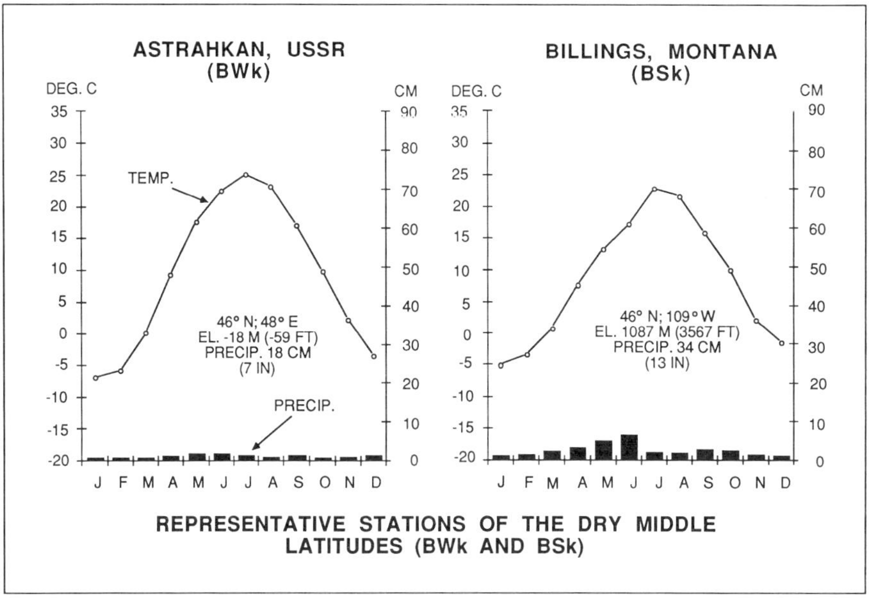

Fig. 2–10. Astrakhan, on the delta of the Volga River at the northern end of the Caspian Sea, is representative of the world's largest middle latitude desert (*BWk*). Billings, in the semiarid western Great Plains (*BSk*), has a warm season maximum of precipitation characteristic of that region. The winters are relatively mild, in spite of the latitude and interior location, making this an excellent area for winter wheat.

original Köppen system classifies these climates as *BWk* and *BSk* because the mean annual temperature is below 18°C (64.4°F), but *k* does not adequately describe these areas. Trewartha suggested they be designated just *BW* and *BS;* thus, omitting *k* indicates they are neither cold nor hot. These cool deserts and steppes are most extensive along the southwest coast of Africa, where the cold Benguela Current influences them, and on the west coast of South America, where the cold Humboldt Current prevails. Although these areas are cool relative to the *Bh* climates, the coolest months average well above freezing, and the annual range of temperature generally is small (fig. 2-11).

The climatic impact of El Niño demonstrates the role of cold ocean currents in extending the range of dry climates equatorward on west coasts. In the latitude of the desert coasts of Peru and southern Ecuador, El Niño causes a warming of the normally cold coastal waters and brings a temporary change from dry to humid tropical climate. Disruption of irrigated agriculture, destruction of transportation facilities, and general dislocation of

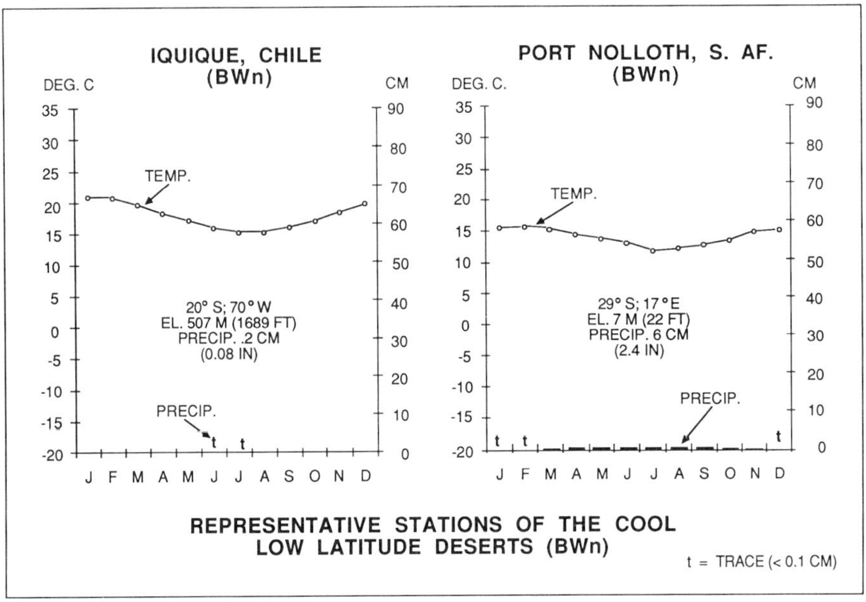

Fig. 2–11. Iquique, Chile, is in the world's driest desert, the Atacama of South America, where temperatures are moderated and aridity is increased by the cold Humboldt Current. Port Nolloth, on the cool, foggy coast of the Namib Desert of South Africa, has an unusually cool and even temperature regime caused by the cold Benguela Current. Other cool low latitude steppes and deserts are found adjacent to cold ocean currents in North Africa (Canaries Current) and Baja California (California Current).

economic activity are direct consequences of torrential rains during an El Niño event. El Niño in 1982–83 was the greatest of this century. Meteorologists and climatologists are now analyzing the possible worldwide climatic effects of this "super" El Niño.

Humid Mesothermal Climates (*Cs, Ca, Cb, Cc*)

The humid mesothermal climates are transitional between the "winterless" tropics, represented by the *A* climates, and the severe winter climates, represented by the humid microthermal (*D*) climates. These *C* climates have a definite but mild winter, with temperatures of the coolest month averaging below 18°C (64.4°F), but above 0°C (32°F). Latitudinal position or marine influences assure mild winters in the *C* climates. For at least part of the year westerlies, with their associated middle latitude cyclonic storms, dominate these climates. The modified Köppen system recognizes three major divisions of the *C* climates: *Cs,* dry-summer subtropical (Mediterranean); *Ca,* humid subtropical; and *Cb* or *Cc,* marine or west coast marine.

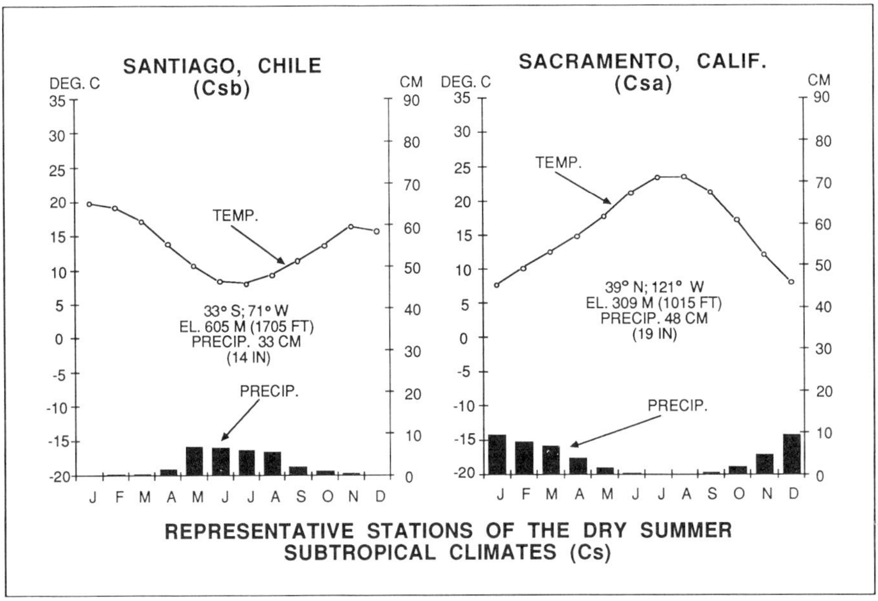

Fig. 2–12. Santiago, near the west coast of Middle Chile, is typical of *Csb* (Mediterranean, cool summer) stations where summer temperatures are modified by cold ocean currents; in this case the Humboldt or Peru Current. Santiago is south of the equator so the seasons are reversed. Sacramento, in the Central Valley of California, represents the *Csa* (Mediterranean, warm summer) climates. Most coastal stations in California have *Csb* climates because of the cold California Current.

Dry Summer Subtropical or Mediterranean Climate (*Csa, Csb*)

Mild, moist winters and dry summers make the Mediterranean climate one of the most clearly perceived of global climates (fig. 2-12). Although the *Cs* climate occupies less than 2 percent of the Earth's land surface, it is justly famous for its pleasant weather and productive and distinctive agriculture. Many of the world's famous resort areas have this climate, from the beaches of California and the French Riviera to the Crimean playgrounds of the Soviet Union.

The dry summer subtropical climate is found mostly on the west coasts of the continents between 30° and 40° latitude. It reaches its greatest extent around the margins of the Mediterranean Sea; thus, the term *Mediterranean climate.* Other type areas include the California coast of the United States, middle Chile, and the southern fringes of Africa and Australia. It extends into the continental interiors only in the Mediterranean area of Europe, North Africa, and the Middle East; principally because of the warm waters of the Mediterranean Sea and the Black Sea, that moderate winter temperatures and create a pressure trough along which winter cyclones move and bring winter rains. Also, the Alps and other mountain ranges protect the Mediterranean Basin from invasions of polar air masses in winter.

It is severe summer drought that sets the *Cs* climate apart from other mild winter climates. Dry summer weather is a direct result of poleward expansion and intensification of the subtropical highs. In summer, west coast locations place these areas of *Cs* climate on the dry side of the great semipermanent anticyclones of ocean basins, in areas of marked atmospheric subsidence and offshore winds (see chapter 1). In winter, the equatorward shift of the westerlies and the accompanying storm tracks bring winter rains.

Winter temperatures are mild, and snow is rare or absent near sea level. Summers in Mediterranean climates vary from cool (*Csb*) to hot (*Csa*). Along coasts bathed by cold ocean currents, summers are cool with warm month average temperatures not exceeding 22°C (71.5°F). Cool summer Mediterranean climates are found on coasts of California, Portugal, Australia, middle Chile, South Africa, and Morocco. Inland from cold water coasts, and at most stations on the shores of the warm Mediterranean Sea, warm month temperatures average above 22°C. This pattern of temperature is apparent if we compare averages for the warmest month at coastal stations in California to those of inland stations, such as San Francisco (16.4°C, or 61.6°F) and Sacramento (23.5°C, or 74.3°F). Similar coastal-inland contrasts exist in other *Cs* areas.

The range in mean annual precipitation within areas classified as *Cs* is rather large, generally falling between 380 mm (15 in) and 635 mm (25 in). At some stations, particularly at higher elevations in mountainous areas, precipitation may exceed 900 mm (35 in), and these probably should be considered in the same category as some stations in the Pacific Northwest of the United States and coastal British Columbia (table 2-5).

TABLE 2-5 Mean Annual Precipitation and Warm Month Mean Temperature at Selected *Cs* Stations

	Mean Annual Precipitation		Mean Warm Month Temperature	
Station	mm	in	°C	°F
San Diego, Calif.	235	10.05	22	71
Santiago, Chile	330	12.99	20	68
Athens, Greece	401	15.79	28	82
Valencia, Spain	578	22.76	25	77
Lisboa, Portugal	705	27.87	23	73
Naples, Italy	874	34.39	25	77
Salerno, Italy	1177	46.35	27	80

Source: Wernstedt 1972.

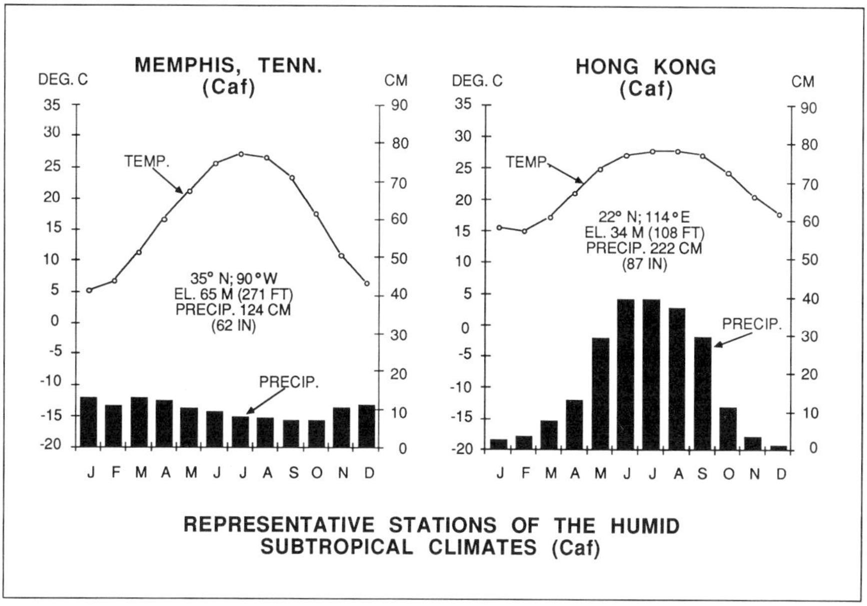

Fig. 2–13. Memphis has a slight concentration of precipitation in the cool half of the year because of increased cyclonic storm activity in winter and its position north of the warm Gulf of Mexico. Hong Kong has a very wet summer and a relatively dry winter, a pattern produced by the strongly developed monsoonal circulation along the eastern margins of Asia. Although Hong Kong lies equatorward of the Tropic of Cancer, its winter temperatures are lowered by the winter monsoon.

Humid Subtropical Climates (*Caf, Caw*)

Humid subtropical climates occupy the east coasts of continents at about the same latitude as the *Cs* climates of the west coasts. In contrast to most areas of Mediterranean climate, major areas of humid subtropical climate extend considerable distances inland from the coasts. East coast location places these climates on the western, or moist, side of the subtropical highs, where monsoonal circulation causes an onshore flow of moist tropical maritime air during summer months. In winter, the equatorward shift of middle latitude cyclonic storm tracks in the westerlies generally brings some precipitation in the cool season. This pattern of atmospheric circulation assures mild, moist or relatively dry winters, and hot humid summers (fig. 2-13).

Because of monsoonal circulation associated with major continents, most humid subtropical climates have a summer maximum of precipitation as represented by Hong Kong (fig. 2-13). An exception is the southeastern United States, where some stations have a slight winter maximum. The Gulf of Mexico plays a role in this pattern, furnishing a source of moisture both for

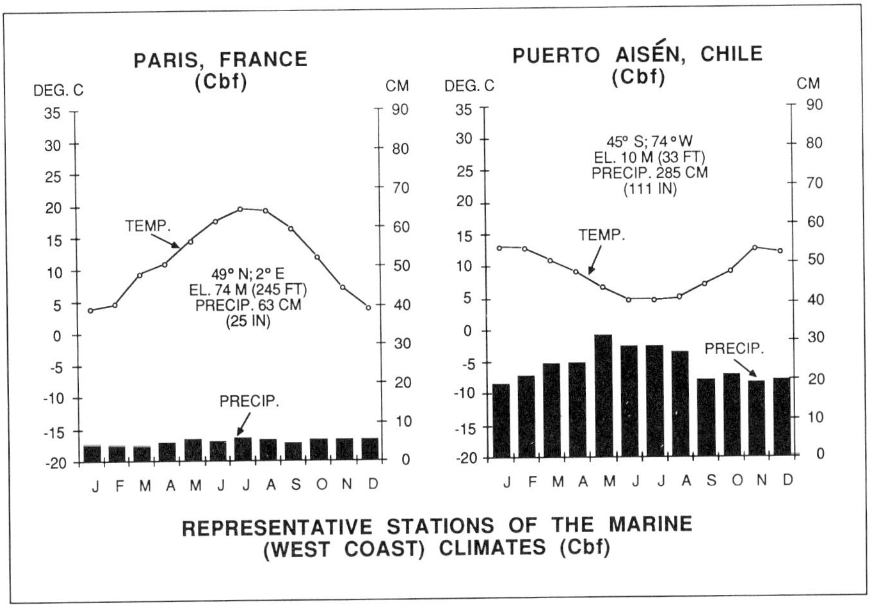

Fig. 2–14. Paris and Puerto Aisén demonstrate the contrast in rainfall between interior and coastal marine climates. Paris, with its moderate and seasonally well distributed precipitation, is rather typical of most *Cb* stations in Western Europe. Puerto Aisén's climate resembles that of other wet marine stations on the coasts of Washington, British Columbia, Norway, the British Isles, and New Zealand, where strong orographic uplift of the westerlies and frequent cyclonic storm activity produce abundant precipitation.

summer monsoonal circulation and for winter cyclonic storms. The climatic regime for Memphis (fig. 2-13) is typical.

Two subtypes of humid subtropical climate are within the classification: *Caf,* with no dry season, and *Caw,* with a dry winter. The *Caf* subtype is best represented in the southeastern United States, where some stations have a slight winter maximum of precipitation, and in southeastern South America, eastern Australia, Japan, and part of China. Less extensive areas of *Caf* climate are found in South Africa, the Po River valley of northern Italy, part of Yugoslavia, and at the eastern end of the Black Sea in Georgia and Abkhazi, the rainiest area of the Soviet Union. The *Caw* subtype occurs mostly on the margins of the wet-dry tropics in Southeast Asia, in northern India, and the interior of China.

West Coast Marine or Marine (*Cb, Cc*)

Marine climates have mild winters and cool summers and, considering their latitude, they have remarkably small annual temperature ranges (fig. 2-14).

Rainfall is moderate at most stations, with mean annual totals between 500 mm (20 in) and 1,000 mm (40 in). Where orographic uplift of westerly winds occurs, however, rainfall may be excessive; for example, many stations in southern Chile have mean annual precipitation in excess of 2,500 mm (100 in). Most areas in Western Europe, Australia, and New Zealand have no real dry season, and the designation *Cbf* is appropriate; in others, including the Pacific Northwest of the United States and the southern part of middle Chile, the seasonal pattern of precipitation resembles the Mediterranean climates. In the original Köppen system, at least some of these areas, designated *Cb* by Trewartha, are classed as *Cbs*.

Major areas of marine climate lie on the west coast of North America, from northern California to southern Alaska; in Western Europe, north of the Alps and west of Berlin; in South America in southern Chile, the east coast of Argentina, and the highlands of southern Brazil; in the highland areas of South Africa; and in southeastern Australia and New Zealand. The *Cc* subtype, with mild winters but exceptionally cool and short summers, occurs in limited areas on the south coast of Iceland, coastal Norway, and the extreme south of Chile. The symbol *c* indicates that only one to three months have mean temperatures above 10°C (50°F).

In west coast locations, the *Cb* or *Cc* climates have a remarkably broad latitudinal range, from about 40°, poleward to 60° latitude and beyond. Some coastal Norwegian stations as far north as 67°N latitude have a *Cc* climate. Prevailing westerly winds and relatively warm ocean currents moderate temperatures, particularly in winter, and the cyclonic storms of the westerlies bring adequate precipitation and cloudy weather. *Cs* climates border the west coast *Cb* on their equatorward margin, and humid continental–cool summer (*Db*) and subarctic (*Dc*) climates fall on their poleward sides.

Marine climates, located on eastern continental margins, generally are found poleward of 40° latitude. In Australia, New Zealand, and eastern Argentina, cool ocean currents moderate the summer temperatures to *Cb* levels, whereas elevation plays a similar role in reducing warm season temperatures in southern Brazil and South Africa. Most of the eastern *Cb* climates are adjacent to *Ca* or *Aw* and have similar rainfall patterns with a summer maximum, and some may be designated *Cbw*.

Humid Microthermal Climates (*Da, Db, Dc, Dd*)

Cold winters with frequent and persistent snow cover distinguish the *D* climates from humid mesothermal climates bordering them on the equatorial and west coast margins. They occupy interiors and eastern margins of the great land masses in the Northern Hemisphere middle latitudes, where marine influences are minimal and great variation in solar radiation occurs from season to season. Since there are no large land masses between 40° and 60°S latitude, the *D* climates are missing in the Southern Hemisphere.

TABLE 2-6 Contrasts in Mean January and July Temperatures Between Marine and Continental Stations in Eurasia Located Between Latitude 50°N and 60°N, and Longitude 6°W and 108°E.

	6°W Dublin		0° Kew		13°E Berlin		38°E Moscow		72°E Sverdlovsk		83°E Novosibirsk		108°E Kirensk		Maximum Temp. Dif. West–East	
	°F	°C	°F	°C	°F	°C	°F	°C	°F	°C	°F	°C	°F	°C	°F	°C
January	40.1	4.5	39.6	4.2	30.7	−0.7	14.2	−9.9	5.7	−14.6	−2.7	−19.3	−17.0	−27.2	57.1	31.7
July	59.5	15.3	63.7	17.6	65.7	18.7	66.2	19.0	64.0	17.8	65.7	18.7	64.9	18.3	7.4	4.1
Annual Range	19.4	10.8	24.1	13.4	35.0	19.4	52.0	28.9	58.3	32.4	68.4	38.0	81.9	45.5		

Note: The January temperature gradient is much steeper than the July gradient, indicating that the moderating effect of the oceans on temperature is much greater in winter than in summer.

TABLE 2-7 Seasonal Distribution of Precipitation for Selected Stations in North America

Station	Mean Annual Precip.		Warm Season Total		Percent in Warm 6 Months	Cool Season Total		Percent in Cool 6 Months
	in	mm	in	mm		in	mm	
Omaha, Neb.	28.05	712	20.99	533	75	7.06	180	25
Chicago, Ill.	33.17	843	20.01	508	60	13.16	335	40
Columbus, Oh.	34.10	866	20.05	510	56	16.24	412	44
Buffalo, N.Y.	37.46	952	17.81	452	48	19.65	500	52
Boston, Mass.	43.32	1100	20.58	523	48	22.74	577	52

Prevailing westerlies extend the continental climates from the interiors to eastern margins of continents where they minimize marine influences; consequently, cities such as New York and Boston on the east coast of the United States have climates more comparable to Des Moines and Minneapolis in the Midwest than to west coast stations at similar latitudes (fig. 2-15). January temperatures at Boston average −1.4°C (29.8°F), whereas at Chicago, at about the same latitude, the January mean is −2.0°C (26.8°F). Crescent City, Calif., at a latitude comparable to Boston and Chicago, has a January mean temperature of 8°C (46.4°F); its relatively warm winter demonstrates the contrast between east and west locations in middle latitudes.

The seasonal range of temperature in the continental climates also exceeds that of most other climates. This is true particularly of the interior of the Eurasian land mass in Siberia, where seasonal ranges of more than 56°C (100°F) have been recorded (table 2-6 and fig. 2-16 and 2-17). Annual temperature ranges in areas such as the American Midwest are significant, although less spectacular than the extreme continentality of Siberia. The range at Chicago is 27°C (48°F), and at Winnipeg it is 38°C (68°F).

In most *D* climates, precipitation peaks in the warmer months, although some stations in eastern North America show a remarkably uniform distribution between winter and summer (table 2-7). Winter precipitation often occurs as snow, with mean annual snowfall of about 300 mm (12 in) on the southern margins of the *D* climates, to more than 2,540 mm (100 in) in more northerly locations. Monsoonal tendencies in the seasonal pattern are apparent in central North America, and they are exceptionally well developed in north China and Siberia.

D climates comprise three major subdivisions: humid continental–warm summer (*Da*), humid continental–cool summer (*Db*), and subarctic (*Dc* and *Dd*). These categories are based principally on the length of the warm season. Where rainfall is highly concentrated in summer, *w* is added to indicate dry winters (*Daw, Dbw, Dcw, Ddw*). Where a dry season is absent, *f* is suffixed, as in *Daf*. The *d* represents extremely cold winters of central and eastern Siberia and indicates a cold month average temperature below −38°C (−36.4°F). North America has no *Dd* climates. Other letters used with *D* have the same values as for the *C* climates.

Humid Continental–Warm Summer Climate (*Da*)

Although this climate is the least severe of the cold winter climates, it is the hot, humid summers that give it its distinctive character (fig. 2-15). July generally is the warmest month, with temperatures averaging above 22°C (71.6°F) along the cooler margins and near 27°C (80°F) on the warmer. The length of the average frost-free season exceeds 150 days throughout most of the *Da* area, and it may reach 200 days along the southern margin in the United States where the *Da* area is adjacent to the *Caf* climates. Temperatures for the coldest month, January, are everywhere below 0°C (32°F), and

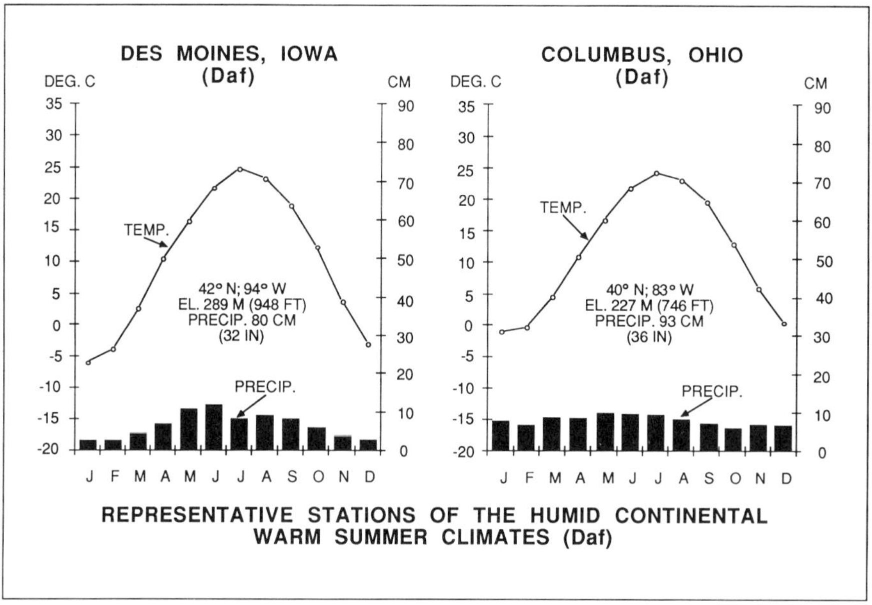

Fig. 2–15. Des Moines and Columbus exemplify the climate of the corn belt of the United States, the most productive corn (maize) growing region of the world. Although Des Moines is designated *Daf,* rather than *Daw,* there is a definite summer concentration of precipitation. At Columbus, precipitation is more evenly distributed between winter and summer, a pattern reflecting more frequent cyclonic storm passage in winter.

they may average as low as −12°C (−10°F) along the poleward margins adjacent to the *Db* climates.

In most areas, summer months have the highest precipitation, and in the *Daw* subtype, the contrast between the dry winter and hot, humid summer is quite pronounced. In North America some stations along the western margins of this climate show a marked monsoonal character, with more than 70 percent of the mean annual precipitation falling in the warm half of the year (table 2-7). In contrast, stations in eastern North America show much less seasonality, with warm season and cool season precipitation more nearly equal.

Spring and summer rainfall comes mostly from thunderstorms, both frontal and local, whereas winter precipitation is almost entirely of cyclonic origin. In both winter and summer, most moisture originates in tropical maritime air masses from the Gulf of Mexico and Western Atlantic. In eastern North America convergence of cyclone paths toward the lower Great Lakes and the St. Lawrence valley (fig. 1-10) accounts for the higher percentage of winter precipitation. Snow furnishes less than 5 percent of the mean annual precipitation in southern areas of the climate, and as much as 20 percent at more northerly stations.

High summer temperatures, long growing seasons, and a warm season concentration of precipitation make the *Da* climate almost ideal for certain crops, such as corn (maize) and soybeans. In North America, the *Daf* climate occupies about the same area as the Feed Grain and Livestock Region, as delineated by the U. S. Department of Agriculture (USDA), better known as the corn belt. Also, the major maize-growing regions of Europe, the lower Danube Basin and the upper Po Valley, have *Daf* climates. Soybeans, the second most important crop in the American corn belt, are the major crop in the Daf area of Manchuria.

Humid Continental–Cool Summer Climate (*Db*)

As this name implies, summers are cooler in the *Db* climate than in the warm summer subtype (fig. 2-16). Also, summers are correspondingly shorter and winters are more severe, but at least four months average above 10°C (50°F). Poleward, the cool summer phase of the *D* climates borders the subarctic (*Dc*) climates, which have very short, cool summers. In North Amer-

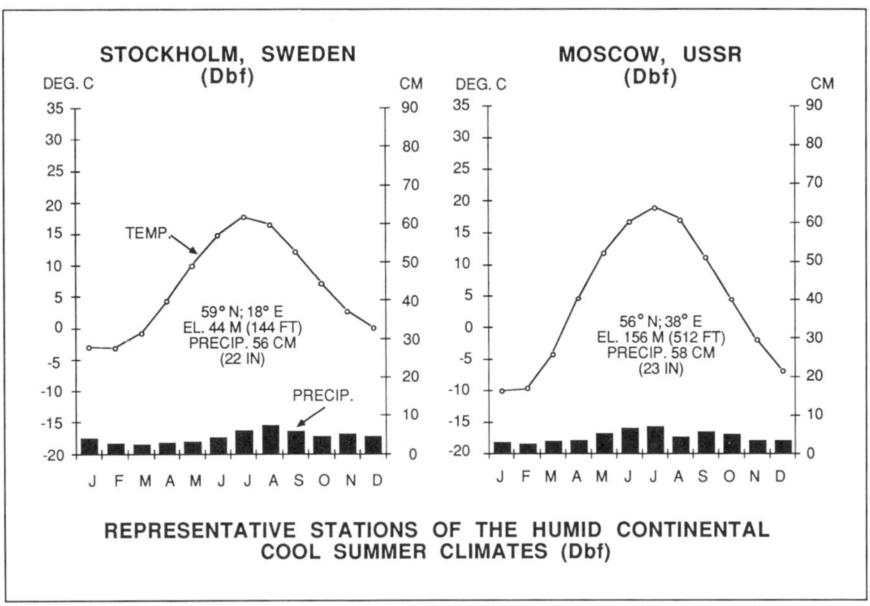

Fig. 2–16. Stockholm has a relatively mild winter typical of *Db* stations on the western margins of the Eurasian land mass. Moscow has a temperature regime representative of *Db* stations in the interior and eastern areas of Eurasia and North America. Rainfall is low at both stations, but it is well in excess of that required for humid climates with these temperature regimes.

ica, *Db* is the climate of the southern part of the prairie provinces of Canada, northeastern North Dakota, northern Minnesota, Wisconsin, and Michigan, southern Ontario and Quebec, and most of New England north of Massachusetts. In Eurasia the *Db* climates occupy most of central and eastern Europe, and southern Scandinavia, except for most of Denmark and marine coastal areas of Norway, and they penetrate eastward in a great wedge between the steppes (*BSk*) to the south and the forested subarctic (*Dc*) to the north and east. *Db* again appears along the eastern margin of Siberia, in north China, and in the northern islands of Japan.

Weather patterns in *Db* are similar to those in *Da* climates, with the same tendency toward a summer maximum of precipitation in most areas, except as noted in eastern North America. In some coastal locations northward from Long Island, particularly in Nova Scotia and eastern Newfoundland, there is a slight winter maximum. A marked summer maximum appears in western areas of North America, where Winnipeg receives 88 percent of its mean annual precipitation in the warm half of the year, possibly justifying a designation of *Dbw*. In contrast, St. Johns, Newfoundland, receives exactly half of its mean annual precipitation in the cool half of the year.

Snow accounts for more than 30 percent of the annual precipitation, except in areas with very dry winters. Some stations in the vicinity of the Great Lakes and in eastern Canada average more than 120 days with a snow cover greater than 25 mm (1 in). Except for some mountainous areas, eastern Canada has the most snow of any region in North America, with Montreal averaging more than 3 m (10 ft) each winter.

Subarctic Climates (*Dc, Dd*)

The subarctic is the most "continental" of climates, reflecting the great seasonal fluctuation in heating and cooling of the continental interiors at high latitudes. Although extremely cold winters are a feature of most areas with subarctic climates, it is the short summer—only one to four months above 10°C (50°F)—that distinguishes them from the other humid microthermal climates (fig. 2-17). The *d* suffix is reserved for the extremely cold winter climates of central Asia where cold month temperatures average below −38°C (−36.4°F). Köppen chose the 10°C (50°F) mean isotherm for the warmest month as the boundary between the subarctic and tundra (*ET*) climates, because it approximates the poleward limit of trees (see chapter 6).

Warm month mean temperatures are in the 10°C to 18°C (50°F to 65°F) range, with interior stations near the top of the range and coastal stations near the bottom. The long summer days of the continental high latitudes, fully twenty-four hours of sun at stations such as Verkhoyansk beyond the polar circle, give surprisingly mild summers. Although some threat of frost exists at almost any time of the year, some unexpectedly high temperatures have been recorded. Yakutsk has an absolute maximum of 38.9°C (102°F), and Verkhoyansk has recorded a high of 34.4°C (94°F).

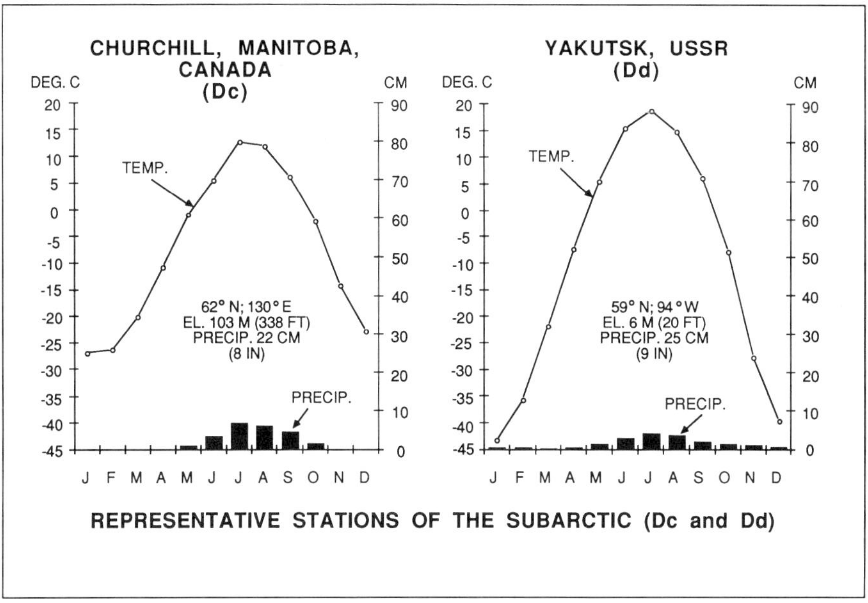

Fig. 2–17. Churchill (*Dc*), on Hudson Bay, and Yakutsk (*Dd*), in eastern Siberia, are representative of the great continental temperature extremes and short summers of the subarctic climates in the interiors of the great land masses at high latitudes. Yakutsk is in an area with the greatest continentality and the lowest reported temperatures on Earth outside Antarctica.

Winter temperatures have much greater geographic variation, with marine influences playing a moderating role near west coasts, whereas extreme continentality prevails in the interiors and toward the eastern margins of Asia and North America (fig. 2-18). We can illustrate the great range of winter temperatures encountered in the subarctic of Eurasia by comparing the January temperature at Narvik, near the Norwegian coast, where the mean is only −3.2°C (26.6°F), to January averages in northeastern Siberia, where mean cold month temperatures are below −45°C (−50°F). Yet mean warm month temperatures are almost 5.6°C (10°F) warmer in Siberia than in coastal Norway.

Scanty precipitation is the rule over large areas of the subarctic, with many stations recording means under 254 mm (10 in); this precipitation falls mostly in the summer months. Winter snowfall is light but, because of the low temperatures, the number of days with snow cover exceeds 240. Again, areas near west coasts are atypical, with many recording mean annual precipitation in the 500–1,000 mm (20–40 in) range. Narvik receives half of its 851 mm (33.54 in) mean annual precipitation during the cool six months of

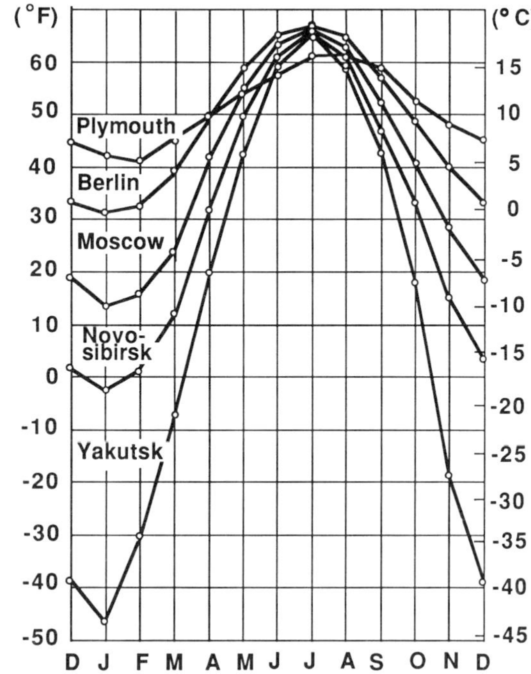

Fig. 2–18. Mean monthly temperature curves for stations in Eurasia, showing increasing continentality from west to east, as indicated by increasing annual temperature range. The winter temperatures decrease with distance from the marine influence of the North Atlantic Ocean. The west-east temperature gradient is much less in summer, indicating that the moderating effect of the ocean is more pronounced in winter.

the year. In Siberia, Yakutsk averages only 212 mm (8.39 in) annually; 74 percent of this is received in the warm half of the year.

Under high barometric pressure during winter and with relatively calm conditions, cold air drains into the topographic low areas and produces pockets of exceedingly cold air below shallow temperature inversions. Most settlements and most weather stations in the subarctic of Siberia and North America are located in river valleys. Because of the low topographic position of these valley stations, isotherms seen on regional temperature maps reflect temperatures in the low areas, and undoubtedly these maps exaggerate the low temperatures for the region as a whole. Since topography has less impact on local temperature variations in summer, the mean annual temperature range is likewise magnified (Kendrew 1953). Since most inhabitants of the subarctic live in the valleys, they experiènce the extremely low temperatures of the frost hollows. Yet, as W. G. Kendrew asserts:

> The Siberian winter is by no means so unpleasant as these low temperatures might suggest. The air in the interior of the continent is bracing and often calm, so that furs give adequate protection. Vegetation hibernates, to flourish luxuriantly again in the summer warmth. [1953, 267]

The relatively warm summers often have temperatures above 27°C (80°F), and temperatures above 32°C (90°F) are not uncommon. The absolute maximum of 38.9°C (102°F) reported at Yakutsk is the highest ever recorded at a station so near the Arctic Circle. Since it has a growing season averaging ninety-eight days, small grains and vegetables can be grown in the vicinity of Yakutsk.

Permafrost is another climate-related phenomenon of the subarctic and the tundra (*ET*). The term *permafrost* refers to permanently frozen ground in which the subsoil remains frozen throughout the year, with only a few feet of the surface materials thawing each summer. In many arctic and subarctic locations, the peculiar problems encountered with permafrost in developing the polar regions make it difficult to construct roads, railroads, buildings, and airfields. Most areas of *Dc* climate in North America have discontinuous or sporadic permafrost, whereas in all areas of *Dd* climate and in large areas of *Dc* climate in the Soviet Union permafrost is continuous.

Polar Climates (*ET*, *EF*)

Most lands lying within or near the polar circles are areas where summer truly never comes. It is appropriate that the equatorward boundary of the summerless *E* climates is marked by the mean isotherm for the warmest month, 10°C (50°F), whereas the poleward boundary of the winterless tropical climates is delimited by the mean temperature of the coolest month. The Köppen system recognizes two types of polar climate: the tundra (*ET*) and the ice cap (*EF*), separated by the 0°C (32°F) mean isotherm for the warmest month (fig. 2-19). Considerable areas of *Dc* and *Dd* climate extend north of the Arctic Circle, where warmer continental summers raise the warm month temperature mean above 10°C, while some areas of *ET* climate, such as coastal Greenland, northeastern Canada, and the eastern Siberian uplands, are located south of the polar circle.

Tundra Climate (*ET*)

Coinciding roughly with the treeless fringe of the Arctic Ocean in North America and Eurasia, the tundra climate takes its name from tundra vegetation, a complex of low-growing plants dominated by mosses, lichens, hardy grasses, and dwarf shrubs. Tundra vegetation is not confined to areas of *ET* climate, but grows in some areas of *Dc* and *Dd* climate, where environmental conditions other than temperature discourage tree growth.

Some small areas of *ET* climate are found in the Southern Hemisphere within the ice-free fringe of Antarctica and on parts of the fjorded coast and islands of the far south of Chile. A similar temperature regime and vegetation cover are found above the tree line in mountains outside the Arctic. As in the subarctic climates, winter temperatures in *ET* climatic areas are not so low, and precipitation is higher, along the western littorals where polar maritime air masses (mP) dominate.

These areas have remarkably high winter temperatures for the latitude, with some stations reporting a mean cold month temperature above 0°C

(32°F). Examples of these marine-dominated *ET* climates are found in Norway, Iceland, Greenland, on islands in the Bering Sea, and on the fjorded coasts and islands in southern Chile, in the Falkland Islands (Islas Malvinas), and other isolated islands in the Southern Ocean.

Ice Cap Climate (*EF*)

This is the climate of the great glacial ice fields of Greenland and Antarctica, the ice-covered Arctic Ocean, and mountainous areas above the snow line. Since ice caps are uninhabited, except temporarily by explorers or at scientific stations, there are few climatic records of long duration. Estimates place the range of mean annual temperatures in the interior of Greenland between −29°C and −34°C (−20°F and −30°F). Eismitte, an expedition base camp maintained at an elevation of 3,030 m (9,941 ft) on the Greenland ice cap from 1929 through 1931, recorded an annual mean for the three years of −30.4°C (−22.4°F). February, the coldest month, averaged −47.3°C (−53.1°F), whereas July, the warmest, had a mean of −10.8°C (12.6°F). A summary of the monthly mean temperatures recorded by the expedition is

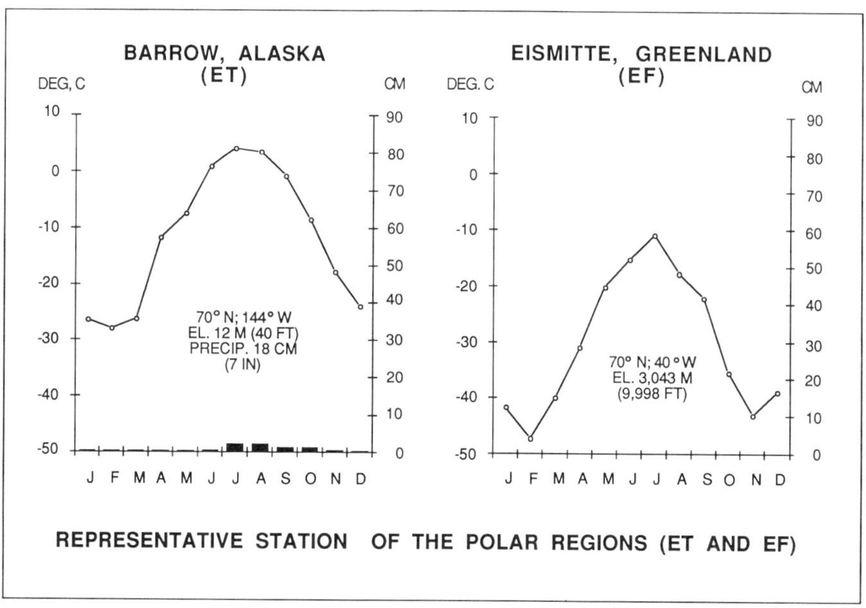

Fig. 2–19. Barrow (*ET,* tundra) is the most northerly weather station in the United States. The mean temperature for the warmest month is well below 10°C (50°F). Eismitte (*EF, ice cap*) was a temporary station in operation between 1929 and 1931 on the Greenland ice cap at an elevation of 3,043 m (9,984 ft). The short record gives some indication of the annual march of temperature.

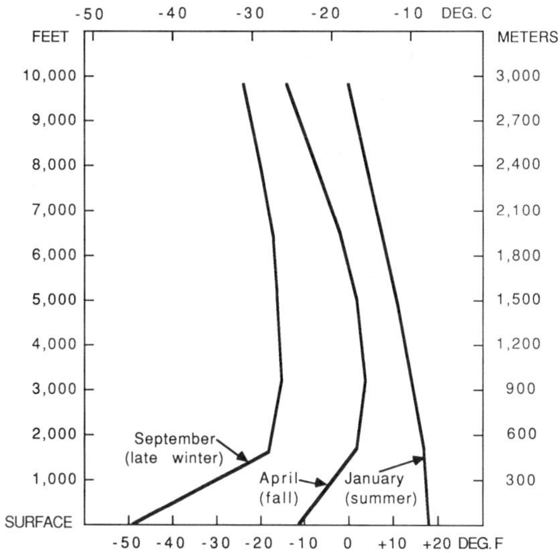

Fig. 2–20. Average lapse rates observed at Little America III, Antarctica, by radiosondes, April 1940 through January 1941. Except for a short period during the brief summer, strongly developed temperature inversions persist throughout the year (U. S. Weather Bureau, *Monthly Weather Review*, Supp. No. 48, 1949.)

shown in figure 2-19. The lowest temperatures observed at Eismitte were −65°C (−85°F), and on eighty-three days, between 10 October and 12 April, temperatures fell below −51°C (−60°F).

Central Greenland is the coldest region of the Northern Hemisphere, but Antarctica has the distinction of being the coldest region on Earth. At Vostok, a Soviet research base and weather station some 1,200 km (800 mi) from the South Pole at an elevation of 3,487 m (11,440 ft), the lowest surface temperature ever recorded, −88°C (−127°F), was noted on 24 August 1960. This record low temperature, and other very low temperature readings, occurred near the end of the long polar night. The highest temperature reported at Vostok during 1958 was −23°C (−9°F). Except for a short period in summer, intense temperature inversions persist throughout the year on the ice cap (fig. 2-20).

Highland Climates (*H*)

Strictly speaking, there is no "highland" climate, but the letter *H* appears on climatic maps because great differences in elevation in relatively short distances make it impossible to show all of the climates in a mountainous area. In tropical mountains, it is often possible to travel in a relatively short distance from the wet tropics near sea level, to tundra above the tree line, or to perpetual snow above the snow line.

In general, highland stations reflect the climatic regime of lowlands at the same latitude. Figure 2-21 shows that Bogota has a typical *Af* temperature pattern, but it is depressed some 12°C to 14°C (22°F to 25°F) below the

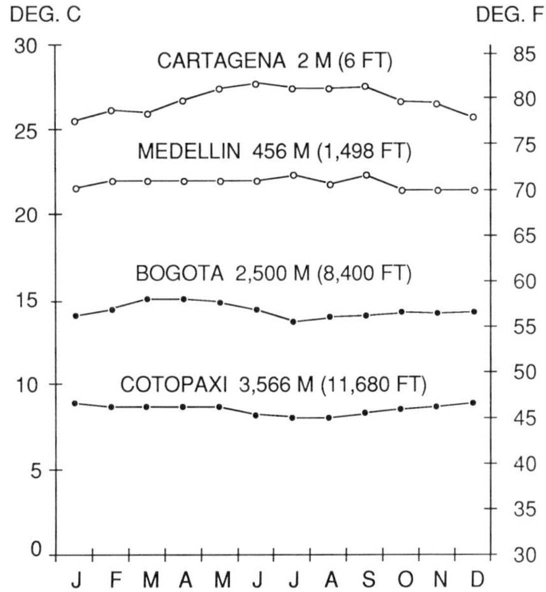

Fig. 2–21. Annual march of mean monthly temperatures at four tropical stations at different altitudes. Note that the lapse rate between Cartagena, at sea level, and Medellin, at 454 m (1,490 ft), exceeds the average lapse rate of 6.4°C in 1,000 m (3.5°F in 1,000 ft), whereas the lapse rate between sea level and the other stations is less than the average.

temperatures experienced at Cartagena, an *Af* station near sea level. This depression amounts to about 4.7°C for each 1,000 m (2.5°F per 1,000 ft), a rate of decrease with elevation that is less than the average lapse rate in the free atmosphere (6.4°C per 1,000 m, or 3.5°F in 1,000 ft).

In the tropics, where vertical zonation of climate is most important in human affairs, several altitude zones are recognized by local people. In Latin America these zones are: *tierra caliente* (hot land), *tierra templada* (temperate land), *tierra fría* (cool land), *páramos* or *puna* (tundra-like areas between tree line and snow line), and *tierra nevada* or *tierra helada* (snow land or frozen land above the snow line). Definitions and elevations of these temperature zones vary from place to place, but those of the tropical Andean region of South America are typical:

Tierra caliente: Sea level to 900 m (3,000 ft); mean annual temperatures 28°C to 24°C (83°F to 75°F); zone of tropical crops, such as bananas, sugar cane, and cacao.

Tierra templada: Elevation 900 m to 1,800 m (3,000 ft to 6,000 ft); mean annual temperatures 24°C to 18°C (75°F to 65°F); zone of coffee.

Tierra fría: 1,800 m to 3,000 m (6,000 ft to 10,000 ft); mean annual temperatures 18°C to 12°C (65°F to 54°F); zone of middle latitude crops, such as potatoes, wheat, and barley, and middle latitude fruits.

Páramos and puna: 3,000 m to 4,300 m (10,000 ft to 14,000 ft); mean annual temperatures 12° C to 0 C (53° F to 32° F); zone of alpine pastures where cultivation is limited by low temperatures and frost.

Tierra nevada and tierra helada: Above 4,300 m (14,000 ft); mean annual temperature below 0°C (32°F).

Of course, in most areas these zones are not sharply defined, and they overlap considerably in vegetation and crop production. As an example, some coffee grown in the *tierra caliente* and some lowland tropical crops, such as bananas, grow in the lower *tierra templada*. In tropical Latin America maize is a significant crop from sea level to above 3,000 m (10,000 ft).

With cooler temperatures, most tropical highlands have a climatic advantage over lowlands, and where there is agricultural land, there is also dense population. Good examples of climatically favored tropical areas are the highland basins in Latin America and higher portions of the East African plateaus. On the other hand, in middle latitudes climates are more favorable at lower elevations and, generally, highlands are sparsely settled.

Other Climatic Classifications

Most of the numerous attempts to classify world climates have followed Köppen's concept of the planet as "a meteorological instrument" that integrates the various climatic elements (Thornthwaite 1948). Some classifications, such as Trewartha's modification of Köppen, merely revise certain values used by Köppen to conform to different ideas concerning desirable boundary location. The widely recognized classifications of C. W. Thornthwaite take a somewhat different approach. He published his first classification scheme in 1931; his second, published in 1948, was modified in 1955. Perhaps the most important contribution of the 1948 classification was the computation and use of potential evapotranspiration for determining the water budget.

Thornthwaite's Second Classification (1948, 1955)

Thornthwaite recognized that the simple evaporation factor contained in his original work was not an adequate expression of the more complex relationships between plants, soil, and climate. He changed Köppen's concept of a plant as a mere "meteorological instrument" that indicates the climate of an area, to one in which "vegetation is regarded as a physical mechanism by means of which water is transported from the soil to the atmosphere; it is the machinery of evaporation as the cloud is the machinery of precipitation" (Thornthwaite 1948,88). Central to Thornthwaite's new system are the concepts of *potential evapotranspiration* and the moisture index.

TABLE 2-8 The Thornthwaite Formula for Computing Potential Evapotranspiration (PE)

$$PE = 1.6\,L\left(10\,\frac{T}{I}\right)^{a}$$

Where: PE = the adjusted potential evapotranspiration for a month in cm
L = adjustment for hours of daylight and days in the month for a particular latitude, as obtained from table 2-9.
T = mean monthly temperature in °C
I = annual heat index derived from the sum of 12 monthly values determined from: $i = (T/5)^{1.514}$
a = an arbitrary exponent derived from:
$a = 6.75 \times 10^{-7}\,I^3 - 7.71 \times 10^{-5}\,I^2 + 1.792 \times 10^{-2}\,I + 0.49239$

Potential Evapotranspiration (PE)

Thornthwaite calculated the evaporating power of vegetation, assuming an unlimited moisture supply, and he called it *potential evapotranspiration* (PE). The PE factor measures the moisture loss from vegetation-covered soil if there is no shortage of soil moisture for the plants to transpire. It is a measure of the ability of a plant to extract water from soil through its root system and transport it to leaf surfaces where it evaporates into the atmosphere. If moisture supply is less than the plant cover's ability to transpire moisture, the actual evapotranspiration will be less than the potential; thus, a water deficit occurs.

Thornthwaite's formula for computing PE is empirical; that is, it is designed to produce values that have already been determined through observations, rather than being based completely on theoretical considerations. The formula for the monthly PE in centimeters, corrected for mean length of daylight period during any particular month, is given in Table 2-8. As is obvious, calculating PE without the aid of a computer or tables and graphs can be time consuming. Fortunately, the task is simplified by computer programs and tables for calculating PE and other factors of the water balance (Thornthwaite and Mather 1957), and published values of PE for many stations throughout the world.

The Moisture Index (I_m)

Thornthwaite defines a moist climate as one where precipitation (P) exceeds potential evapotranspiration ($P > PE$), and a water surplus exists; climate is dry if PE exceeds P and there is a moisture deficit ($P < PE$). At most stations in the eastern United States, winter months have a moisture surplus, whereas, summer months generally show a deficit; this is in spite of the seasonal pattern of rainfall showing a summer maximum at many stations. On

TABLE 2-9 Duration of Sunlight Correction (L) for Use in Computing PE.

Latitude	Jan.	Feb.	Mar.	Apr.	May	Jun.	Jul.	Aug.	Sep.	Oct.	Nov.	Dec.
50°N [a]	.74	.78	1.02	1.15	1.33	1.36	1.37	1.25	1.06	.92	.76	.70
45°N	.80	.81	1.02	1.13	1.28	1.29	1.31	1.21	1.04	.94	.79	.75
40°N	.84	.83	1.03	1.11	1.24	1.25	1.27	1.18	1.04	.96	.83	.81
35°N	.87	.85	1.03	1.09	1.21	1.21	1.23	1.16	1.03	.97	.86	.85
30°N	.90	.87	1.03	1.08	1.18	1.17	1.20	1.14	1.03	.98	.89	.88
20°N	.95	.90	1.03	1.05	1.13	1.11	1.14	1.11	1.02	1.00	.93	.94
10°N	1.00	.91	1.03	1.03	1.08	1.06	1.08	1.07	1.02	1.02	.98	.99
0°	1.04	.94	1.04	1.01	1.04	1.01	1.04	1.04	1.01	1.04	1.01	1.04
10°S	1.08	.97	1.05	.99	1.01	.96	1.00	1.01	1.00	1.06	1.05	1.10
20°S	1.14	1.00	1.05	.97	.96	.91	.95	.99	1.00	1.08	1.09	1.15
30°S	1.20	1.03	1.06	.95	.92	.85	.90	.96	1.00	1.12	1.14	1.21
35°S	1.23	1.04	1.06	.94	.89	.82	.87	.94	1.00	1.13	1.17	1.25
40°S	1.27	1.06	1.07	.93	.86	.78	.84	.92	1.00	1.15	1.20	1.29
45°S	1.31	1.09	1.07	.91	.82	.73	.80	.90	.99	1.17	1.24	1.34
50°S [a]	1.37	1.12	1.08	.89	.77	.67	.74	.88	.99	1.19	1.29	1.41

Source: Thornthwaite 1948, 93 (slightly modified).
Note: Value of 1.00 is for a standard month of 30 days, each with 12 hours of potential sunlight.
[a] Poleward of 50° use values for 50°N for Northern Hemisphere and 50°S for Southern Hemisphere.

an annual basis, all stations east of the Mississippi River, and the first tier of states west of the river, show a water surplus, and these are considered to be *humid* or *moist subhumid.*

The moisture index (I_m) is expressed as a percentage of P over PE. If P is greater than PE, the index is positive; if P is less than PE, the index is negative. The moisture index is calculated as follows:

$$I_m = 100 \frac{P - PE}{PE}$$

On the basis of the moisture index, we recognize five moisture regions:

Climatic Type	*Moisture Index (1955)*
A Perhumid	100 and above
B_4 Humid	80–100
B_3 Humid	60–80
B_2 Humid	40–60
B_1 Humid	20–40
C_2 Moist Subhumid	0–20
C_1 Dry Subhumid	−33.3–0
D Semiarid	−66.7 to −33.3
E Arid	−100 to −66.7

Humidity and Aridity Indexes

Thornthwaite's second classification uses two additional indexes: the *humidity index* (I_h) and the *aridity index* (I_a). Both are an expression of rainfall compared to need, expressed as surplus or deficiency. When rainfall exceeds need, as expressed by PE, and there is a surplus of moisture, the humidity index is:

$$I_h = 100\frac{\text{surplus}}{\text{need}} \text{ or } 100\frac{s}{n}$$

When rainfall is less than PE there is a deficit, and the aridity index is:

$$I_a = 100\frac{\text{deficit}}{\text{need}} \text{ or } 100\frac{d}{n}$$

These two indexes indicate seasonal variation in effective moisture supply. The magnitude of the aridity index in a humid climate, or the humidity index in a dry climate, indicates whether the humid climate experiences a marked dry season, or the dry climate a marked season of water surplus. The symbols used in the 1955 revision and their definitions follow:

Moist Climates (A, B, C_2)		*Aridity Index*
r	little or no water deficiency	0–10
s	moderate summer water deficiency	10–20
w	moderate winter water deficiency	10–20
s_2	large summer water deficiency	20+
w_2	large winter water deficiency	20+

Dry Climates (C_1, D, E)		*Humidity Index*
d	little or no water surplus	0–16.7
s	moderate winter water surplus	16.7–33.3
w	moderate summer water surplus	16.7–33.3
s_2	large winter water surplus	>33.3
w_2	large summer water surplus	>33.3

Note that the meaning of *s* and *w* depends on the first letter of the classification: If the climate is moist (A, B, or C_2), *s* indicates a moderate summer water deficiency; if dry (C_1, D, or E), *s* indicates a moderate winter water surplus. Thus, the symbols *s* and *w*, although defined differently, have the same meaning both in moist and in dry climates; they indicate the season of water deficiency. Note also that a subscript 2 added to *s* or *w* indicates a large deficit or surplus.

Thermal Efficiency

Thornthwaite believed that evapotranspiration is the best measure of thermal efficiency, since it is a function of temperature and duration of daylight;

consequently, potential evapotranspiration determines the thermal efficiency index directly. Thermal efficiency is represented by the following categories:

Climatic Type	*Annual PE (T-E Index)* (mm)
A' Megathermal	>1,114
B'_4 Mesothermal	1,114–997
B'_3 Mesothermal	997–855
B'_2 Mesothermal	855–712
B'_1 Mesothermal	712–570
C'_2 Microthermal	570–427
C'_1 Microthermal	427–285
D' Tundra	285–142
E' Frost	< 142

Thornthwaite recognizes the seasonal concentration of thermal efficiency by adding a fourth element. This lower-case letter symbol indicates the degree of T-E concentration in the warm half of the year:

Symbol	*Percent Summer Concentration of T-E*
a'	<48.0
b'_4	48.0–51.9
b'_3	51.9–56.3
b'_2	56.3–61.6
b'_1	61.6–68.0
c'_2	68.0–76.3
c'_1	76.3–88.0
d'	>88.0

Table 2-10 classifies selected stations according to the Thornthwaite system. An extremely wet climate such as Kribi, Cameroon, classifies as *AA'ra'*, and it is described as perhumid (*A*), megathermal (*A'*), with little or no water deficit (*r*), and no high concentration of thermal efficiency in summer (*a'*). Note that *a'* means essentially that there is no winter in the sense of temperature. In the Köppen scheme, Kribi would be *Af*. Many stations classified as Af in the Köppen scheme, however, are humid rather than perhumid, and are designated as *BA'ra'*. A humid continental (*Dbf*) station, such as Muskegon, Mich., classifies as $B_1B'_1rb'_1$; it is described as first-order humid (B_1), first-order mesothermal (B'_1), with little or no water deficiency (*r*), and a marked concentration of thermal efficiency in the summer months of between 61.6 and 68 percent (b'_1).

The moisture index is of paramount importance in the distribution of natural vegetation and in agriculture, and most maps using the Thornthwaite system are based on this factor alone (fig. 2-22). Researchers using

TABLE 2-10 Selected Climatic Stations Classified According to Thornthwaite's Second Classification as Modified in 1955

Station	PE mm	Summer need %	Annual Precip. mm	Surplus mm	Deficit mm	I_h	I_a	I_m	Thornthwaite Type	Köppen Type
Kribi, Cameroon	1,385	28.2	2,949	1,568	4	113.2	0.3	112.9	AA′ra′	Af
Gandajika, Zaire	1,298	26.1	1,384	229	143	17.6	11.0	6.6	$C_2A'wa'$	Aw
Alice Springs, Australia	1,162	43.8	245	0	917	0.0	78.9	−78.9	EA′da′	BWh
Memphis, Tenn.	923	52.5	1,232	404	95	43.8	10.3	33.5	$B_1B'_3sb'_3$	Caf
Athinai, Greece	913	51.5	382	0	531	0.0	58.2	−58.2	$DB'_3db'_4$	Csa
Birmingham, U.K.	607	49.1	674	99	32	16.3	5.3	11.0	$C_2B'_1rb'_4$	Cbf
Huron, S.D.	633	62.2	487	0	146	0.0	23.1	−23.1	$C_1B'_1db'_1$	Daf
Muskegon, Mich.	591	63.1	810	259	40	43.8	6.8	37.0	$B_1B'_1rb'_1$	Dbf
Irkutsk, U.S.S.R.	464	73.7	369	0	95	0.0	20.5	−20.5	$C_1C'_2dc'_2$	Dcf

Source: Carter and Mather 1966.

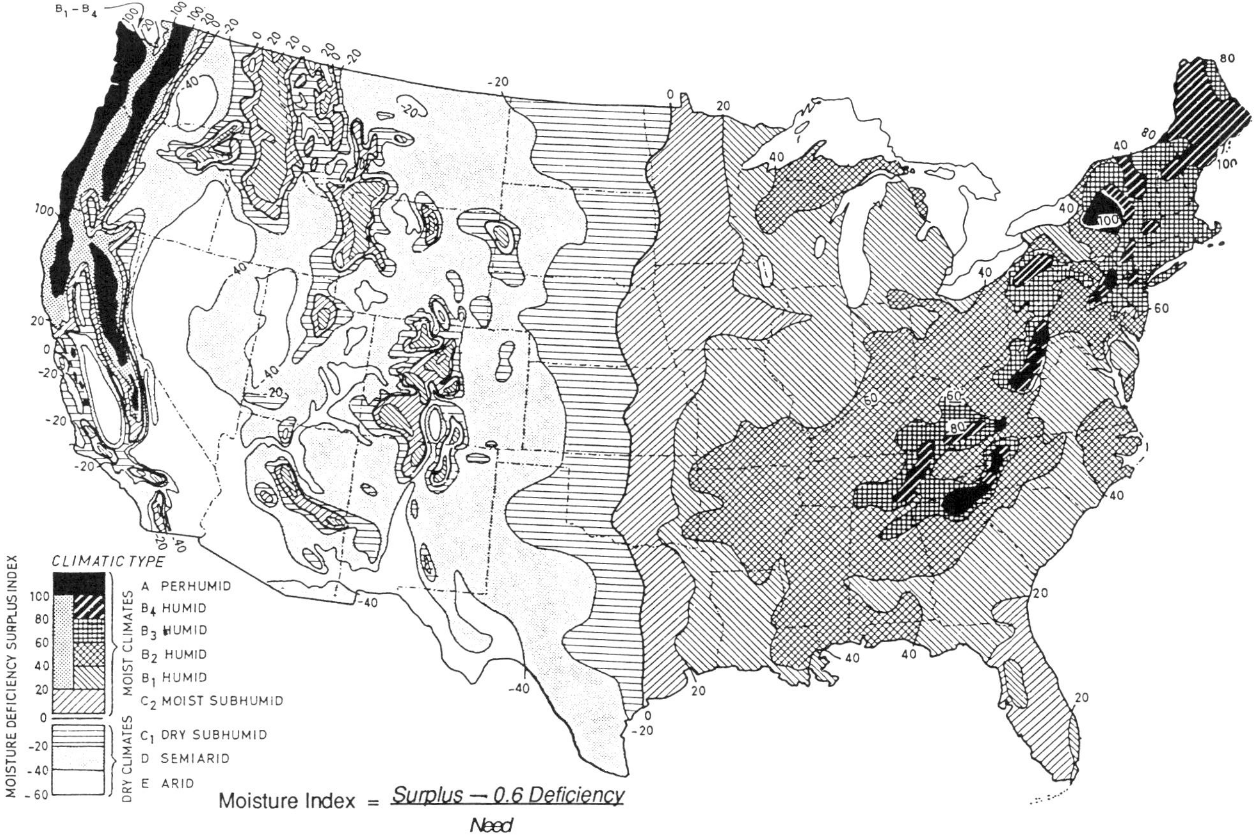

Fig. 2–22. Climates of the United States according to the 1955 Thornthwaite classification. (From Thornthwaite 1948, plate 1. Copyright © 1948 by the American Geographical Society. Reproduced with permission.)

the work of Thornthwaite have emphasized potential evapotranspiration and have computed data for numerous stations throughout the world. These data are available in a series of volumes produced by the Laboratory of Climatology (see part one of the bibliography). In chapter 3 we will study the relationship between PE and other factors in the water budget.

This wealth of waters might but seem to draw
From yon dark cave, but, son, the source is higher,
Yon summit half-a-league in air—and higher,
The cloud that hides it—higher still, the heavens
Whereby the cloud was moulded, and whereout
The cloud descended. Force is from the heights.

TENNYSON, *The Ancient Sage*

Chapter 3 THE WATER BUDGET

Withdrawal and replenishment of water at its source may be compared to debits and credits in a bank account, but unlike a money manager, the water budget manager must take into account quality as well as quantity. The earth's gross water budget has changed very little since the Earth-atmosphere-ocean system came into existence billions of years ago. Volcanic activity and geochemical reactions have added minor amounts to the system—the so-called *juvenile water*—and some has been lost through dissociation of water molecules in the upper atmosphere. Hydrogen atoms have escaped into space, leaving behind the O_2 molecules. But, essentially, these two processes are in balance, and the total quantity of water on Earth has varied little over most of geologic time.

If the total amount of water is constant, why the present concern with water shortages and surpluses? Why is the water budget often "in the red"? The answer lies in the distribution, both in time and in space, and increasingly in quality as well. Where water is plentiful, quality may be the primary concern but in dry areas, both limited quantity and quality may present problems. Of course, in both humid and dry areas excess water may cause flooding, waterlogging of soils, and crop damage.

Life requires large volumes of water. Water is essential for crops and animals, wildlife and recreation. It removes and renders innocuous the water-transported wastes of cities and towns, supplies the needs of industry, provides energy for hydroelectric power, and cools the generators that produce electricity from fossil fuels.

Water use is either *consumptive* or *nonconsumptive*. Changing water into other compounds by chemical reactions—for instance, *photosynthesis* of carbohydrates by plants (see chapter 5)—it is a consumptive use in the strictest sense. But most so-called consumptive uses do not destroy water, they just remove it from one hydrologic system into another. If we remove water from a stream to generate steam for driving a turbine, the water becomes vapor. Some of it enters the atmosphere and becomes lost to the stream from which it was taken; thus, the use is consumptive. But some of the steam condenses and returns to its original source, and this represents a nonconsumptive use.

By far the largest consumptive use of water is to grow crops, either by irrigation or rainfall; all other consumptive uses are minor by comparison. Generally, municipalities and industries are the largest nonconsumptive users. They rely on water to remove their transportable and degradable wastes. A study of water demand in Iowa shows that municipalities and farmsteads in the Des Moines River Basin used a quantity of water in one year equivalent to only 1.27 mm (0.05 in) of the average annual precipitation of 762 mm (30 in), prorated over the entire basin. A large share of this water was "used" but not used up, since it was returned to the streams as wastewater. In contrast, water demand for growing principal agricultural crops in the basin was about 579 mm (22.8 in) of precipitation during a single growing season (Iowa Natural Resources Council 1953, 23–25). Wild vegetation uses water at rates comparable to cultivated plants. This use is truly consumptive. Cycled from soil moisture storage through the plants, the water evaporates from leaves; thus, it is "lost" in the sense that it is not returned to the soil moisture supply from which it came. Only a very small part of the transpired water is retained in the plant tissues (see chapter 5).

With the exception of the amount needed for crops, livestock, and rural population, water demands mostly are concentrated at certain points within cities and towns, and at industrial sites. Although nonagricultural uses

TABLE 3-1 Distribution of Water in Earth Environments

Location	Volume in km^3 (mi^3)	Percent of Total
World ocean	1,322,000,000 (317,000,000)	97.2
Icecaps and glaciers	29,200,000 (7,000,000)	2.15
Groundwater (below water table)	8,400,000 (2,000,000)	0.62
Freshwater lakes	125,000 (30,000)	0.009
Saline lakes and inland seas	104,000 (25,000)	0.008
Soil moisture (above water table)	67,000 (16,000)	0.005
Atmosphere	13,000 (3,100)	0.001
Stream channels	1,250 (300)	0.0001
Total liquid water in land areas	8,630,000 (2,070,000)	0.635
World total	1,360,000,000 (326,000,000)	100.00

Source: U.S. Geological Survey, reported in Strahler 1975.

account for a relatively small percentage of total use, concentrated withdrawals at specific points may place heavy pressure on limited sources, and critical water shortages may develop even in humid areas. Of course it is in semiarid and arid regions that scarce water resources may severely limit human settlement. A world population map shows vast expanses of very sparse population in dry regions; where deserts and steppes do have dense human populations, as does the valley of the Nile in Egypt, dependable water supplies are available in spite of the dry climate.

Table 3-1 summarizes the distribution of water in earth environments. Oceans contain more than 97 percent of the water on earth, and glaciers contain another 2 percent. Although fresh water on land and in the atmosphere accounts for only 0.635 percent of the total, it is vital for land plants, animals, and humans. In this chapter, our primary concern will be with this component of the earth's water budget.

The Hydrologic Cycle

Water is moving constantly from one to another of the earth's great storage systems in the *hydrologic cycle* (fig. 3-1). A particular water molecule may require hundreds, thousands, even millions of years to move through all the systems, particularly if it should be removed temporarily from the cycle and

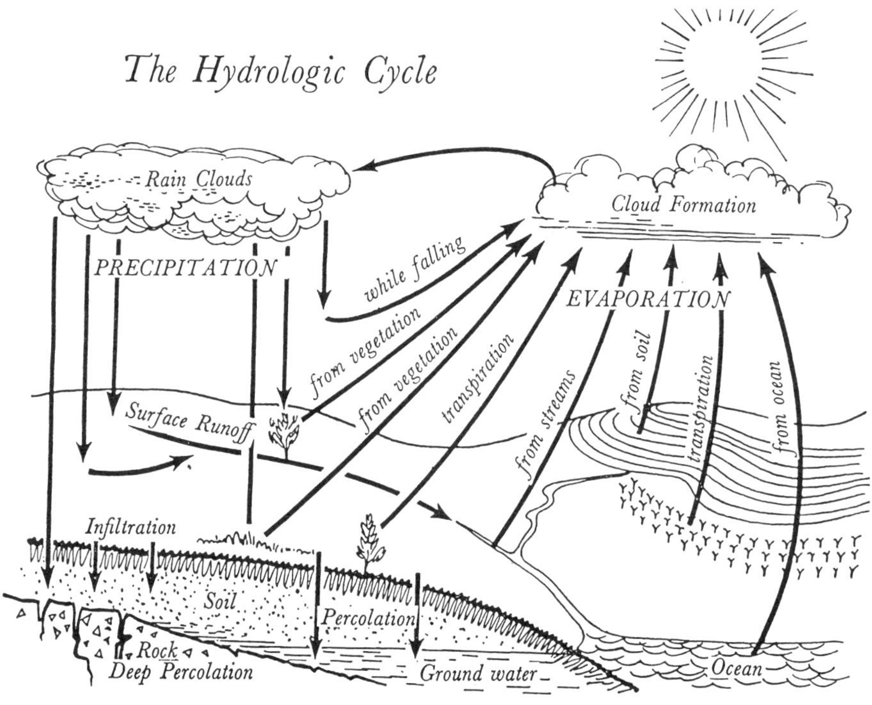

Fig. 3–1. Diagrammatic model of the hydrologic cycle. (From USDA, 1955, 42.)

stored in glacial ice or in deep groundwater reservoirs, or if it becomes a part of the slow-moving deep ocean bottom water. Other molecules may move from ocean to atmosphere to ocean again in a relatively short time, measured in weeks, days, or even hours.

Water on the Land

Water falling on land surfaces as rain or snow may infiltrate the ground and become a part of the *soil moisture* or *groundwater* reservoir, or it may run off over the surface and become part of the stream flow (*surface runoff*). Some of it infiltrating below the land surface eventually may be discharged into streams where it becomes a part of surface runoff. Although all surface and subsurface water systems are interconnected in a relatively complex way, we shall consider them separately.

Infiltration and Subsurface Water

If the land surface is relatively dry, rainfall or melting snow may infiltrate below the surface. If rainfall is prolonged, some of the infiltrating water will percolate downward toward the water table (*gravity water*) and temporarily saturate the soil by filling all pore spaces between soil particles. When precipitation ceases, the gravity water drains to the groundwater zone, leaving some water adhering to soil particles by molecular and capillary attraction (fig. 3-2). The moisture retained in the upper zone is the *soil moisture*.[1] It does not occupy all of the empty space between the soil particles, but shares this space with air—thus, the soil moisture zone sometimes is called the *zone of aeration*. This water from the zone of aeration supplies the soil moisture needs of most plants.

The *water table* marks the upper boundary of the saturated zone; here, groundwater occupies all empty space. This zone provides water for wells and discharges water to streams, springs, and lakes. Most plants cannot use water from the groundwater zone, but must depend on soil moisture (see chapter 5).

Soil Moisture

Although the amount of water stored as soil moisture at any one time represents a relatively small part of the total water budget, unquestionably, it is the most widely used water source on the land surface. Even irrigation is just a means for creating optimum soil moisture conditions for growing crops. At any one time the stored soil moisture is equivalent to a layer 117 mm (4.6 in) over the earth's total land surface (Ackerman and Löf 1959), an amount insufficient for abundant plant growth. Of course, soil

[1] *Soil moisture*, for convenience, is defined as moisture available for plant growth; this definition includes some of the gravitational water used by plants called *phreatophytes*. Most soil moisture, however, is *capillary water*, which is held in a continuous film around soil particles and moves in response to capillary forces.

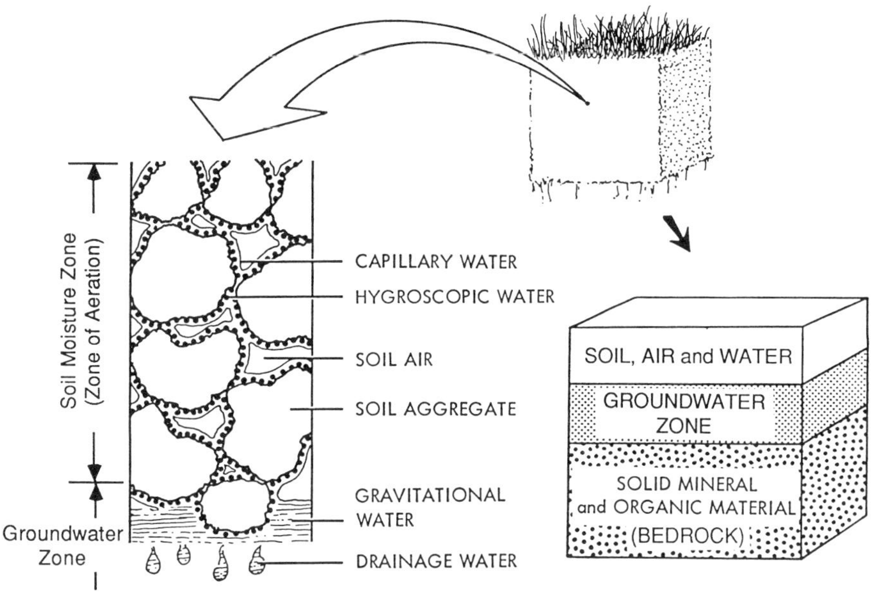

Fig. 3–2. Subsurface water zones. The soil moisture zone, sometimes called the zone of aeration, contains capillary and hygroscopic water, in addition to space occupied by air. In the groundwater zone, gravitational water (moving under the pull of gravity) occupies all spaces between soil particles, and air is excluded. Most plants use capillary water from the aeration zone. In a good silt loam, illustrated in the diagram, solid material occupies only half the volume, with air and water each occupying about one-fourth. (From K. C. Berger, *Sun, Soil, and Survival,* Norman: University of Oklahoma Press, 1972, 8.)

moisture availability varies greatly from place to place and from time to time, depending on the amount and frequency of replenishment and the time of year it is available for plant growth.

The soil moisture component of the water budget of a particular climatic station is calculated from data for precipitation, evaporation, transpiration, runoff, and soil moisture storage. These procedures are described in a later section of this chapter. Soil moisture availability for plants depends on precipitation, solar radiation, soil and air temperature, and wind. Different plants have different moisture requirements, but optimum soil moisture conditions are seldom present throughout the growing season. In an average year the natural supply of soil moisture is less than optimum for 95 percent of the cultivated area of the United States (Ackerman and Löf 1959, 16). The success of supplemental irrigation for increasing crop yields in humid areas, particularly on droughty soils, demonstrates the inadequacy of precipitation for maintaining optimum soil moisture levels.

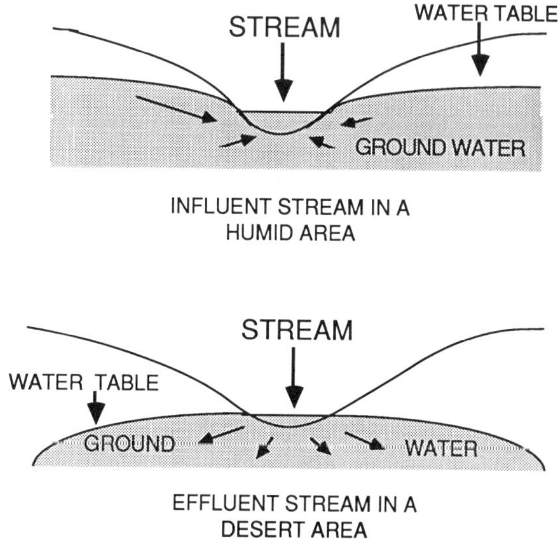

Fig. 3–3. Configuration of the water table in humid and dry areas in relation to streams. In humid areas the water table slopes toward the stream and helps maintain base flow in periods of low surface runoff; in dry areas it slopes away from the stream and feeds the groundwater reservoir. If the water table drops below the level of the stream bed, the stream will cease to flow.

Groundwater

Groundwater accounts for nearly 98 percent of all the fresh water in the water budget, except for water stored in icecaps and glaciers. Most is in rocks within 3,800 (12,500 ft) of the surface, potentially available for human use. The most accessible groundwater is at depths less than 760 m (2,500 ft), representing about 44 percent of fresh water stored in the Earth-atmosphere system.

The upper boundary of the groundwater zone is the *water table*, defined as the upper limit of the saturated zone—the level to which water will rise in a well, without pumping. The water table intersects the surface at the margins of lakes and flowing streams, in springs, and in swamps and marshes (fig. 3-3). Below the water table all open spaces, or voids, are filled with water. Earth materials that contain such voids are *permeable*, and the amount of empty space available for water storage represents their *porosity*. The ease with which water can move through materials depends on *permeability*, which in turn reflects the size of the pore spaces and the degree to which they are interconnected. *Aquifers* are waterbearing subsurface formations that have high porosity and high permeability.

The capacity of subsurface materials for storing and releasing water varies greatly among geological formations. Aquifers may be classified broadly as *unconsolidated*—loose materials, such as gravel, sand, and silt—and *consolidated,* or "solid" rocks. The most productive unconsolidated aquifers are coarse gravel; ranking next are uniformly sized coarse sands; fine sands and silts are less productive. Of consolidated aquifers, weakly cemented sandstones may have large storage capacities and relatively free movement of water within; these constitute some of the best aquifers. Other sandstones are so tightly cemented that they yield little water. Shales are almost impervious to water movement and release very little water, but they may form cap-rock seals (*aquitards*) for some artesian aquifers (fig. 3-4). Geologic formations composed of rocks having inherently low porosity and permeability may nonetheless contain large quantities of stored groundwater. These include limestones and many igneous and metamorphic rocks with joints, fractures, faults, and solution channels containing water.

Origin, Discharge, and Recharge of Groundwater

The origin of water stored below the surface of the ground is precipitation that infiltrated through the soil and subsoil to accumulate in water-bearing formations. If soil is dry or only partially saturated, precipitation infiltrates by molecular and capillary forces analogous to the action of a sponge soaking up water. If soil is saturated, gravity governs the penetration of water through the soil-water interface. This addition to the groundwater system is termed *recharge*.

Streams may replenish a water-bearing formation if the adjacent water table is lower in elevation than the stream, as is often the case in dry areas. Such streams are *influent streams*. In humid areas, however, most perennial streams receive water from groundwater runoff and are *effluent* (fig. 3-3). A stream that under natural conditions is aquifer fed may become influent in certain reaches if the adjacent water table is lowered by the operation of wells or infiltration galleries. Such aquifer recharge is called *induced infiltration*.

Direct Groundwater Use by Plants (Phreatophytes)

Although most plants depend on soil moisture stored above the water table, certain plants have the ability to extract water from the groundwater zone below the water table, or from the capillary fringe just above it. These are the *phreatophytes* (Greek "well plants"), which are "natural wells" with the ability to lift water, sometimes from great depths, to the surface and transpire it through their leaves. In desert regions, phreatophytes consume great volumes of scarce water, abstracting it from alluvial aquifers along streams and drainage ways. Most common of these in the western United States are saltgrass (*Distichlis spicata*), greasewood (*Sarcobatus vermiculatus*), mesquite (*Prosopis juliflora*), salt cedar (*Tamarix gallica*), cottonwood (*Populus* spp.), and willow (*Salix* spp.). Some cultivated plants, such as alfalfa, also are phreatophytes.

A study by the U.S. Geological Survey (Meyers 1962) estimated the annual loss from fifty reservoirs and lakes with the highest evaporation

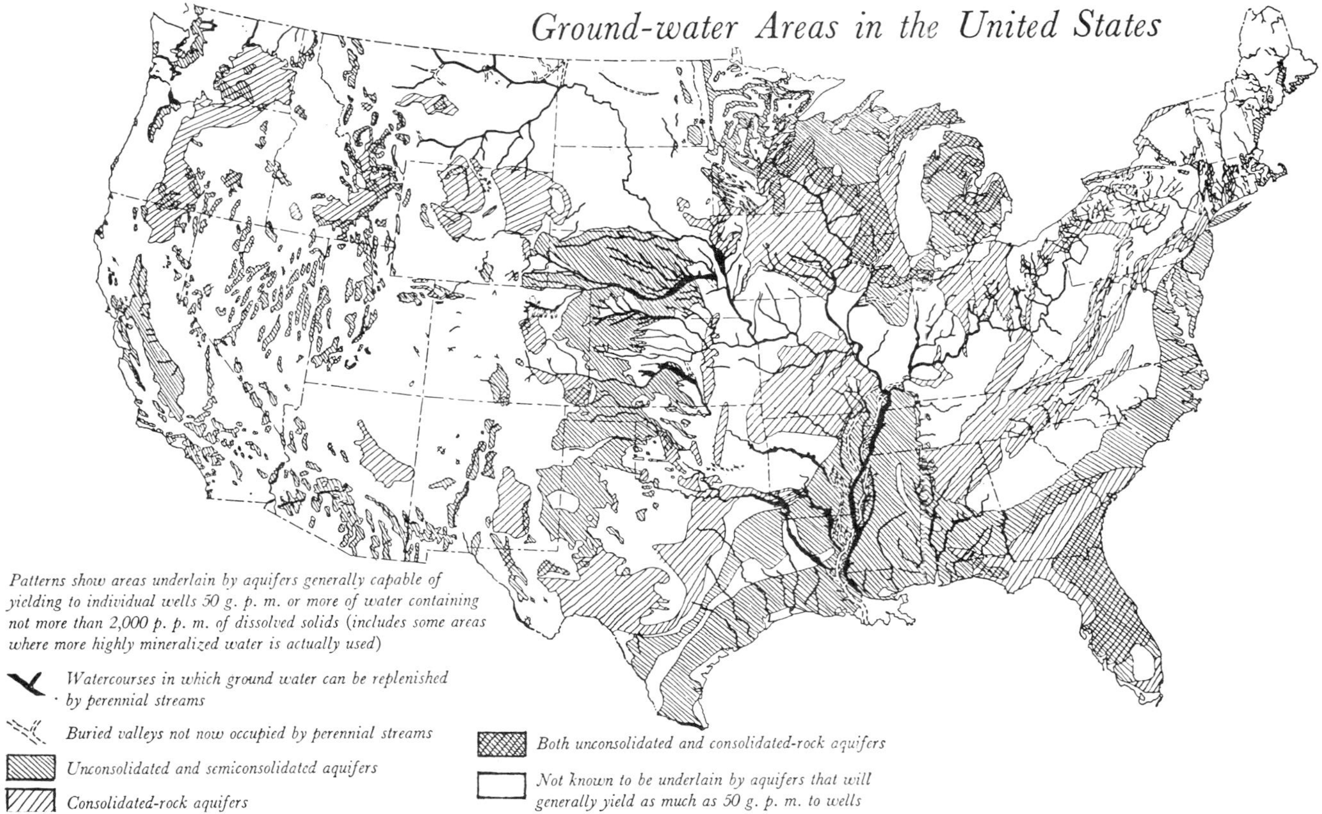

Fig. 3–4. Groundwater areas of the United States. (From USDA, 1955, 66.)

at slightly more than 9 million *acre feet*.[2] The same survey estimated that transpiration by phreatophytes in the arid West was 25 million acre feet. Phreatophytes may be beneficial in some areas because they protect the soil from erosion, thereby reducing sedimentation of reservoirs. For the most part they are undesirable because they are viewed as "wasting" a valuable resource.

Springs

A *spring* discharges groundwater at the surface in sufficient quantity to produce a well-defined channel. Lesser surface discharge may produce wet spots (*seepage*). In the past, the origin of springs was considered mysterious, and the curative powers of their waters were legendary. Even in the twentieth century, health spas have been established at the site of springs (the term *spa* comes from the name of a town in Belgium that was famous for its medicinal mineral springs). The water in some springs is hot, since it is heated by contact with underground thermal sources, sometimes producing a special and spectacular type of spring, called a *geyser*.

Springs often are small and intermittent, supplying only limited and undependable amounts of water. However, a number of areas, with underlying formations of cavernous limestones or basaltic lavas, have springs with enormous flows. In the United States, the U.S. Geological Survey (USGS) identifies sixty-five springs with average yields of 100 cubic feet per second (cfs) or more, several exceeding 500 cfs. A discharge of 100 cfs equals about 64,600,000 gallons a day, an amount sufficient to supply the water needs of a large city (fig. 3-5). These mammoth springs occur mostly in northern Florida, the Ozark region of Missouri, central Texas, the Upper Snake Plains of Idaho, western Oregon, northern California, and central Montana. Most large springs in the eastern United States occur in limestone bedrock, whereas in the northwestern portion of the country they are associated mostly with porous basaltic lavas.

Wells

Except in a few areas where large springs yield significant quantities, wells supply groundwater for human use. Wells are classified according to the hydrologic conditions of their aquifers. These aquifers fall into two general categories: (1) *surface aquifers* consisting of unconsolidated materials that mantle the bedrock; and (2) *bedrock aquifers*. The principal surface aquifers are the following: sands and gravels associated with major streams (*alluvial aquifers*); sands and gravels buried in preglacial bedrock valleys (*buried channel aquifers*); and localized sand and gravel bodies within glacial drift (*glacial drift aquifers*). Bedrock aquifers consist principally of sandstones and limestones. Figure 3-4 locates the major aquifers in the United States.

[2] One acre foot is the amount of water required to cover an acre of land to a depth of 1 ft; therefore, it is equal to 12 in (304.8 mm) of precipitation; it consists of 43,560 ft^3 (1,233.483 m^3).

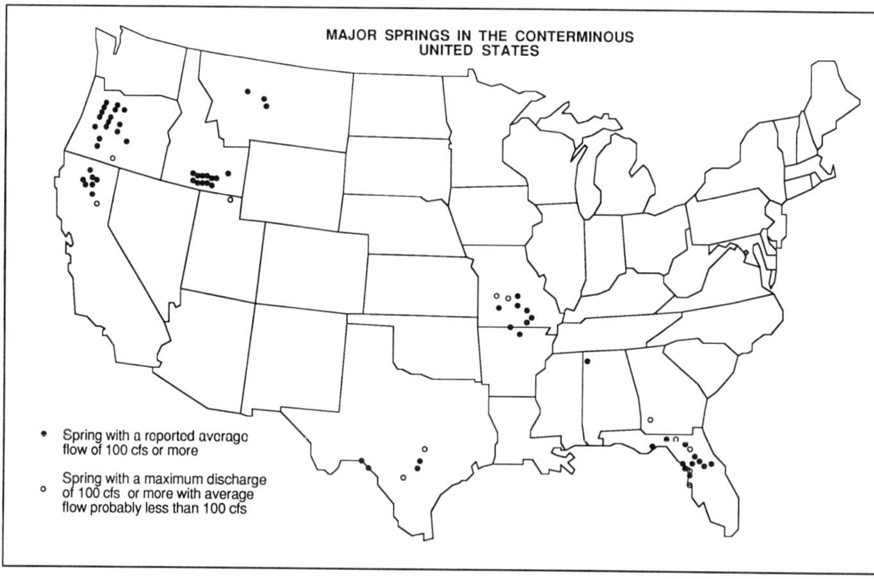

Fig. 3–5. Major springs in the conterminous United States. In eastern states, most large springs occur in karst topography where many caverns and extensive joint systems in limestone provide channels for underground drainage. In the northwestern states, caverns and tubes in basaltic lava flows serve the same purpose. (After Meinzer 1927.)

Groundwater is stored and transmitted under two hydraulic conditions—*artesian* (confined) and *water table* (unconfined). Overlying, impermeable formations (*aquitards*) confine water under pressure (*hydrostatic pressure*) in an artesian aquifer. When a well penetrates such an aquifer, water will rise above the bottom of the confining bed (fig. 3-6). Water enters confined aquifers at a distant point where the aquifer first dips below the surface, or where the confining bed ends underground. The *piezometric surface* is the level to which water under hydrostatic pressure will rise in wells. If this imaginary surface is above ground level, a flowing artesian well results. An unconfined aquifer is one in which a water table exists, and water levels in wells correspond to the undulating form and position of the water table. In most areas, the water table generally follows the land surface, rising beneath hills, declining toward stream valleys, and intersecting the surface in streams, lakes, and ponds (fig. 3-3).

When the pumping rate from an unconfined aquifer exceeds the rate at which water enters the well—and this is usually the case—the water level in the well drops (*drawdown*), and a cone-shaped depression (*cone of depression*) is created in the water table around the well (fig. 3-7). More rapid

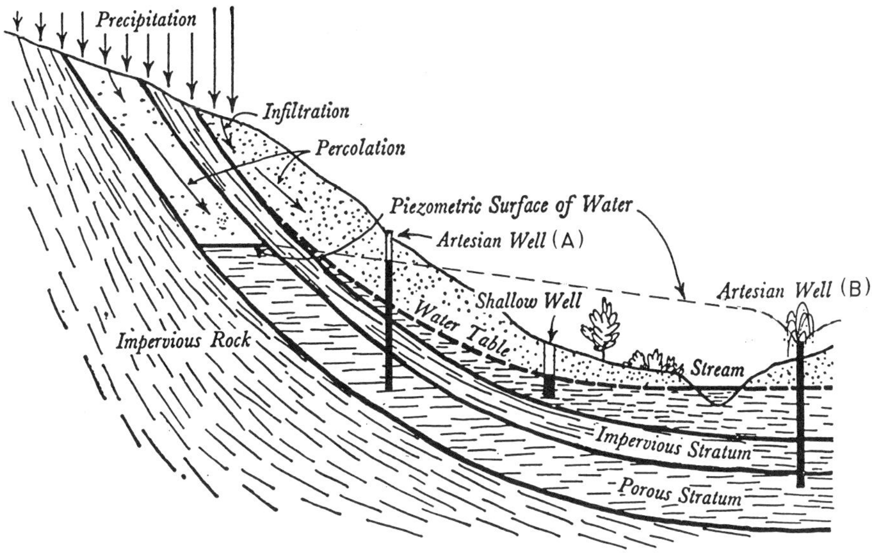

Fig. 3–6. Artesian and water table aquifers. In the shallow well, water rises to the water table if the well is not pumped. In the artesian well *A,* water pressure causes water to rise to the piezometric surface which lies below the surface. In artesian well *B,* the piezometric surface is aboveground and the well flows, creating a cone of depression in the piezometric surface. (Modified from USDA, 1955, 50.)

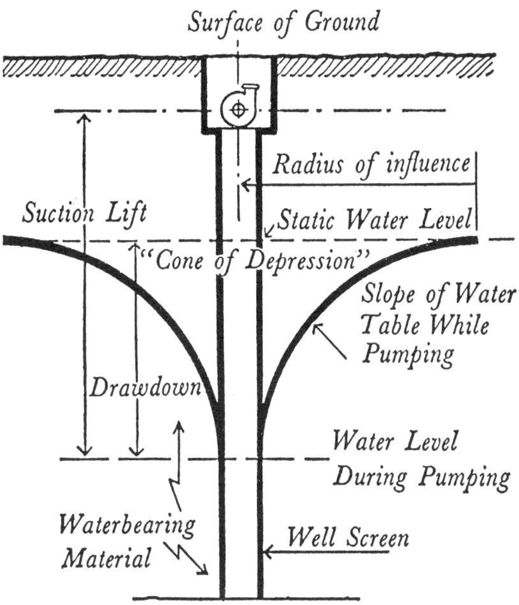

Fig. 3–7. Hydraulics of wells. During pumping a cone of depression forms in the water table or piezometric surface around the well; thus, the hydraulic gradient is increased toward the well and inflow of water increases. If the pumping rate increases, the radius of influence and slope toward the well increases, until the cone of depression intersects the bottom, and the well is pumped dry. If pumping ceases, the well recovers, and water returns to the static level. (From USDA 1955, 336.)

pumping produces a steeper gradient toward the well, until a maximum-yield pumping rate is reached. Further drawdown will fail to increase yield, and continued withdrawal at an excessive rate could pump the well dry. Pumping from artesian wells produces a similar cone of depression in the piezometric surface. Cones of depression around heavily pumped municipal, industrial, or irrigation wells may extend as far as 16 km (10 mi) from a single well. Where a number of wells are being pumped in the same area, their cones of depression may intersect and lower the general level of the water table.

When pumping ceases, the cone of depression gradually disappears, and water returns to the static level. Since water moves very slowly through permeable materials—usually a few feet per day—recovery may be slow, particularly where overlapping cones of depression have lowered the level of the regional water table. In the southern Great Plains, where the famous Ogalalla Formation has furnished irrigation water at a rate exceeding the annual aquifer recharge, water table levels have fallen drastically (fig. 3-8). Such gross overpumping is called "mining" the aquifer, as opposed to pumping at the *sustained yield* level.

Stream Flow

Precipitation flowing across the land and collecting in natural and artificial drainage channels is *storm runoff*, which occurs only during and after periods of heavy precipitation. This runoff, supplemented by discharge of groundwater, supplies water that maintains stream flow. Although stream flow is largely beneficial, flooding may cause severe damage when surface runoff exceeds channel capacities.

A considerable portion of *perennial* (permanent) stream flow derives from shallow groundwater reservoirs. This flow, although relatively small in late summer when groundwater levels are low, is the main source of stream flow during periods of minimal surface runoff. *Ephemeral* streams, which carry water only after heavy rainfall, and seasonally *intermittent* streams, which are dry during periods of minimum runoff, receive little or no groundwater outflow. In some deserts, water actually flows away from streams and recharges ground water supplies (fig. 3-3).

Stream flow records, collected at strategic points over many years, reflect a stream's discharge characteristics and provide essential data for water use planning and control. Most records are made at stream gaging stations operated by the U.S. Geological Survey in cooperation with other federal and local agencies.

Stream flow is measured in terms of water volume passing a gaging station in cubic feet per second (cfs) or cubic meters per second (cms). This quantity is called *discharge* and appears in equations in hydrology as Q. If a

Fig. 3–8. Air travelers often are intrigued by great circles of green on the landscape of the Great Plains: fields of corn, alfalfa, and other crops irrigated by automatic center-point irrigation systems. Each circle has a shallow well in its center and a rotating sprinkler system that delivers water to the field. Water comes from the Ogalalla Formation, a vast aquifer extending from Texas north to the Dakotas. Each circle in the photo has a diameter of one-half mile and an area of about 51 hectares (126 acres). Overpumping of the Ogalalla aquifer has lowered water levels and increased the pumping costs to a prohibitive level, causing farmers in some areas to revert to dry farming. *U.S. Soil Conservation Service photo.*

stream channel had vertical sides and a flat bottom, giving it a box-like cross section, the discharge could be calculated by:

$$Q = wdv$$

Where Q is the discharge in *cfs*, w is the width of the stream; d is the depth; and v is the velocity at which the stream flows. If a stream were 10 ft wide and 1 ft deep, and water were moving at a rate of 1 ft per sec, the discharge would be:

$$Q = 10 \times 1 \times 1 = 10 \textit{ cfs}$$

Most stream channels are rather irregular, and the simple linear relationships used in the formula just given would be a very rough approximation.

Fig. 3–9. A stream gaging station in North Carolina. The artificial weir controls the channel cross section, and water level is measured by instruments in the gage house. Water height and velocity of flow determine the discharge of water from the watershed above the gage. (*U.S. Forest Service photo by L. J. Prater, August 1952.*)

A more accurate determination of discharge is:

$$Q = AV$$

where Q is the rate of flow or discharge, A is the cross-sectional area of the stream, and V is the average velocity of the stream. Not only does the cross-sectional area of a stream channel change with water depth, but velocity increases with depth and volume.

Permanent gaging stations are located at points on a stream where the channel cross section is likely to be stable. Such places include solid bedrock channels, bridge sections, or artificial weirs (fig. 3-9). A *rating curve* must be developed for a gaging station by repeated measurement of stream velocity and the cross-sectional area of the channel at different water levels. These measurements determine the discharge of water flowing past the station when the water level, as indicated by the gage, is at a certain height.

Discharge records of a stream over time are a good indicator of the water balance of the drainage area above the gaging station. Ultimately, water received on the watershed as precipitation—less the amount lost through evapotranspiration or to deep underground storage—must flow past the gaging station. Annual stream discharge usually is expressed in equivalent inches or millimeters of runoff spread over the entire drainage basin. Runoff sometimes is given in millions of gallons per day per square mile (mgd/mi^2), or in millions of liters per day per square kilometer (mld/km^2). Figure 3-10 shows a map of average annual stream runoff in inches for the United States.

Stream runoff characteristics reveal much about the climate and geology of a drainage basin. Knowing the maximum and minimum flows one can expect is important in water resources planning. Maximum flows are a measure of flood potential, and minimum flows indicate limitations on beneficial use of streams for irrigation, hydroelectric power generation, and other uses. The *flow duration curve* is a graphic technique for depicting the flow characteristics of a stream. This type of graph plots discharge against the percentage of time when any given flow rate was equaled or exceeded during the period of record. By plotting discharge in volume per second per square mile, it is possible to compare flow characteristics of river basins of different sizes.

Figure 3-11 plots flow duration curves for four stream gaging stations in northeastern Iowa. Here, the high flows are remarkably similar, whereas periods of minimum runoff (*base flow*) diverge. The upper end of the curve indicates flood potential, and the lower end indicates reliability of stream flow during dry periods. The low flow characteristic of the Maquoketa River is the best of the four streams, and that of the Wapsipinicon River above Independence is the poorest. Farther downstream at De Witt, the lower Wapsipinicon basin has much better base flow than the upper. The good low flow characteristics of many northeastern Iowa streams depend on large groundwater discharge from limestone bedrock supporting a high base flow. Upstream from Independence, the Wapsipinicon flows through glacial till where limited supplies of shallow groundwater maintain a low base flow.

Evapotranspiration and the Water Budget

Of the average 762 mm (30 in) of precipitation received annually in the conterminous United States, about 546 mm (21.5 in), or 72 percent, returns to the atmosphere by evaporation from water surfaces and wetlands, and by transpiration from vegetation. The amount of loss, however, and its percentage of the precipitation received vary greatly from area to area, as

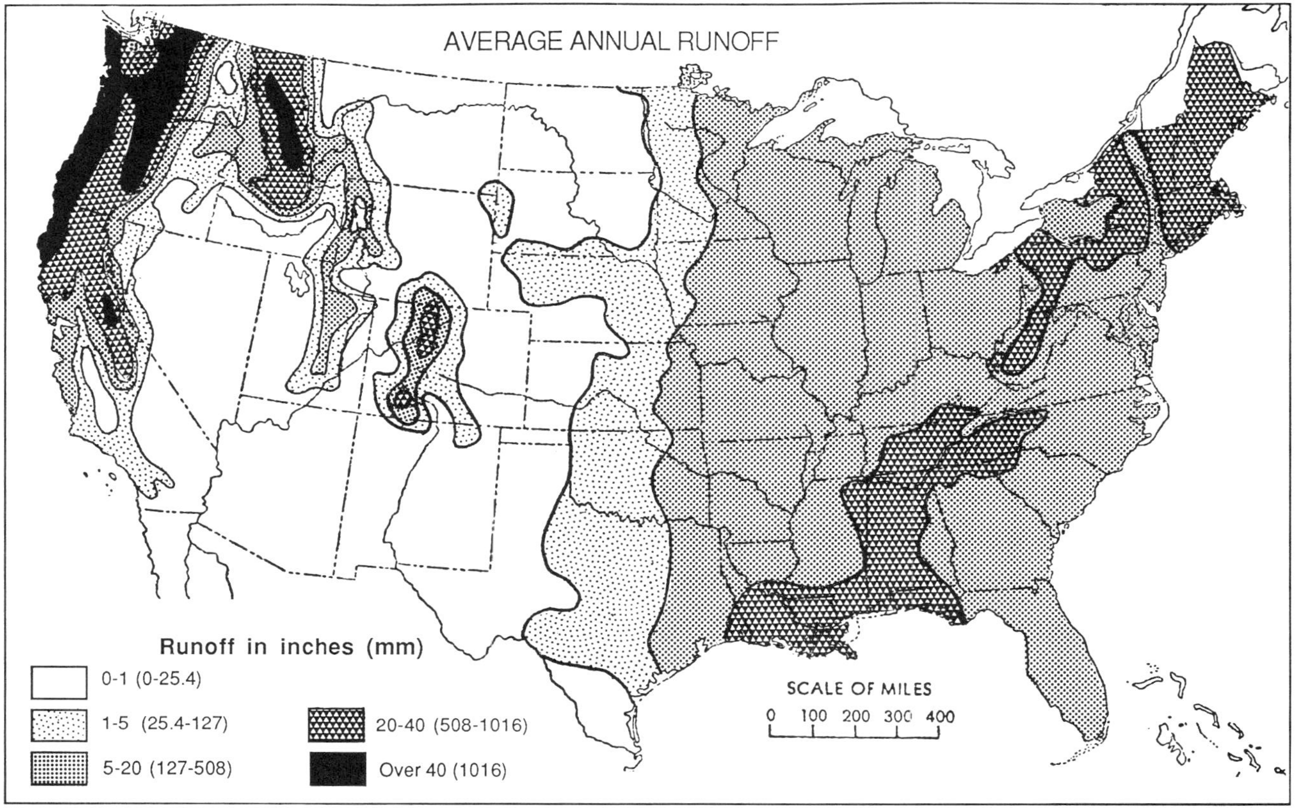

Fig. 3–10. Average annual runoff in the United States. (Redrawn from USDA 1981, 192.)

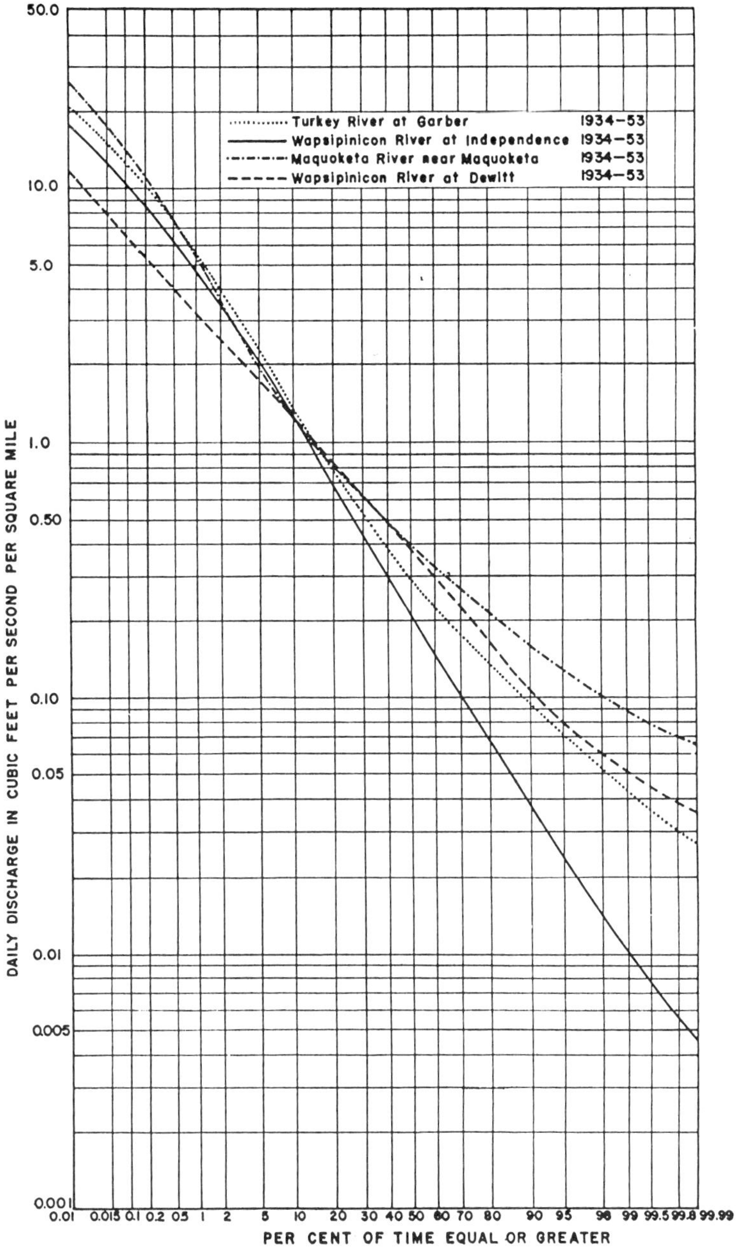

U. S. Geological Survey Data

Fig. 3–11. Flow-duration curves for four gaging stations on streams in northeast Iowa. (Iowa Natural Resources Council, *Water Resources and Water Problems, Northeastern Iowa River Basins, Des Moines: INRC,* 1955, 27.)

TABLE 3-2 Ratios of Potential Evapotranspiration (PE) to Precipitation (P) at Selected Stations

Station	PE (mm)	P (mm)	PE/P (%)
Northeast			
Portland, Me.	593	1,077	55
Southwest			
Miami, Fla.	1,308	1,547	85
Midwest			
Decatur, Ill.	740	931	79
Middle South			
Memphis, Tenn.	923	1,232	75
E. Margin Great Plains			
Fargo, N. D.	601	538	112
Dallas, Tex.	1,035	879	118
Desert Southwest			
Yuma, Ariz.	1,103	89	1,239
Mediterranean W. Coast			
Los Angeles, Calif.	840	378	222
Pacific Northwest			
Seattle, Wash.	683	840	81
Aberdeen, Wash.	622	2,073	30

Source: J. R. Mather, *Average Climatic Water Balance Data of the Continents, Part VII: United States.* Publications in Climatology, XVII, 3. Centerton N.J.: C. W. Thornthwaite Associates Laboratory of Climatology, 1964.

shown in table 3-2. In chapter 2, the concept of potential evapotranspiration (PE) was used to help describe and classify climates; here, we will consider it as an important element of the water budget.

PE is a measure of estimated water loss through evapotranspiration from land surfaces, assuming an unlimited moisture supply. PE depends on the energy available for effecting change of state from liquid water to water vapor, a factor closely related to air temperature and solar radiation.

Reliable values for evapotranspiration are much more difficult to obtain than are measurements of precipitation and stream flow. Evaporation loss causes falling water levels in lakes and streams, but inflow and outflow of surface water and groundwater makes them unreliable evaporation indicators. Instead, other methods and devices are used to estimate loss of water through evapotranspiration, including direct measurements using *evaporation pans* and *lysimeters*, and indirect computations based on available meteorological data. Estimates of PE may be made by empirical methods based on a limited number of actual measurements of PE correlated with more easily measured climatological elements, such as temperature and solar radiation. Thornthwaite's method, described in chapter 2, uses mean monthly temperature, hours of daylight, and an annual heat index based on mean monthly temperatures—all factors influencing the energy available for evapotranspiration.

Evaporation Pans

Hydrologists base estimates of evaporation from free water surfaces, such as reservoirs, on measurements of evaporation loss from shallow metal pans. The most widely used in the United States is the Class A Land Pan, employed by the National Weather Service. Pan evaporation, as a general rule, will give reasonable estimates of evaporation loss from lakes when multiplied by a factor of 0.70 to 0.80. Table 3-3 compares pan and lake evaporation based on pan data with PE values calculated by Thornthwaite's equations. Pan evaporation is greater than calculated PE in all parts of the country, and estimated lake evaporation is fairly close to PE values in humid areas, although it is much greater in drier areas. Only in the Pacific Northwest does computed PE exceed estimated evaporation loss from lakes.

Lysimeters

Lysimeters are soil-filled containers in which plants grow. They can be used to monitor additions of water by irrigation and precipitation and loss through runoff and evapotranspiration. Change in soil moisture content is measured by weighing the lysimeter. Even small changes can be measured; for example, the lysimeters at the Coshocton Watershed experimental area in Ohio can measure and record automatically additions and losses of moisture to the plant-soil system of as little as 0.24 mm (0.01 in). Lysimeter records at Coshocton indicate that crops transpire 80 percent of the moisture infil-

TABLE 3-3 Comparison Between Mean Annual PE, as Calculated by the Thornthwaite Method, and Mean Annual Class A Pan Evaporation and Mean Annual Lake Evaporation at Selected Stations in the United States

Station	PE (mm)	Class A Pan (mm)	PE/Pan (%)	Lake (mm)	PE/Lake (%)
Portland, Me.	593	762	78	610	97
Baltimore, Md.	746	1,143	65	914	82
Miami, Fla.	1,308	1,615	81	1,320	99
Madison, Wis.	647	1,016	64	762	85
Memphis, Tenn.	923	1,397	66	1,014	91
New Orleans, La.	1,118	1,651	69	1,219	92
Fargo, N.D.	601	1,016	59	737	82
Omaha, Neb.	734	1,448	50	1,016	72
Oklahoma City, Okla.	888	2,159	41	1,524	58
Fort Worth, Tex.	1,037	2,023	50	1,448	72
Yuma, Ariz.	1,103	3,226	34	1,981	56
Los Angeles, Calif.	840	1,524	55	1,168	72
Seattle, Wash.	683	762	90	584	117

Source: For PE data, same as for table 3-2; for pan and lake data: U.S. Department of Commerce (ESSA), *Climatic Atlas of the United States,* Washington, D.C.: U.S. Government Printing Office, 1968.

trating the soil, while the remaining 20 percent percolates down and out of the soil (USDA 1955, 685).

Empirical Computations of PE

Numerous attempts have been made to devise empirical formulas for calculating PE by comparing actual measurements by evaporation pans and lysimeters with meteorological conditions, particularly with air temperature and solar radiation. The concept of energy or heat balance is the basis of this approach. The rather complex balance between incoming and outgoing radiation is *net radiation,* which we express as:

Net radiation = insolation - reflection and reradiation

At any given time, net radiation will vary greatly with atmospheric and surface conditions, but during daylight hours it averages between 60 and 70 percent of the insolation. Distribution of the net radiation within the environment is:

Net radiation = evapotranspiration + soil heat + air heat + photosynthesis + miscellaneous factors

When unlimited water is available, evapotranspiration uses about 80 percent of net radiation, but if water is not available, a larger share of the net radiation raises the soil and air temperature. Notwithstanding, its great importance to life photosynthesis uses very little net radiation. Net radiation, therefore, is a good relative measure of actual evapotranspiration if water is available, or of potential evapotranspiration if water is in short supply (Shaw and Burrows 1966).

A number of methods exist for computation of PE, all giving values that may differ quantitatively but are of the same order of magnitude. Because data are available from a large number of stations throughout the world, we will use the Thornthwaite method for determining the water budget.

Computing a Water Budget

Table 3-4 illustrates two water budgets from which it is possible to determine moisture deficiency and moisture surplus, and give an estimate of the average runoff. PE is computed by the Thornthwaite formula given in chapter 2; P is from weather records; and the capacity of the soil to store moisture is estimated from local soil characteristics and vegetation cover. Figure 3-12 displays graphically the water budgets of the two stations.

Soil moisture storage (ST) records, unlike other factors considered, are not readily available. The storage capacity of soils (field capacity) will vary from a few millimeters in shallow sandy soil to more than 400 mm in deep, well-aerated silt loam (tables 3-5 and 3-6).

TABLE 3-4 Water Budget Computations in Millimeters for Berkeley, Calif. and Seabrook, N.J. (300 mm depth of water stored in soil layer at field capacity)

	Month												
	J	F	M	A	M	J	J	A	S	O	N	D	Year
Berkeley, Calif.													
Potential Evaporation	26	32	45	56	71	84	88	82	75	63	43	28	693
Precipitation	130	112	94	37	24	5	1	1	13	31	62	106	616
Difference	104	80	49	−19	−47	−79	−87	−81	−62	−32	19	78	−77
Storage change	104	22	0	−19	−41	−56	−47	−33	−19	−8	19	78	
Moisture storage	278	300	300	281	240	184	137	104	85	77	96	174	
Actual evaporation	26	32	45	56	65	61	48	34	32	39	43	28	509
Moisture deficit	0	0	0	0	6	23	40	48	43	24	0	0	184
Moisture surplus	0	58	49	0	0	0	0	0	0	0	0	0	107
Runoff[a]	0	29	39	19	10	5	3	1	1	0	0	0	107
Moisture detention	278	329	339	300	250	189	140	105	86	77	96	174	
Seabrook, New Jersey													
Potential Evaporation	1	2	16	46	92	131	154	136	97	53	19	3	750
Precipitation	87	93	102	88	92	91	112	113	82	85	70	93	1,108
Difference	86	91	86	42	0	−40	−42	−23	−15	32	51	90	358
Storage change	0	0	0	0	0	−38	−35	−17	−10	32	51	17	
Moisture storage	300	300	300	300	300	262	227	210	200	232	283	300	
Actual Evaporation	1	2	16	46	92	129	147	130	92	53	19	3	730
Moisture deficit	0	0	0	0	0	2	7	6	5	0	0	0	20
Moisture surplus	86	91	86	42	0	0	0	0	0	0	0	73	378
Runoff[a]	61	76	81	61	31	15	8	4	2	1	1	37	378
Moisture detention	361	376	381	361	331	277	235	214	202	233	284	337	

Source: Thornthwaite and Mather 1955.

[a] Assuming 50% of the water available for runoff in any month is held over until the following month.

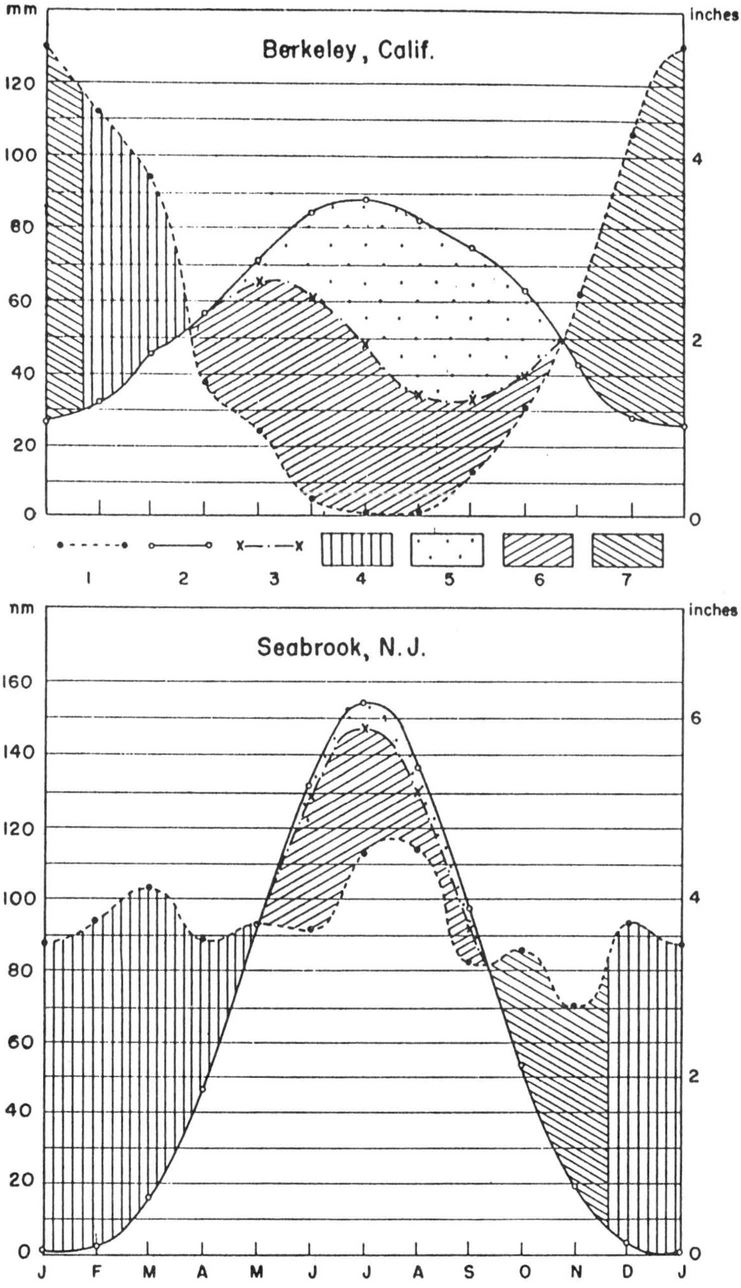

Fig. 3–12. Thornthwaite water budget graphs of data in table 3-4 for Berkeley, Calif., and Seabrook, N.J. Key: Lines show mean monthly (1) precipitation, (2) potential evapotranspiration, and (3) actual evapotranspiration. Patterns indicate periods of (4) water surplus, (5) water deficit, (6) soil moisture utilization, and (7) soil moisture recharge. (From C. W. Thornthwaite, "Modification of Rural Microclimates," in Thomas 1956, 575. Copyright © 1956, 1984 University of Chicago Press. Reproduced with permission.)

TABLE 3-5 Estimated Water-Holding Capacity of Soils According to Texture

Soil Texture	Water-Holding Capacity (mm/m)	(in/ft)
Fine sand	100	1.2
Fine sandy loam	150	1.8
Silt loam	200	2.5
Clay loam	250	3.0
Clay	300	3.6

Source: J. R. Mather, *The Climatic Water Balance,* Publications in Climatology, vol. 14, no. 3, Centerton, N.J.: Laboratory of Climatology, 1961, 255.

TABLE 3-6 Water-Holding Capacity in Root Zones of Various Types of Vegetation Cover Growing in a Silt Loam

	Depth of Root Zone		Total Water-Holding Capacity in Root Zone	
Capacity Vegetation	m	ft	mm	in
Shallow Rooted (vegetables)	.62	2.08	125	5.0
Moderately Deep-Rooted (corn, cotton, small gr.)	1.00	3.33	200	8.0
Orchards	1.50	5.00	300	12.0
Mature Closed Forest	2.00	6.66	400	16.0

Source: J. R. Mather, *The Climatic Water Balance,* Publications in Climatology, vol. 14, no. 3, Centerton, N.J.: Laboratory of Climatology, 1961, 255.

Actual Evapotranspiration (AE) vs. Potential Evapotranspiration (PE)

Potential evapotranspiration and actual evapotranspiration are equal as long as precipitation equals or exceeds PE. This is the case at Seabrook, N. J. from October through May (table 3-4 and fig. 3-12). In fall, winter, and spring the excess of P over PE recharges soil moisture, or it is lost in surface and ground water runoff. From June through September at Seabrook, PE exceeds P, and vegetation must draw on moisture stored in the soil. As the soil becomes drier, however, plants have more difficulty in transpiring additional water, and AE becomes less than PE. Measurements show that the rate of evapotranspiration is proportional to the amount of available water remaining in the soil. For example, if soil moisture content is only 60 percent of capacity, the rate of evapotranspiration will be about 60 percent of the potential rate, and AE will be about 60 percent of PE.

Water Surplus and Water Deficiency

During an average climatic year, as summer progresses at Seabrook, plants use soil moisture at a rate less than the PE, resulting in a deficiency in June,

July, August, and September. From October to May, P exceeds PE, resulting in a surplus of water (P—PE is positive). From September through November, excess precipitation recharges soil moisture that was depleted during summer months. By December, soil is again recharged to 300 mm capacity, and the surplus from December through April contributes to surface and groundwater runoff. Although Seabrook has a period of moisture deficiency, its annual precipitation of 1108 mm (43 in.) exceeds its PE of 750 mm (29 in.), giving the station a humid climate.

Berkeley's climatic regime (table 3-4 and fig. 3-12) is dry summer subtropical (*Cs*), quite different from that of Seabrook. From November through March, the weather is moist, and there is a water surplus. Water deficiency is considerably greater during the dry summer than at Seabrook, where summer is the season of maximum precipitation. At Berkeley, the annual PE (693 mm, or 27.28 in) exceeds P (616 mm, or 24.25 in).

The Climatic Water Budget of the United States

Figure 3-13 compares evaporation from shallow lakes and reservoirs in the United States with average precipitation and with the Thornthwaite bound-

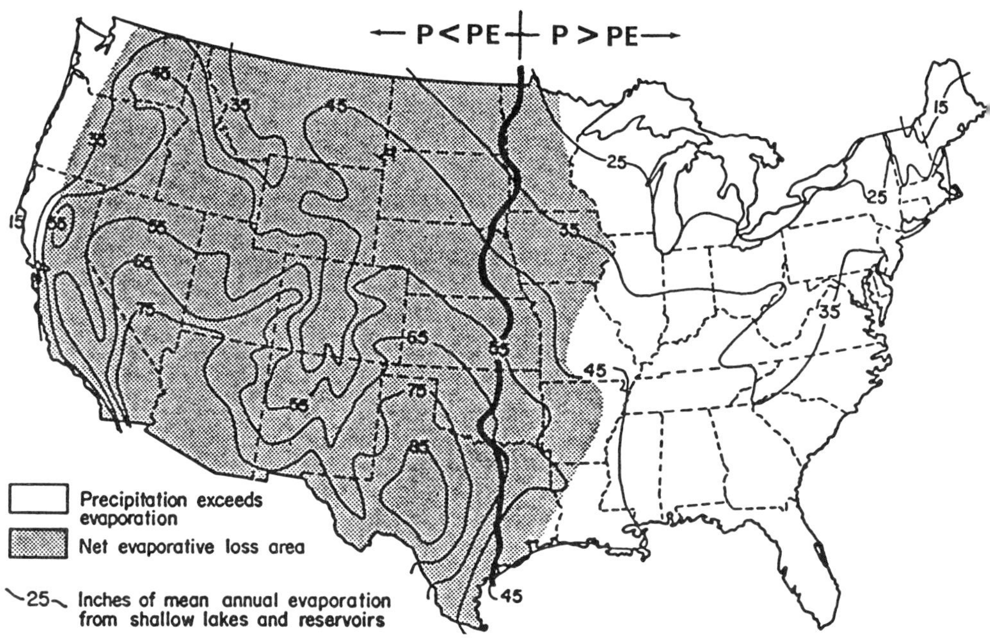

Fig. 3–13. Evaporation and net evaporative loss areas in the conterminous United States, based on corrected pan data. The heavy line separates moisture deficiency areas from moisture surplus areas, according to Thornthwaite's calculations (Thornthwaite and Mather 1955). (Modified from USDA, 1981, 194.)

ary between humid and dry climates. As the figure shows clearly, large areas of the West have evaporation loss exceeding precipitation, and water deficiency reaches particularly high levels in the arid Southwest. Areas of potential water surplus occur only in higher mountains and along the Pacific coast and in valleys north of San Francisco Bay. Nearly all perennial stream flow of the West originates in these limited areas. Typical of these permanent streams are the Rio Grande and the Colorado River, whose headwaters are located in the southern Rocky Mountains, in Colorado and northern New Mexico. Except for southern Florida, the eastern United States is an area of water surplus in which average annual precipitation exceeds PE. Consequently, the major climatic pattern in the United States is one of humid East and arid West.

Water Resources of the United States

The Second National Water Assessment for the United States, made by the U.S. Water Resources Council, was published in 1979 as *The Nation's Water Resources, 1975–2000.*[3] This report estimates water resources for an assumed statistical year based on estimates of water supply and use under average conditions and at 1975 population levels. The year 1975 in the report, thus, does not necessarily represent the actual conditions for historical 1975. The projected water uses and demands for 1985 and 2000 were made by projecting base year data on the basis of historic trends. The report identified ten problem categories:

1. Inadequate surface-water supply
2. Overdraft of ground water
3. Pollution of surface water
4. Pollution of ground water
5. Quality of drinking water
6. Flooding
7. Erosion and sedimentation
8. Dredging and disposal of dredged material
9. Wet-soils drainage and wetlands
10. Degradation of bay, estuary, and coastal waters

The report examined each problem with respect to public health, environmental quality, economic efficiency, resource conservation, international relations, pervasiveness, and urgency. It is obvious that all these problems are interrelated. For example, pollution of surface water and groundwater reduces usable water supplies as effectively as an actual water shortage; erosion and sedimentation have an adverse effect on reservoir storage capacity and contribute to the degradation of bay, estuary, and coastal waters; drain-

[3]Unless otherwise indicated, the sources for data used in this section are: U.S. Water Resources Council, *The Nation's Water Resources, 1975–2000,* Vol. 2: *Water Quantity, Quality, and Related Land Considerations* (Washington, D.C.: U.S. Government Printing Office, 1978); and USDA, *1980 Appraisal: Soil, Water, and Related Resources in the United States: Parts 1 and 2* (Washington, D.C.: USDA,1981).

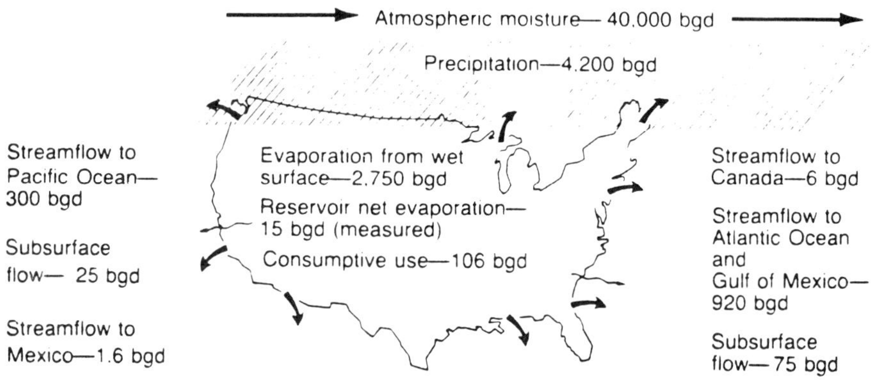

Fig. 3–14. Water budget for the conterminous United States. Quantities are in billions of gallons per day (bgd). (From USDA, 1981, 190.)

age of wetlands, although it adds to available agricultural land, reduces wildlife habitat. Similar interrelationships exist between other problem areas on the list. Consequently, because any solution for one problem has an impact on the others—favorable or unfavorable—effective planning for preservation, development, and use of water resources must be broad enough to encompass these interrelationships.

Although we are concerned here primarily with physical aspects of water resource problems, we should not lose sight of related legal and institutional aspects. Space does not permit a thorough treatment of water rights, land-use planning, water conservation, policy and organization, data collection and analysis, research, and public participation in addressing the problems. The 1978 report noted that "this Nation has already faced and is currently facing problems of water management. In view of the long lead-time needed for planning, research, education, and implementation of programs to resolve these conflicts, a comprehensive, coordinated program is vital to long-term national well-being."

Contributions to the U. S. Water Budget

About forty thousand billion gallons of water in the form of atmospheric water vapor passes over the conterminous United States each day (fig. 3-14).[4] Most of this water vapor, about 90 percent, originates in the Gulf of Mexico and the Atlantic and Pacific Oceans; the remainder derives from evaporation and transpiration from inland sources such as lakes, rivers, and vegetation. About one-tenth precipitates, mostly as rain or snow, for a total of about 762 mm (30 in.) of equivalent rainfall in an average year.

[4]Conterminous United States consists of the forty-eight mainland states without Alaska or Hawaii.

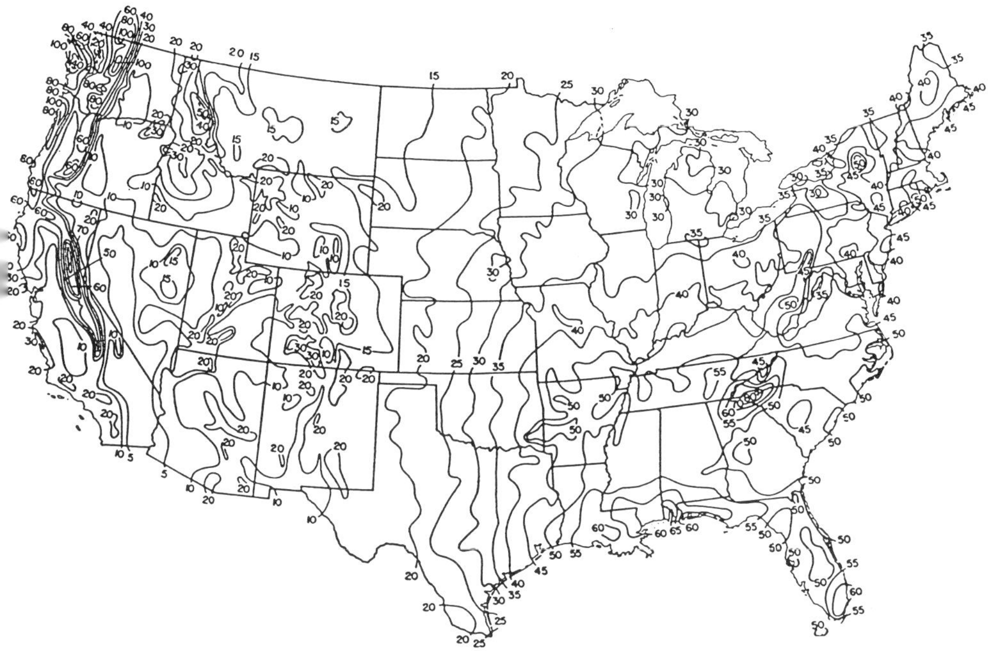

ig. 3–15. Average annual precipitation for the conterminous United States. Values of *isohyets* ines of equal rainfall) in inches. (From USDA, *Climate and Man, Yearbook of Agriculture, 1941,* Vashington, D.C.: USDA, 1941, 711.)

This precipitation is not distributed uniformly in either space or time. Average amounts range from less than 100 mm (4 in.) in the arid Southwest, to more than 3,080 mm (120 in.) in some areas of the Pacific Northwest (fig. 3-15). Of the 762 mm (30 in.) of average annual precipitation, about 510 mm (21 in.) returns to the atmosphere through evapotranspiration. Most of the remaining 228 mm (9 in.) eventually returns to the ocean as stream flow supplied directly by surface runoff or indirectly through groundwater runoff, although about 76 mm (3 in.) replaces groundwater lost through previous discharge.

Ground Water Supplies

The volume of groundwater stored beneath the land surface of the United States is far greater than the total capacity of all its lakes and reservoirs. It equals approximately thirty-five years of surface runoff. In spite of this great volume, intensive use in some areas results in mining of groundwater, leading to diminishing artesian pressure, declining stream flow, land subsidence, and intrusion of salt water (fig. 3-16). Most excessive groundwater mining

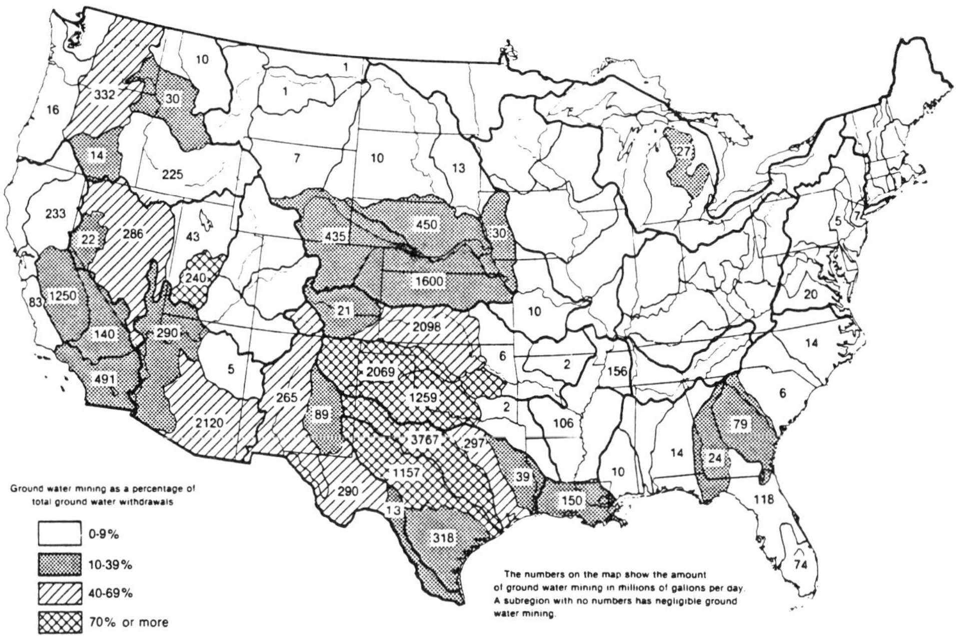

Fig. 3–16. Areas of groundwater mining in the United States, based on average water use du ing a "normal" year, 1975. Mining implies that withdrawals exceed natural recharge of aquife On the southern Great Plains, 70 percent of the water withdrawn is in excess of natural r charge. Water levels in wells are falling rapidly in those areas, and pumping costs are increa ing. (From USDA 1981, 194.)

takes place in the dry areas of the West, where heavy demand for irrigation exceeds the capacity of limited surface supplies. Some areas of groundwater mining, however, occur in the humid East; here, demands on aquifers by rapidly growing urban population may lead to excessive pumping levels.

The removal of large amounts of groundwater without adequate recharge of aquifers will not only lead inevitably to substantial declines in water levels in wells, but will produce other undesirable side effects. One of these is land subsidence which may cause ground elevations to drop several feet over large areas. Such subsidence may cause damage to buildings, highways, and railroads, and, near the coast, inundation by higher tides.

In coastal areas and on islands (as in Hawaii), fresh water often "floats" on the denser salt water. If pumping is in excess of recharge, salt water will migrate into zones formerly occupied by fresh water, and aquifers may be permanently ruined. Even in interior areas, saline aquifers may underlie and be connected to freshwater aquifers, and excessive pumping may cause saline water to migrate into freshwater zones.

In the Houston-Galveston area of the Texas Gulf Coast, both saltwater intrusion and land subsidence cause excessive losses in property values. Other areas with real or potential problems from mining of aquifers include the Gila River basin of New Mexico and Arizona; the high plains region in Nebraska, Colorado, Kansas, Oklahoma, New Mexico, and Texas; the San Joaquin Valley of California; the upper Susquehanna River basin in New York and Pennsylvania; southeast Georgia, in the Savannah and Brunswick areas; and the island of Oahu in Hawaii, particularly Pearl Harbor.

Surface Water Supplies

Surface water supplies are much more dependent on annual fluctuations in precipitation than are ground water sources. Consequently, stream flow varies widely from region to region and from year to year. In some regions, variations in annual runoff may range from less than 30 percent of normal in a dry year to more than 200 percent of normal in a wet one. Stream flow also may vary from season to season even in a normal year; for example, in the dry Southwest, as much as 75 percent of annual runoff occurs while the spring snow is melting.

Adequacy of Surface Water Supplies

Serious surface water supply problems already exist, or will exist by 2000, in many of the water resources regions located west of the Mississippi River. Here, withdrawals for irrigation place heavy demands on both surface and ground water supplies. High probabilities of drought and increased water losses from evaporation exacerbate water problems.

Typical surface water supply problems are those plaguing the lower Colorado River region in Arizona and California. More than half the population of the western United States depends on the inadequate and poorly distributed water supply of the Colorado River basin. In addition to demands made by urban areas of southern California on surface water supplies of the region, about half the irrigated acres depend on surface sources. Also, the Mexican Water Treaty of 1944 provides for delivery of 1.5 million acre feet annually; this obligation will be difficult to fulfill in the future.

Since demands on water supplies of the Lower Colorado Region of 4.6 bgd in the assumed base year of 1975 exceeded available renewable surface supplies by about 2.4 bgd, groundwater resources have had to make up the difference. Consequently, because of groundwater mining, water levels have declined at an average 2.4 to 3 m (8–10 feet) per year. In addition to the problem of land subsidence, pumping in excess of annual recharge decreases stream flow and increases pumping costs.

Water Quality

Water quality is difficult to define. In part, it depends on the intended use of the water. Even saline water serves as a coolant in industrial processes, and low levels of dissolved minerals in surface water and groundwater may not interfere with most uses. In fact, many supplies are unsuitable because of natural pollution, even without human contribution to the problem.

Water quality may be adversely affected by a number of environmental factors, both natural and human. Pollution is especially serious when it occurs in groundwater. Once polluted, aquifers are not as easily restored to a healthful condition as are surface waters because of slow groundwater movement. Surface sources, because of rapid movement of water through them, are more easily renovated than are underground aquifers.

Pollutants from natural and human sources include infectious agents, nutrients, dissolved solids, suspended solids, pesticides, and industrial chemicals. If the exact point of origin and entry into the water supply, such as discharge from a sewage treatment plant or seepage from a known landfill, can be identified, it is called *point source pollution*. Dispersed sources, such as runoff from agricultural land or fallout from the atmosphere, cause *nonpoint source pollution*.

Infectious Agents

Infectious agents—bacteria, fungi, and viruses capable of spreading diseases to plants, animals, and humans—are widespread in polluted streams and shallow aquifers. Water supplies are monitored for *fecal coliform bacteria*. Although not harmful in themselves, these bacteria are an indicator of contamination by fecal matter, which can transmit diseases, including typhoid fever, hepatitis, brucellosis, encephalitis, poliomyelitis, prittacosis, and tuberculosis.

Infectious agents may originate from both point sources (effluents from municipal treatment plants and feedlots) and nonpoint sources (sewer overflows, septic tank fields, urban and agricultural runoff, and even manure from migrating waterfowl). Polluted waters also may contribute indirectly to the spread of disease by serving as breeding areas for mosquitoes and other insect vectors.

Nutrients

Plant nutrients, both those naturally occurring and those applied as fertilizer, leach and infiltrate shallow aquifers. Also, runoff can transport them in solution or attached to sediment from eroding farmland. In groundwater, they may pose a health threat both to humans and animals, whereas in surface waters they also over-stimulate aquatic growth, often to the detriment of fish habitat. When nutrients concentrate in water bodies in relatively large quantities, aquatic plants proliferate and deplete oxygen at night when oxygen-producing photosynthesis is dormant. This process, called *eutrophication*, occurs naturally in lakes and eventually converts them to bogs, but it accelerates when excessive nutrients are added from human activities.

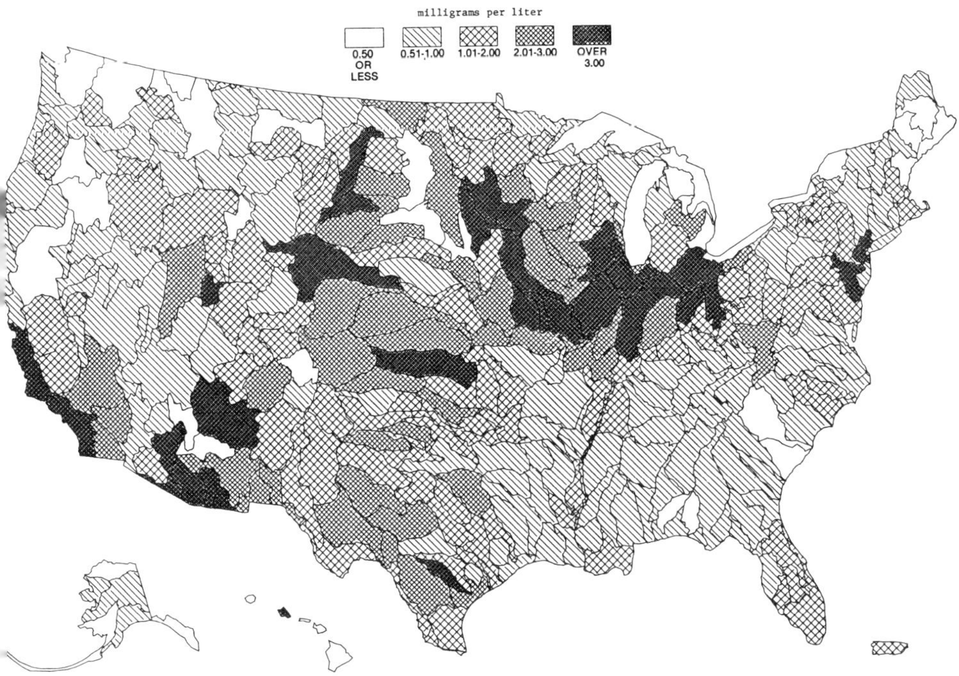

g. 3–17. Total nitrogen levels in streams of the United States. (From U.S. Water Resources ouncil 1978.)

Nutrient pollutants consist principally of nitrogen, phosphorus, and potassium compounds (fig. 3-17). Naturally occurring nitrates may originate in the atmosphere, and also in soils through the action of nitrogen-fixing bacteria. Major sources of nutrient pollutants, originating from human activity, are artificial fertilizers, animal wastes, septic tank drainage, discharge from sewage treatment plants, and certain industrial wastes such as those from food processing plants.

Dissolved Solids

The most common dissolved solid is salt—both sodium chloride (NaCl) and other mineral salts. Calcium and magnesium salts produce "hardness" in water—a condition that may cause problems in use but seldom imposes health risks. Other minerals in naturally hard water may be bicarbonates or sulphates, chlorides of calcium, magnesium, iron, or other minerals. Commercial softening processes are available for removing or reducing hardness.

As mentioned previously, salinity is a particularly serious problem in coastal areas because of saltwater intrusion into overpumped aquifers. In dry areas, salinity is a special problem for irrigated agriculture because application of water with high salinity causes accumulation of salts in the soil and may eventually make the soil useless for agriculture. Salt concentrates because some of the applied water is transpired by plants, leaving salt in the soil or increasing salinity of water returning to streams. In the western half of the United States all drainage systems, except for the Columbia River, have high levels of salinity in their tributary rivers. A major part of this salinity derives from natural sources, but human activity may exacerbate its detrimental effects. Of 10 million tons of salt carried annually in the Colorado River, two-thirds comes from natural geological sources, whereas one-third is from return flows from irrigated land.

Suspended Solids

Suspended solids include soil particles, organic matter, plankton, and other substances in suspension in water. These particles reduce light transmission through water and thus inhibit growth of underwater vegetation, causing deterioration of habitat for aquatic life. Suspended solids may settle and form accumulations of oxygen-demanding sludge, or they may act as transport agents for other pollutants. By far the largest part of suspended solids in surface waters consists of sediment.

Most suspended solids derive from nonpoint sources; accelerated erosion from crop land is the major contributor. Agricultural sources are of special concern because nutrients and pesticides adhere to particles carried into streams. Other significant sources of sediment are dredge and fill operations, wave turbulence, and urban runoff from new construction sites.

Pesticides (Herbicides and Insecticides)

Current agricultural practices depend heavily upon applications of herbicides and insecticides, a potential health threat. The problem is of special concern in the Midwest and California, where pesticide applications are heavy and where residues appear in low concentrations in shallow groundwater and surface water supplies. Table 3-7 gives pesticide use data for Iowa, drawn from a 1986 program that monitored groundwater quality in the state.

A study by the Iowa Department of Natural Resources for a groundwater protection strategy for the 1987 Iowa General Assembly (Hoyer, et al. 1987, 16–17) offers the following assessment of the pesticide problem:

> The implications to human health from the long-term exposure to pesticides are unclear. And it is because of our lack of knowledge that concern has been expressed by both the scientific community and the general public. Certainly, various epidemiological studies suggest cause for concern. For example, mortality from some cancers is significantly higher in rural farm families. Aggregate county studies reveal that excessive mortality from leukemia, multiple myelonia and non-Hodgkin's lymphoma is consistently associated with area herbicide usage.

TABLE 3-7 Summary of All Pesticide Data from Groundwater Quality Monitoring in Iowa, 1986

Common Name of Active Ingredient	Maximum Concentration (μg/liter)	Percentage of All Detections	Month(s) in Which Detected
Herbicides			
Alachlor	16.6	15	Jan.–Dec.
Atrazine	13.0	72	Jan.–Dec.
Chloramben[a]	1.7	< 1	July
Cyanazine	13.0	13	Jan.–Dec.
Dicamba[a]	2.3	2	Mar., June, July
Metolachlor	9.0	9	Jan.–July, Nov., Dec.
Metribuzin	4.4	10	Jan.–Dec.
Trifluralin	0.2	1	June, July
2.4-D	0.2	< 1	Apr.
Insecticides			
Fonofos	0.9	2	Apr., June, Aug.
Sulprofos[b]	1.4	< 1	May
Terbufos[b]	12.0	5	May

Source: R. D. Kelley et al., *Pesticides in Ground Water in Iowa,* (NWWA Agricultural Impacts of Ground Water Conference, Omaha, Neb., 1986), reported in Hoyer et al. 1987.
[a] Analyzed by different methods, thus N not the same as for other herbicides.
[b] Detected in only one study.

Because we are only just beginning to look for and find pesticides in groundwater and drinking water and chronic health disorders may require twenty years or more to manifest themselves, the impacts to public health and the resulting economic ramifications may not be known for years. However, a substantial part of the state's population may be at risk. . . .

Current pesticide management practices have resulted in the problems presently being experienced in Iowa. Without changes in these practices, it is reasonable to expect the problem to continue and expand both in geographic area and into deeper aquifers.

It seems logical to assume that similar problems may exist in other areas with comparable high levels of agricultural chemical use.

Industrial Wastes

Many industrial processes require the use of toxic and carcinogenic chemicals. They are present, for example, in fuels, cleaning compounds, industrial solvents, and materials used in the manufacture of plastics. These chemicals may enter the environment through spills, leaks, runoff from improper aboveground storage, or disposal of waste products (table 3-8). They may contain potentially health-threatening substances such as lead, cadmium, zinc, arsenic, barium, chromium, chlorides, copper, cyanides, fluorides, mer-

Table 3-8 Synthetic Organic Compounds Other than Pesticides Found in Iowa Groundwater.

Organic Solvents
Tetrachloroethene
Trichloroethene
1,1,1-Trichloroethane
1,1-Dichloroethane
Methylene Chloride
Carbon Tetrachloride
1,1,1-Dichloroethyene
1,2-Dichloroethane
1,1,2-Trichlorethane
Aromatics
Benzene
Toluene
Styrene
Ethylbenzene
Trihalomethanes
Chloroform
Bromoform
Bromodichloromethane
Dibromochloromethane

Source: R. D. Kelley, Synthetic Organic Compound Sampling Survey of Public Water Supplies (Iowa DWAWM Report 1985), reported in Hoyer et al. 1987.

cury, nickel, phenols, selenium, and silver. Included in industrial sources of contamination is the storage and transport of agricultural chemicals.

Natural Pollution of Water Supplies

Natural pollution is especially serious in some groundwater sources because of high mineralization from such naturally occurring chemicals as magnesium, iron, sulfur, nitrates, phosphorus, calcium, fluoride, arsenic, and chloride. Natural radioactivity may occur in groundwater in Florida, Maine, New York, Illinois, Iowa, and Missouri. Natural pollution may not make groundwater totally unusable in all cases, but it certainly makes it less desirable.

Groundwater pollution from natural sources also may be the result of human activities. Decline in water levels in overpumped aquifers in coastal areas leads to saltwater intrusion from the ocean. Similarly, drilling of oil and gas wells may cause intrusion of brine, oil, or gas into high quality aquifers lying above the oil-producing strata.

Pollution of Ground Water

Fifty percent of the population of the United States derives its drinking water from underground sources; thus, groundwater pollution poses a serious health threat. Continued overdraft of groundwater, as we have seen, causes saltwater intrusions, and may induce infiltration of contaminated surface water into aquifers. Groundwater pollution also occurs from surface infiltration of industrial contaminants and fecal coliform bacteria from human and animal sources, particularly from septic tanks in rural and recently urbanized areas, and from commercial feedlots.

Other sources of aquifer pollution include (a) seepage from landfills containing municipal and industrial wastes, some of it highly toxic; (b) leaching of salts and agricultural chemicals from the soil zone by irrigation water carrying them downward into aquifers; and (c) leaching of herbicides, insecticides, and fertilizer from agricultural lands by infiltrating rainwater. Of particular concern is contamination of shallow groundwater supplies by nitrates, mostly from commercial fertilizers or animal wastes. Even small amounts of nitrates in water supplies may be toxic to cattle and cause methemoglobinemia (blue baby syndrome) in human infants. Because of increased and often excessive application of nitrogen to crops, water from many farm wells has high nitrate concentrations.

Figure 3-18 indicates some relationships between geological conditions and potential groundwater contamination. Shallow alluvial aquifers, shallow glacial drift aquifers, and bedrock aquifers lying directly beneath very shallow soils are particularly susceptible to contamination from surface infiltration. In areas of limestone *karst*,[5] where surface water has direct access to aquifers by way of sinkholes, the potential for contamination from both infiltration and surface "run in" of sediment, pathogens, and chemicals is extremely high.

The Economic Value of Water

It is beyond the scope of this book to detail the economics of water supply and demand, and the priority of use. Many water problems, however, stem not from actual water shortage but from traditional attitudes of people living in humid areas. Because water usually is plentiful in such areas, people tend to regard it as a free good, comparable to air. Charges seldom appear for actual water withdrawn from a source and costs are due primarily to the expense of distribution. Charges for water seldom reflect the costs of putting water to one use at the expense of another.

In a foreword to a study by Resources for the Future, Inc., on the economic value of water (Gibbons 1986, vii), Kenneth D. Frederick used the proposed 500,000-acre expansion of the Columbia Basin Irrigation Project to illustrate how instream water values often are ignored. He expressed concern for offstream supplies developed without regard to costs and to the prospective values derived from the water. This proposal would divert 8.3 million acre feet of water each year (11,500 cfs) from the Columbia

[5] A *karst* is an area with limestone bedrock in which most of the drainage is underground through sinkholes, solution channels, and caverns. The name comes from the Karst district of Yugoslavia.

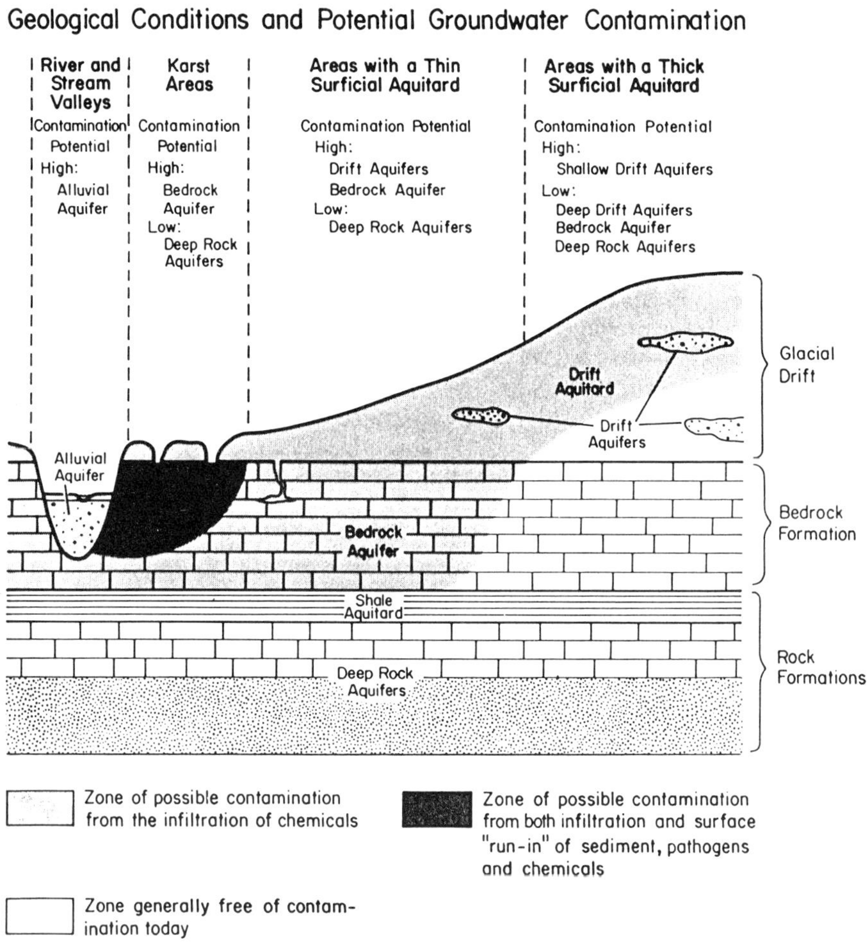

Fig. 3–18. Geological conditions and potential groundwater contamination in glaciated areas of the upper Midwest. (From Hoyer et al. 1987, 6.)

River above Grand Coulee and ten other hydropower dams for irrigating already productive wheat lands. The estimate for investment cost is $4,000 per irrigated acre, and each acre irrigated would add about $200 a year to the cost of supplying electricity to the region. Frederick pointed out that the federal government would have to subsidize this heavily, and that such a project would increase scarcity of water in nonfavored sectors. This proposed project illustrates the traditionally favorable position of agriculture in establishment of water use priorities in the United States.

Water Law

Laws for regulating water use vary widely from state to state, and between surface water and groundwater. Water laws often have been based on outdated and erroneous information about water supplies, particularly those found below the surface. In the past, the interrelationship between surface and groundwater sources was seldom recognized in formulating water laws. In the eastern states where, until recently, ample rainfall and numerous rivers assured an adequate supply for most users, water laws have been based on the *common law doctrine*; in the West, with limited water supplies, the *prior appropriation doctrine* predominates. In recent decades many eastern states have found the old common law doctrines inadequate for handling increasing and conflicting demands. After considering and rejecting appropriation doctrine, they have turned to permit systems for regulation (fig. 3-19 and 3-20). California law allocates water according to the concept of *correlative rights*, and the state enforces reasonable use and some system of agreed sharing.

Under common law the *English Rule* governs groundwater, granting land owners absolute rights to all they can pump. It does not recognize the fact that pumping water from beneath one property may deplete supplies below others lying above the same aquifer. In some states, water is not "owned" until it is "captured and reduced to possession." In others, the so-called *American Rule* of "reasonable use" takes into account the impact on water rights of others and on the common good.

Under common law, the ownership of land adjacent to a body of water (*riparian ownership*) includes the right to use that water. In humid areas, early use often involved the development of water mill sites along streams, and an assured flow of water in the stream had priority. This led to the *natural flow* theory, which stipulates a riparian owner's responsibility to leave any water flowing past his or her land unimpaired in quality and undiminished in quantity. A less restrictive common law theory applies the test of reasonable use to the riparian owner's right to withdraw and use water from a stream or lake. Of course, "reasonable use" is open to many interpretations and these might well change with time and circumstances.

After 1850, most western states broke away from the English concepts of water rights, and prior appropriation doctrine replaced the concept of riparian and other common law rights. This doctrine holds that a property right is created by diverting water from its source and applying it for beneficial use. Furthermore, the right is not limited to a riparian owner; rather, water may be used at any location without regard to the location of the source. This concept also bases priority of use on the date of first use: "first in time is first in right." When water is in short supply, the last user to divert and make use of the water will be the first to have his or her supply terminated.

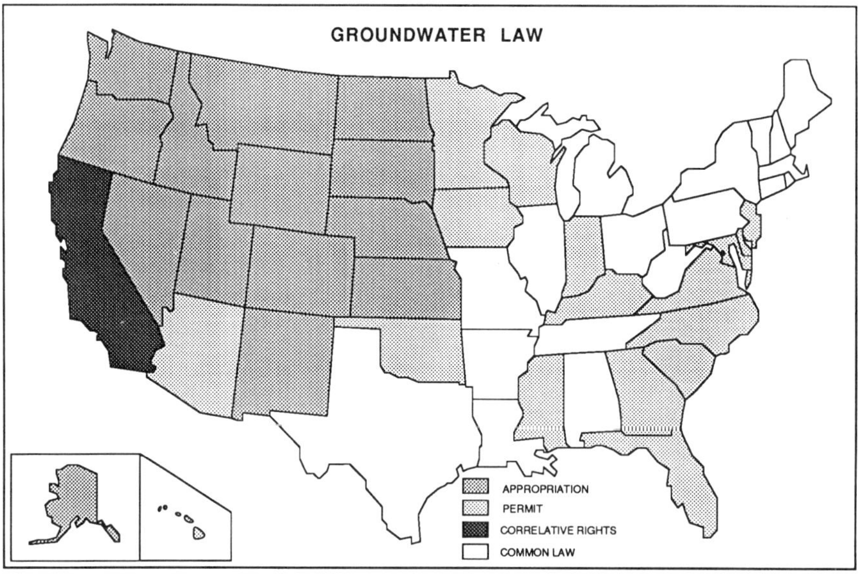

Fig. 3–19. Groundwater law in the United States. (From Matthews 1984; U.S. Water Resources Council 1978.)

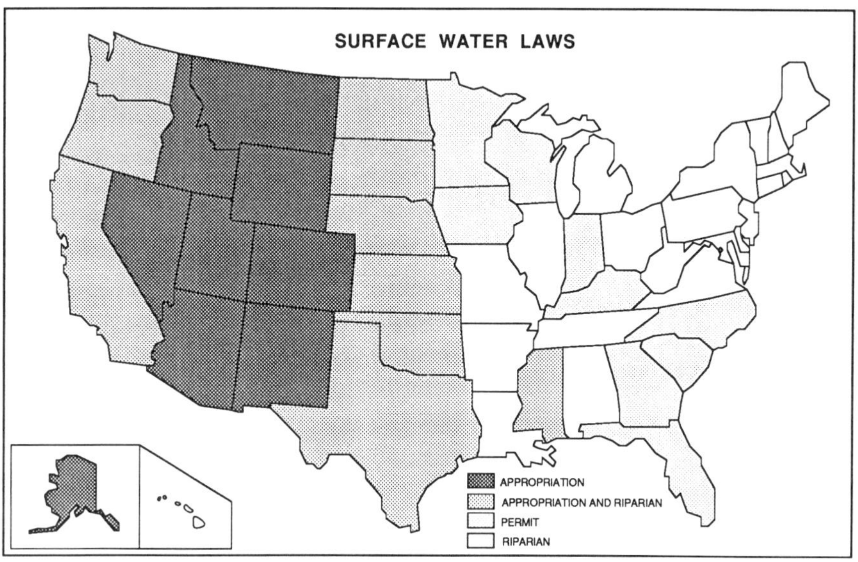

Fig. 3–20. Surface water law in the United States. (From Matthews 1984; U.S. Water Resources Council 1978.)

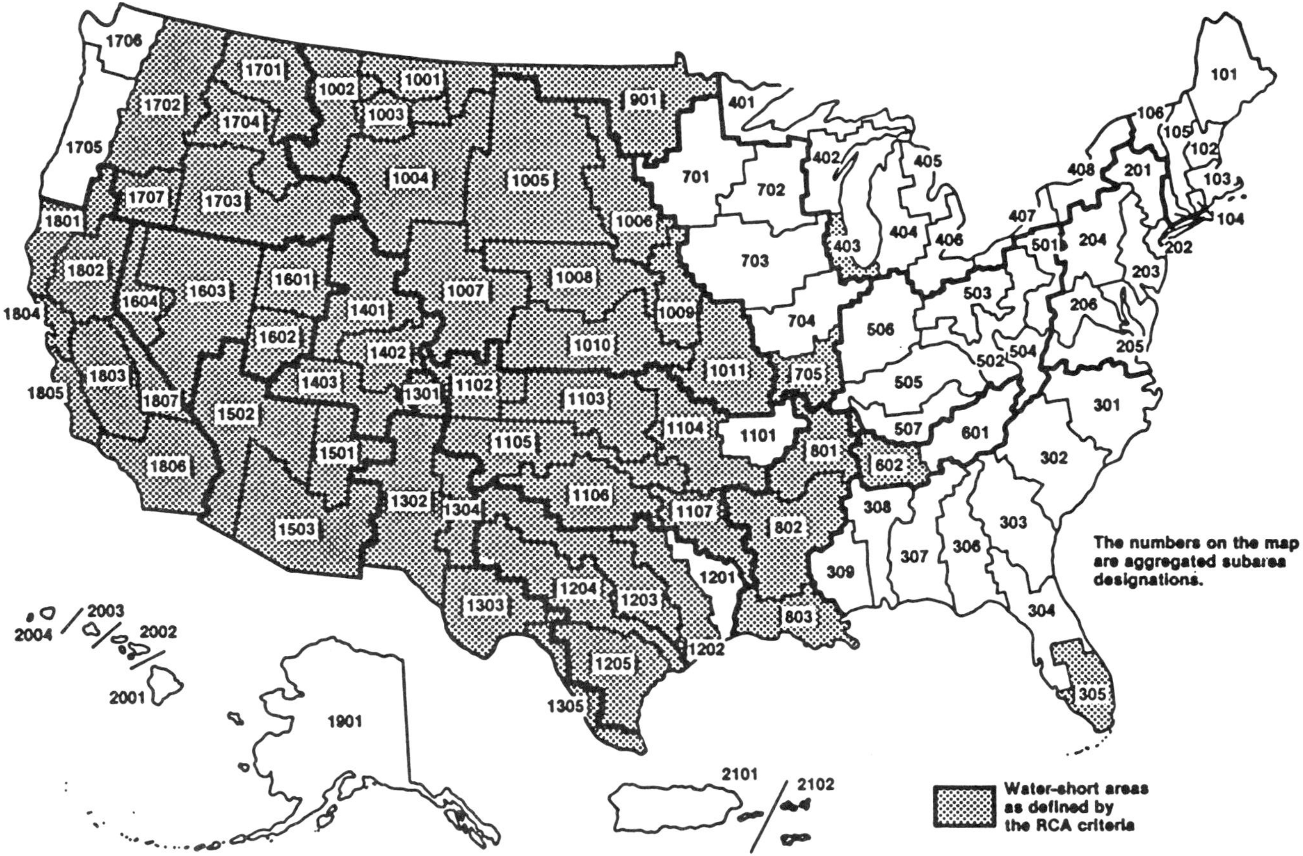

Fig. 3–21. Hydrologic units in the United States designated as water-short by the Resources Conservation Act (RCA) of 1977. (From USDA 1981, 94.)

Prior appropriation doctrine developed in the West in response to special circumstances related to mining. Establishment of a water right was analogous to filing a claim on public land for mining rights. Many miners diverted streams to process ores, a practice prohibited under riparian doctrine. Since the land was public, no one could claim to be a riparian owner, and a new system of water allocation was necessary. The result was the prior appropriation doctrine. Although established for mining, this same doctrine today governs the use of water for irrigation.

A Water Budget for the Future

Little can be done to increase the absolute quantity of water on the Earth. Future availability of this major resource depends on many factors, some already recognized, others not. Unquestionably, population increases and expanding industrialization will place additional demands on water resources, and many human activities will adversely affect the quality of existing supplies (fig. 3-21). Changes in water storage and distribution must be expected. Society will base some decisions concerning use of this vital resource on careful physical and economic planning, while other decisions will be made on the basis of political power or expediency.

On a more optimistic note, technological advances certainly will result in more efficient use of the water. Already, progress has been made toward more efficient irrigation, less water-demanding processes in industry, better control of pollution, and desalination of sea water. Perhaps just as significant as technological advances is general awareness of water problems among state and federal legislators, the news media, environmentalists, and the general public. This awareness may lead to vital changes in the economic, political, legal, and institutional approaches to water conservation and use.

Finally, an area of future uncertainty is the prospect of climatic changes that could alter radically the world pattern of precipitation. Such changes could cause some areas now dry to become moist, and moist areas to become dry. Obviously, they would cause vast dislocations in agriculture and would have profound geopolitical and geoeconomic ramifications. Climatic change will be the subject of the next chapter.

Observe constantly that all things take place by change, and accustom thyself to consider that the nature of the Universe loves nothing so much as to change the things which are, and to make new things like them.

MARCUS AURELIUS, *Meditations, Book IV*

Chapter 4

CLIMATIC CHANGE

Past, Present, and Future

During the decade of the 1970s, a series of climate-related events focused unprecedented media attention on climatic fluctuation and change. In the United States, there were record-breaking blizzards in the East, severe droughts in California, devastating flash floods in the Rocky Mountains, and severe tornado outbreaks in many states. Outside the United States, climatic events included droughts and famine in the Sahel of West Africa, crop failures because of erratic behavior of the Asian monsoon, drastic declines in the Peruvian fishing catch (in part the result of changes in sea temperatures, the so-called *El Niño* event), and unusual heat waves and drought in the British Isles and Western Europe. These were the highly publicized events; there were others as well.

Many severe economic and political dislocations occurred, in part because of these and other erratic climatic events following several decades of relatively benign weather patterns. In 1972, for example, there was a serious agricultural crisis in the Soviet Union. Severe drought in the Moscow area and lack of winter snow cover in the wheat lands of the Ukraine caused crop shortages. The Soviets purchased vast quantities of wheat and other grains from the United States and, as a result, American food prices rose dramatically, an important factor in the phenomenal inflation that followed.

During this era, newspapers and magazines were filled with accounts of these climatic disasters and their human, economic, and political impacts, and many writers speculated on the "changing climate." After an unusually cold winter or cool summer, "the coming Ice Age" was a common theme;

after an unusually hot summer or mild winter, the subject of numerous articles was likely to be whether the climate was growing warmer.

Also in the 1970's the scientific community, and to a lesser degree the public, was becoming concerned with human-induced changes in the composition of the atmosphere, which threatened to increase acid rain, deplete the ozone layer, and change future earth temperatures. No longer was air pollution just an urban problem for Los Angeles and other urban-industrial centers, it was also the concern of farmers, fishermen, foresters, and an informed public in general.

Concerns of the 1970's continue, reinforced by such climatic events as the warming of the Western Pacific Ocean in 1982–83, which produced the most severe and prolonged El Niño on record, and by Northern Hemisphere temperatures that have been the warmest in a century—certainly a contrast to the cold winters of the late 1970's in the eastern United States (Kates, Ausubel, and Berberian 1985). Furthermore, there was the specter of a *nuclear winter,* a climatic phenomenon that could result from a major exchange of nuclear weapons, and with it the possibility that most life on earth could become extinct. While the probability and severity of a nuclear winter is still being debated, concern about the depletion of the atmospheric ozone has been rekindled by the discovery in 1987, and again in 1989, of a "hole" in the ozone layer in the stratosphere over Antarctica.

During the 1980s, world temperatures averaged higher than during any previous decade of the instrumental climatic record (Hansen and Lebideff 1988). In some areas of the United States 1988, the warmest year of the decade, was the driest, with drought conditions equaling or exceeding those of the 1930s. Again, public and media attention focused on climate, and questions were raised about the *greenhouse effect* and the possibility of a warmer and perhaps drier future. As of fall 1989, the rains of August and September already had dampened this concern about the impending greenhouse warming, and media and public attention had turned away from climate to other concerns. It is difficult to tell whether the momentum toward solving the atmospheric problems related to human intervention will increase or decrease in coming years.

In this chapter, we will trace the record of climatic change from the remote geologic past, through the Ice Ages of the Pleistocene, and into the present geologic epoch, the *Holocene.* We will base our interpretation of the past climatic record on evidence and data from paleoclimatology and historical and instrumental climatology, as described in chapter 2. As we approach the present, we will encounter evidence of the increasing impact of humans on the climatic environment. Although humans have been ecologically dominant for only a minute part of geologic time, their potential for changing the earth's climate may be second only to that exercised by green plants, which produce Earth's oxygen-rich atmosphere. Unfortunately, human impact may, in large part, be detrimental both to humans and to other living things.

Following this excursion into the past, we will consider various theories for explaining the mechanisms of climatic change, how these mechanisms might be affecting the present climate, and how they might be involved in future changes. Finally, we will consider the possible climatic futures—a warmer climate or a new Ice Age—and weigh some of the possible economic, social, political, and geopolitical consequences of each.

Climates of the Past: Are They the Key to the Future?

Perhaps the long-accepted tenet of geology, "The present is the key to the past," should be modified to, "The past is the key to the future." Since climate certainly has changed, sometimes drastically, future changes are certain to occur. By understanding past climatic patterns, and by formulating acceptable theories about them, it may be possible to forecast future trends, but rapid growth of world population and related industrial, economic, and land use changes, add a new, and perhaps ominous, dimension to the problem.

Evidence for Past Climatic Change

Earth's climates are viewed against the geologic time scale summarized in figure 4-1, which indicates significant climatic events of past geologic ages. On the scale of hundreds of millions of years, the best proof that climates vary is the geologic and paleontologic evidence of major ice ages that interrupted long periods of warmer climate. The glacial record of the late Cenozoic Ice Age of the Pleistocene epoch is the most complete existing; this epoch ended only some ten thousand years ago.

Alongside glacial evidence, other research fields contribute valuable data to reconstruct past climates. To be of value, the data must answer two questions: What happened? When did it happen? Table 4-1 gives the characteristics of sources of paleoclimatic data and climatic inferences that can be drawn from them. A brief description follows, covering some but by no means all of these climatic data sources, and the research tools used to develop them.

Glaciology

The advance and retreat of glaciers are obvious indicators of climatic change. Glaciers form when more snow falls in winters than melts in summers. Given sufficient time, ice formed from this snow will accumulate to a depth sufficient to flow under its own weight, thus becoming a glacier. Cold alone is not enough to trigger a glacial event. Winters with heavy snowfall and cool summers seem to be most favorable for the accumulation and advance of glaciers. A case in point is Alaska: here, the most active glaciers are in the

GEOLOGIC TIME

TIME IN MILLIONS OF YEARS	EONS	ERAS	PERIODS	EPOCHS	PERCENT OF GEOL. TIME (APPROX.)	LIFE FORMS
	PHANEROZOIC	CENOZOIC	QUATERNARY	RECENT	1.3	MODERN HUMANS
2				PLEISTOCENE		LARGE MAMMALS
			TERTIARY	PLIOCENE		EARLY HOMINIDS
				MIOCENE		
				OLIGOCENE		MAMMALS INCREASE IN NUMBERS AND SIZE
				EOCENE		
				PALEOCENE		MAMMALIAN EXPANSION
62		MESOZOIC	CRETACEOUS		3.7	LARGE DINOSAURS
			JURASSIC			FIRST BIRDS
			TRIASSIC			FIRST MAMMALS
230		PALEOZOIC	PERMIAN		9	FIRST REPTILES
			PENNSYLVANIAN			AMPHIBIANS AND COAL FORESTS
			MISSISSIPPIAN			
			DEVONIAN			FIRST LAND PLANTS
			SILURIAN			FIRST VERTEBRATES
			ORDOVICIAN			INVERTEBRATES (MARINE)
			CAMBRIAN			
			EDIACARIAN			
640	PRECAMBRIAN	PROTEROZOIC	UPPER		43	MULTICELLED MARINE LIFE
			MIDDLE			
			LOWER			
2,600		ARCHEAN			22	ALGAE - OLDEST LIFE FORMS OBSERVED TO DATE
3,600		NO STRATIGRAPHIC RECORD			22	NO RECORD OF LIFE
4,600	FORMATION OF EARTH					

Fig. 4–1. Geologic time scale based on major units of sedimentary rocks and distinct fossils associated with them. Absolute time scales are supplied by radioactive isotope dating techniques.

south, where winters are milder but snowfall heavier, whereas the Brooks Range, in the cold north, has limited glaciation.

Glaciers may vary in size from small valley glaciers in mountainous areas to great ice sheets of continental proportions, such as the ice sheets of Greenland and Antarctica, remnants of the last Ice Age. These ice sheets, and the many mountain glaciers still existing, furnish a fascinating opportunity to study glacial behavior and to view evidence of glaciation in postglacial landscapes.

Familiarity with the physical nature of recent glacial deposits, alteration of the shape of valleys by glacial action, glacial striations on bedrock, and formation of unique glacial landforms makes it possible for geologists to recognize evidence of ancient glaciation. Also, as glaciologists and climatologists study the climatic regimes of modern glaciers, it should be possible to apply that knowledge in reconstructing past climatic regimes.

Paleobotany and Paleoecology

The natural vegetation and animals occupying a region are among the best indicators of its climate. Chapter 5 deals with the present distribution of the world's natural vegetation in relation to climate and other environmental factors. From present climate-vegetation correlations, it is possible to evaluate the past climatic conditions under which certain types of vegetation flourished. Certain plants and animals have a very narrow range of tolerance for temperature or moisture or for other climatic variables. By examining fossil remains of these indicator plants, it is possible to deduce the climate that prevailed at the time they grew there.

Plants in particular are good indicators because they are fixed—they cannot migrate with changing climate as do animals. If a vegetation assemblage grows in an area, it becomes an integrator of the range of climatic variation within which it can survive. If climate varies so much in one direction, say from moist to dry, that the vegetation can no longer adjust, the former assemblage will be replaced by more drought-tolerant species. Consequently, fossil remains furnish some of the best clues of past climates. A good example is the luxuriant coal forests of the Pennsylvanian period of late Paleozoic time: these forests undoubtedly grew in a warm, moist environment with vigorous photosynthesis and rapid plant growth, perhaps aided by an atmosphere rich in CO_2.

Palynology (Pollen Analysis). Wind disperses plant pollen over large areas where it settles on the surface as the *pollen rain,* a good sample of the regional vegetation. A palynologist must take into account such variables as the amount of pollen produced by each type of plant, the frequency with which the plant blooms (some forest trees do not bloom every year), the dispersal mechanism of the pollen, plus other correction factors. The key to pollen rains of the past is pollen rain produced by present vegetation (Huntley and Prentice 1988, 1,007–1,008).

Under certain environmental conditions, such as the acid environment of a peat bog, pollen is almost indestructible. Usually it can be identified at the genus level (see chapter 5); that is, it is possible to distinguish pine pollen from those of oak or grass, but not to separate individual species of pine from one another.

TABLE 4-1 Characteristics of Paleoclimatic Data Sources

Proxy Data Source[a]	Variable Measured	Continuity of Evidence	Potential Geographical Coverage
Layered ice cores	Oxygen isotope concentration, thickness (short cores)	Continuous	Antarctica, Greenland
	Oxygen isotope concentration (long cores)	Continuous	Antarctica, Greenland
Tree rings	Ring-width anomaly, density, isotopic composition	Continuous	Midlatitude and high-latitude continents
Fossil pollen	Pollen-type concentration (varved core)	Continuous	Midlatitude continents
	Pollen-type concentration (normal core)	Continuous	50°S to 70°N
Mountain glaciers	Terminal positions	Episodic	45°S to 70°N
Ice sheets	Terminal positions	Episodic	Midlatitude to high latitudes
Ancient soils	Soil type	Episodic	Lower and midlatitudes
Closed-basin lakes	Lake level	Episodic	Midlatitudes
Lake sediments	Varve thickness	Continuous	Midlatitudes
Ocean sediments (common deep-sea cores, 2–5 cm/1000 yr)	Ash and sand accumulation rates	Continuous	Global ocean (outside red clay areas)
	Fossil plankton composition	Continuous	Global ocean (outside red clay areas)
	Isotopic composition of planktonic fossils; benthic fossils; mineralogic composition	Continuous	Global ocean (above $CaCO_3$ compensation level)
(rare cores, >10 cm/1000 yr)	As above	Continuous	Along continental margins
(cores, <2 cm/1000 yr)	As above	Continuous	Global ocean
Marine shorelines	Coastal features, reef growth	Episodic	Stable coasts, oceanic islands

Source: Reproduced with permission from National Research Council 1975.

[a] Proxy data furnish evidence of climatic conditions by indirect means.

Period Open to Study (YBP)	Minimum Sampling Interval (Yr)	Usual Dating Accuracy (Yr)	Climatic Inference
10,000	1–10	±1–100	Temperature, accumulation
100,000+	Variable	Variable	Temperature
1,000 (common) 8,000 (rare)	1	±1	Temperature, runoff, precipitation, soil moisture
12,000	1–10	±10	Temperature, precipitation, soil moisture
12,000 (common) 200,000 (rare)	200	±5%	Temperature, precipitation, soil moisture
40,000	—	±5%	Extent of mountain glaciers
25,000 (common) 1,000,000 (rare)	—	Variable	Area of ice sheets
1,000,000	200	±5%	Temperature, precipitation, drainage
50,000	1–100 (variable)	±5% ±1	Evaporation, runoff, precipitation, temperature
5,000	1	±5%	Temperature, precipitation
200,000	500+		Wind direction
200,000	500+	±5%	Sea-surface temperature, surface salinity, sea-ice extent
200,000	500+	±5%	Surface temperature, global ice volume; bottom temperature and bottom water flux; bottom water chemistry
10,000+	20	±5%	As above
1,000,000+	1000+	±5%	As above
400,000	—	±5%	Sea level, ice volume

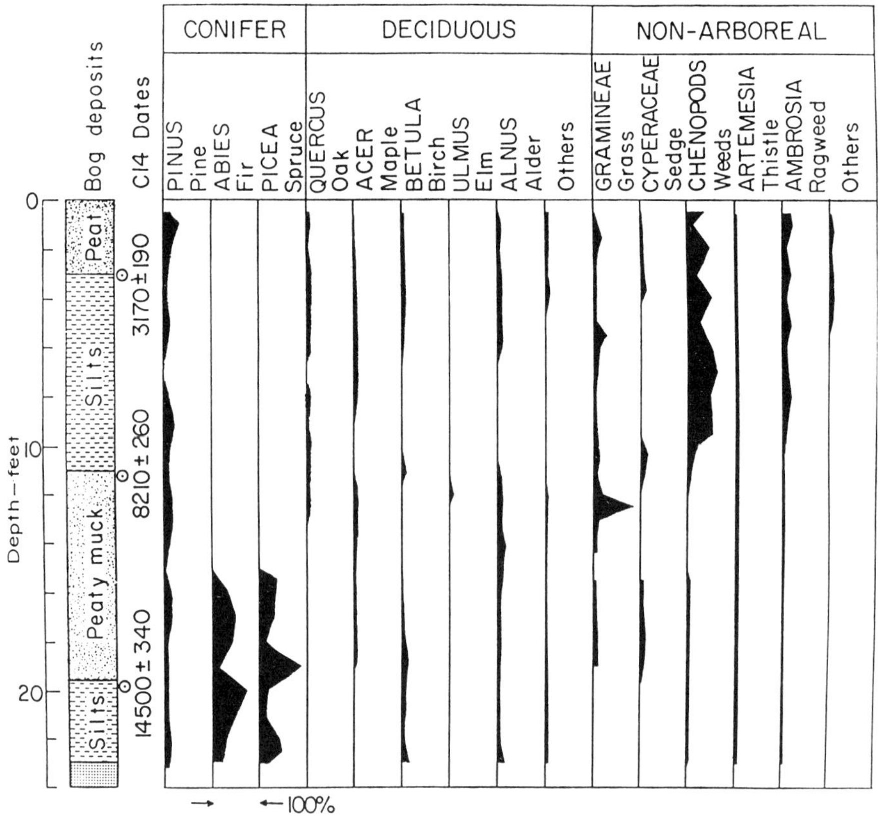

Fig. 4–2. Pollen profile diagram of McCulloch bog in southern Hancock County, Iowa. (Reprinted by permission from Robert V. Ruhe, *Quaternary Landscapes of Iowa,* © 1969 by Iowa State University Press, Ames, Iowa 50010.)

Pollen can be identified from other deposits, such as soil, silt, clays, and sands. As an example, analysis of pollen contained in the pulp cavities of mastodon tusks gives a clue to the vegetation of the western Ozarks in Missouri some 16,500 to 25,000 years ago (Mehringer, King, and Lindsay 1970). A pollen profile diagram is shown in figure 4-2. Since other organic matter often occurs with pollen, it is possible to add a time scale to the pollen profile by radiocarbon (C^{14}) dating (see below).

Pollen analysis is an excellent tool for reconstructing former vegetation cover because the small size of the grains permits very small samples of material to yield enough pollen grains to permit statistical analysis. Also it is possible to sample very thin layers of sediment, whereas macrofossils (large fossils) may penetrate several layers, sometimes making it difficult to determine their exact stratigraphic position.

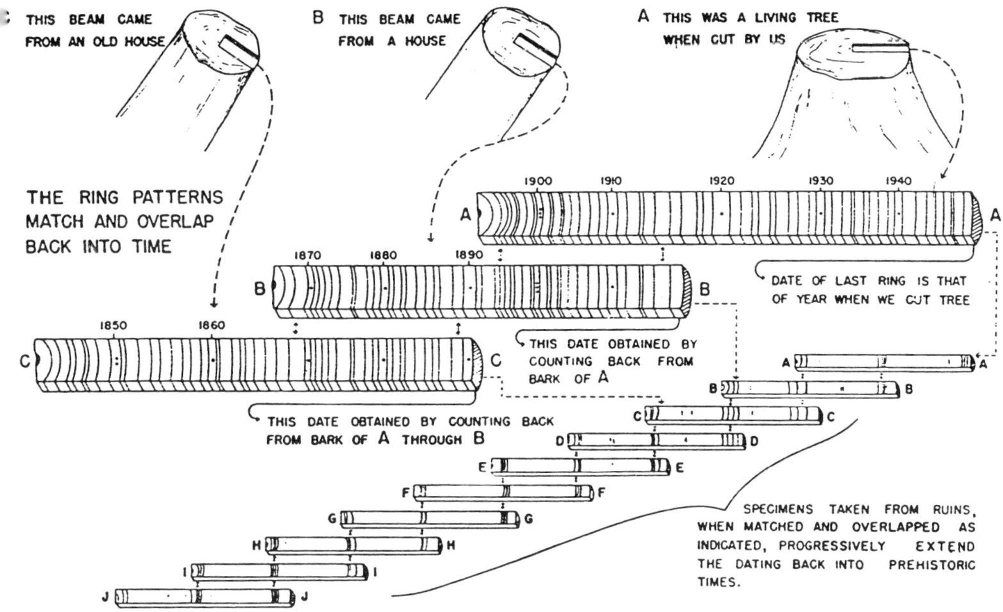

Fig. 4–3. Illustration of tree-ring dating technique of building chronologies by matching overlapping sequences of rings from successively older trees—from living trees to remains of trees from prehistoric times. (From W. Stallings, Jr., *Dating Prehistoric Ruins by Tree Rings,* Santa Fe, N.M.: Laboratory of Anthropology, School of American Research, General Series Bulletin 8, 1939. Reproduced with permission of Tree-Ring Laboratory, University of Arizona.)

Dendrochronology (Tree-Ring analysis). Trees that grow during only part of each year add annual growth layers; when a tree is cut down, these layers appear as rings in the trunk.[1] In favorable years, rings are thick, in unfavorable years, thin. Generally, in areas of plentiful rainfall, the controlling factor in tree ring growth is temperature; in dry areas, it is precipitation. It is possible to identify the year a particular ring formed in a living tree by counting its annual growth rings, or by locating a particular ring in a sequence of thick and thin rings, characteristic of trees in a particular area. By cross dating—overlapping ring sequences from near the center of living trees with the outer rings of the remains of dead trees—it is possible to extend the tree-ring dates back several thousand years (fig. 4-3). By this method, in the southwestern United States, tree-ring chronologies for the bristlecone pine (*Pinus longaeva*) extend back 8,200 years. (The oldest living

[1] It is not necessary to cut down a living tree to view its tree-ring record. A precision instrument called a Swedish increment borer removes a small core from a living tree without harm. This small core—smaller in diameter than a pencil—is easy to transport, examine in the laboratory, and store for future reference.

bristlecone pines known are 4,600 years old, which makes them probably the oldest living things on Earth.)

Not only can individual years and groups of years be identified by a dendrochronologist, but also variations in widths of rings provide clues to climatic conditions at the time the rings formed. Best results are obtained where trees grow near the limits of their range, as on the Colorado Plateau of the Southwest; here, precipitation is the limiting climatic factor. Careful selection of samples is necessary to ensure that some environmental factor other than climate did not cause the narrow or wide rings. As an example, a tree growing on a river flood plain may produce a narrow ring in a wet year because poor drainage inhibits its growth, whereas a tree growing on a slope above the flood plain may produce a wide ring in the same year. Today, computer statistical analysis has replaced the visual matching of tree-ring sequences.

Isotope Analysis

Most elements are a mixture of *isotopes*, some stable, others radioactive and changing into other elements (*daughter elements*). For example, we know that spontaneous decay of both carbon^{14} (C^{14}) and rubidium^{87} (Rb^{87}) occurs when a neutron in the nucleus of an atom splits into an electron (beta particle) and a proton. The electron leaves the nucleus as a beta particle, and the proton remains in the nucleus, thus changing the atomic number but not the mass (an electron has practically no mass). The equations for these decay processes are:

$$_{6}C^{14} - \beta \rightarrow {}_{7}N^{14}$$
$$_{37}Rb^{87} - \beta \rightarrow {}_{38}Sr^{87}$$

Isotopes of the elements uranium (U^{235} and U^{238}) and thorium (Th^{232}) decay into the daughter elements lead and helium by emitting particles, each equivalent to the nucleus of a helium (He^4) atom, as follows:

$$U^{235} - 7\alpha \rightarrow Pb^{207} + 7He^{4}$$
$$U^{238} - 8\alpha \rightarrow Pb^{206} + 8He^{4}$$
$$Th^{232} - 6\alpha \rightarrow Pb^{208} + 6He^{4}$$

The time required for half the atoms in a sample of a radioactive *parent element* to change into a daughter element is the *half-life* of the isotope. Half-lives vary greatly, from a small fraction of a second to billions of years. C^{14}, for example, has a half-life of about 5,870 years, whereas the half-life of Rb^{87} is 47 billion years (47×10^9). Half-lives for the uranium and thorium isotopes and other isotopes such as potassium40 amount to hundreds of millions or billions of years; thus, they are the choice for dating very old rocks. With a shorter half-life, radiocarbon is preferred for dating events of the last thirty thousand years.

Radiocarbon dating is particularly valuable for determining the age of or-

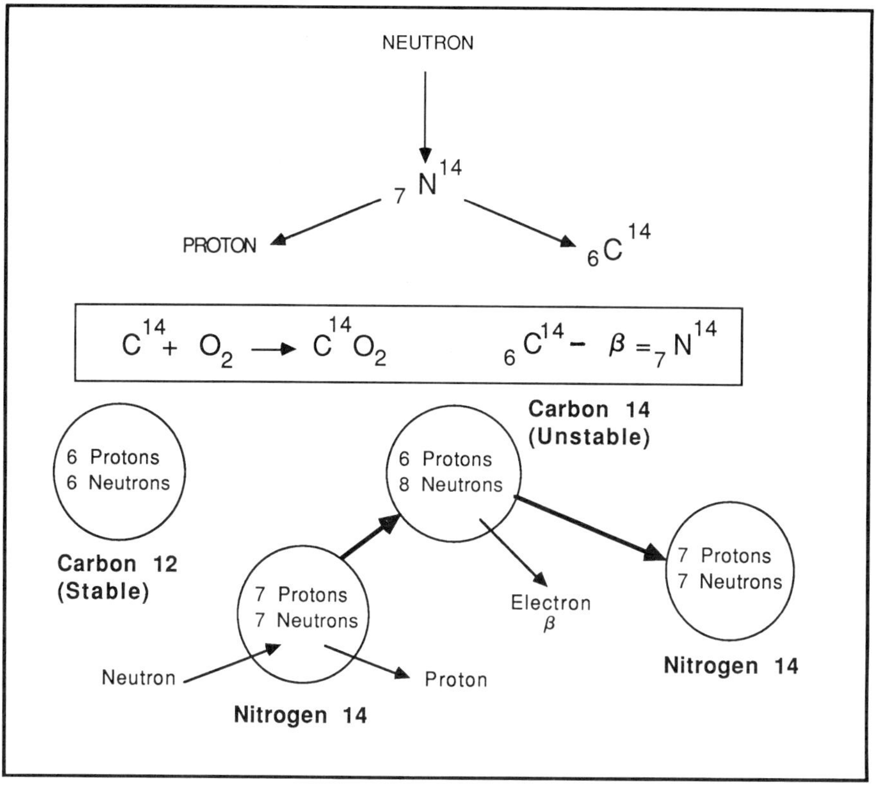

Fig. 4–4. Formation of C^{14} in the upper atmosphere.

ganic matter. C^{14} forms in the upper atmosphere when a neutron strikes an N^{14} atom, displacing a proton in the nucleus, and thereby changing the atomic number from 7 to 6, but leaving the mass unchanged (fig. 4-4). The displaced proton picks up an electron and becomes a hydrogen atom. The reaction can be written as:

$$N^{14} + n \rightarrow C^{14} + H^{1}$$

The former nitrogen atom now behaves chemically as a carbon atom, combines with oxygen to form CO_2, and enters the life chain through photosynthesis.

Living things, both plants and animals, contain a known ratio of C^{14} to C^{12} atoms in their organic compounds. After a plant or animal dies, of course, no new carbon is added, and the C^{14} begins to diminish in relation to C^{12}. If a sample of the remains of a tree contains only half the expected C^{14} in relation to the total carbon, the specimen is assumed to be about 5,870 years old—the half-life of C^{14}. Also, radiocarbon dates must be corrected to account for variations over time in the amount of C^{14}. Some of this variation may reflect such factors as changing intensity of cosmic radiation; variations in the earth's magnetic field, which alter the pattern of cosmic radiation reaching the atmosphere from space; increase or decrease in the amount of CO_2 in the atmosphere; and, more recently, testing of nuclear devices. Dendrochronological dates are used to verify and correct radiocarbon dates encompassing the last eight thousand years.

Isotope analysis in paleoclimatology is not restricted solely to the use of radioactive isotopes for supplying time markers for climatic events. One of the most valuable indicators of past climatic change is variation over time in the ratio of O^{18} to O^{16}. The ratio between the stable oxygen isotope (O^{18}) to common oxygen (O^{16}) in the calcium carbonate ($CaCO_3$) shells of marine animals, especially foraminifera, is a sensitive indicator of the water temperature when the shells formed. Shells from different layers in ocean bottom sediment cores show warming and cooling trends correlated with glacial advances and retreats on land. The O^{18}/O^{16} ratios in ice cores from Greenland and Antarctic ice sheets indicate the temperature of snow that formed ice in different layers. These temperatures aid in estimating the age of the layers by correlating them with dates and temperature profiles obtained from sea bottom cores.

The Paleomagnetic Record

Use of magnetic compasses in navigation for the last four hundred years provides a lengthy historical documentation of the geomagnetic field, undoubtedly the longest continuous instrumental record in the earth sciences. This record demonstrates that *declination* (variation in the north-south orientation of the compass needle), *inclination* (dip of the compass needle from horizontal), and strength of the Earth's magnetic field all vary from place to place and change over time. Today, it is possible to extend the geomagnetic record far into the geologic past by measuring fossil magnetism (*paleomagnetism*) in rocks and sediments.

When lavas solidify, the orientation of the Earth's magnetic field is imprinted in particles of *magnetite* (magnetic iron oxide) contained in the rocks (*thermoremanent magnetization*). Some sediments and sedimentary rocks also may show remanent magnetism if they contain magnetite particles. As the sediments settle through water or air, magnetic particles react like tiny compass needles and become aligned with the Earth's magnetic field, thus preserving the geomagnetic orientation at the time of deposition (*depositional remanent magnetization*).

This paleomagnetic orientation at the time the lavas and sediments formed or were deposited can be analyzed in the laboratory. Analysis gives clues as to their former geographic location in relation to the earth's magnetic poles. This is an important tool in *plate tectonics* for measuring "continental drift" or, more accurately, plate movements (see below). Paleomagnetism also furnishes evidence of numerous reversals in polarity of the Earth's magnetic field; these magnetic reversals may serve as valuable time markers in some geological strata.

Plate Tectonics, Continental Drift, and Paleogeography

The idea that the arrangement of continents on the Earth has changed through time is not new. By the end of the sixteenth century, cartographers had drawn fairly accurate maps showing both coastlines of the Atlantic Ocean, and geographers had recognized the jigsaw puzzle fit between the African and South American shores. In 1620, the English natural philosopher, Francis Bacon, commented on the parallelism of the two sides of the Atlantic Ocean.

In the late nineteenth century, geologists noted similarity among flora and fauna of the Gondwana strata in India and other Southern Hemisphere areas; on this basis, they postulated a vast continent in the Southern Hemisphere, which they named Gondwanaland. This supercontinent consisted of parts of India, Africa, Australia, South America, and supposed intervening continuous land in what is now the South Atlantic and Indian oceans. Geologists continued invoking these ancient land bridges as an explanation for the distribution of flora and fauna into the middle of the twentieth century.

In 1915, Albert Wegener, a German meteorologist, published *The Origin of Continents and Oceans,* citing geological and paleontological evidence to support his hypothesis of a supercontinent he called *Pangaea,* composed of all the present continents (fig. 4-5). Wegener believed that Pangaea began to break up some 200 million years ago, and that the continents slowly drifted into their present positions (fig. 4-6). Although Wegener had some strong supporters, his ideas were not widely accepted by geologists because of his failure to develop a plausible theory to explain the mechanism of continental drift.

Only after convincing evidence began to emerge from extensive oceanographic exploration following World War II did the idea of drifting continents re-emerge. By the end of the 1960s, most earth scientists accepted Wegener's general theory of the break-up of Pangaea and the movement of continents to their present position, but now the mechanism—*plate tectonics*—was understood. Plate tectonics modified the theory of continental drift to include movement of a number of semirigid plates which form a mosaic on the Earth's surface.

Plates consist of both oceanic and continental crust and lithosphere; the continents are carried along by plate movement. Plate tectonics helps explain radical climatic changes of the past represented today by evidence of glaciation in rocks now lying near the equator and by fossils of warm climate flora and fauna in areas that today have cold climates. It is no longer

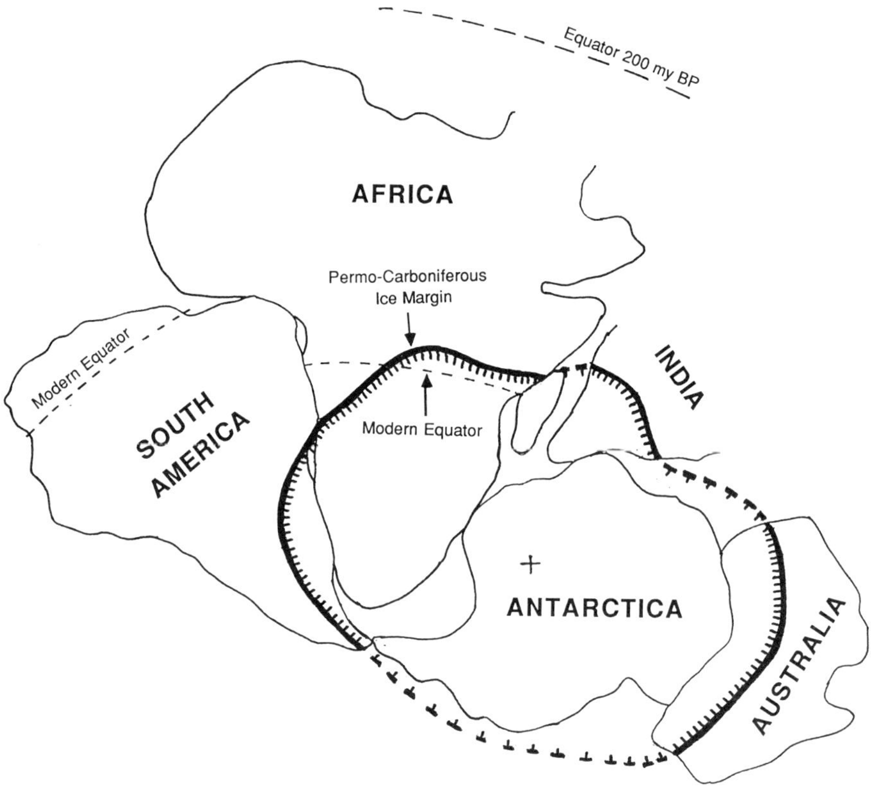

Fig. 4–5. Theoretical configuration of Pangaea showing Permo-Carboniferous glaciation. Possible position of the South Pole near the center of the polar icecap is indicated by a cross. This position is not far from the present pole, an indication that Antarctica has drifted little from its ancient position.

necessary to postulate that the Earth's climate underwent sufficient changes in temperature to produce polar conditions near the equator, or tropical conditions in polar regions.

The Historical Record

Historical records were seldom written with the specific aim of recording climatic data, but fortunately they often contain valuable information for interpreting past climate and related phenomena. These climatic clues may include such obvious information as comments on severity of a winter, success or failure of crops, presence of vineyards in areas where the present climate is unsuitable, and records of seasons of ice-free navigation on major rivers. Of particular value are accounts of Viking settlements in Iceland and

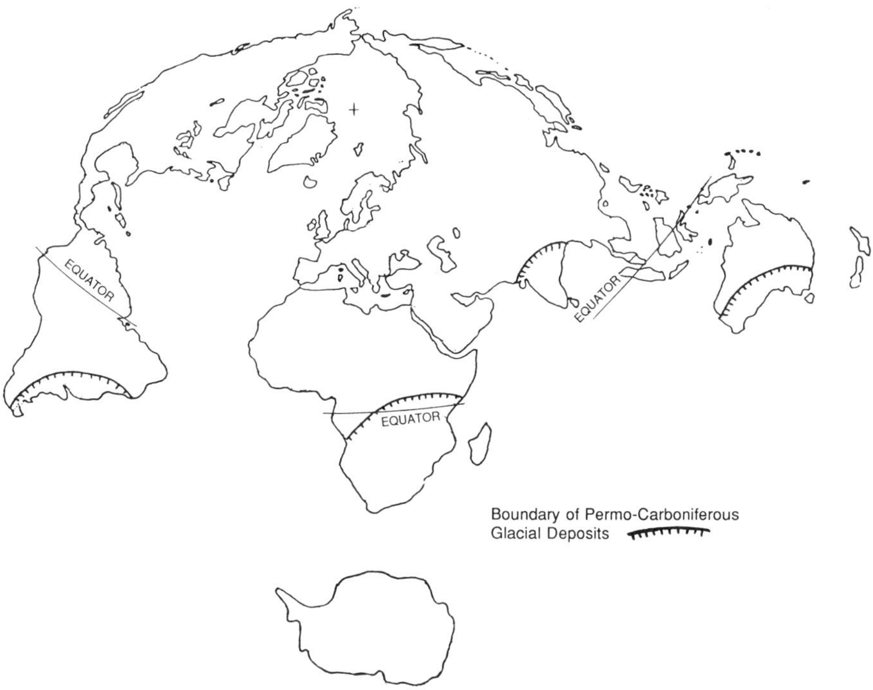

Fig. 4–6. Present position of continents indicating paleoglacial deposits that formerly were under the polar icecap.

Greenland preserved in the medieval Icelandic sagas and verified by recent archaeological investigations.

Still, most climatologists are not trained historians, and many historians fail to recognize the possible role of past climates in the shaping of historical events. Lack of communication between scholars in the physical and the social sciences often leads to gross misinterpretation of history by some climatologists, while many historians fail to recognize significant climatological information contained in historical documents.

Historical records contain indirect evidence of what might be climate-related stress on human and animal populations over the last two thousand years. Certain epidemic and pandemic diseases could be related at least indirectly to climate. Malnutrition, partly a product of poor crop years, undoubtedly creates higher susceptibility to diseases; also, the movement of wild animals and insects in response to climatic events, such as floods or drought, could spread diseases to humans or contribute to crop damage.

For example, consider the Black Death, perhaps the greatest pandemic of all time, which swept out of Asia into Europe in the mid-fourteenth century

(1348–50), killing approximately one-fourth of the European population. Bubonic plague was the primary disease of the Black Death, but two other related diseases, pneumonic plague and septicemic plague, undoubtedly played a role (Ziegler 1969). The favored vector spreading the plague bacillus, *Pasteurella pestis,* is the flea, *Xenopsylla cheopsis*. This flea prefers to live in the hair of rodents, including marmots and the common black rat (*Rattus rattus*). Throughout recorded history, plague has been endemic in the rodent population of Central Asia; it spreads to human victims only when some environmental disturbance causes the rodent population to migrate. Here is a possible climatic connection, since available evidence suggests that floods and prolonged droughts may have precipitated mass movement of rodents, particularly the black rat, at the time of the Black Death (Ziegler 1969).

Instrumental Records

In parts of Europe quantitative climatic records based on actual weather observations with meteorological instruments extend back to about 1650, when the first standardized thermometers became available. In 1643, a pupil of Galileo, Evangelista Torricelli, invented the barometer. Since 1700, some European stations have kept continuous rainfall records, and some of these records are early enough to give us an intriguing but incomplete glimpse of the "Little Ice Age"—that period between 1430 and 1850 during which occurred the most severe weather since the end of the glacial period some ten thousand years ago.

Until the twentieth century, instrumental records were largely confined to temperature, precipitation, and barometric pressure, although observations of winds and humidity and the production of continental and regional climatic maps progressed in the later part of the nineteenth century. Upper-air observations began at mountain stations, using manned balloons and instrument-carrying kites (Benjamin Franklin was one of the earliest investigators). The U.S. Weather Bureau began systematic kite observations in 1893, and in the same year Australians obtained kite data to a height of over 9,000 m (30,000 ft).

Daily upper-air data on pressure, temperature, and humidity are now received by *radiosonde,* a lightweight instrument package carried aloft by balloon and capable of transmitting data by radio to ground stations. The first use of radiosondes was in the Soviet Union in 1928; the first ascent in the United States occurred in 1935 (Mather 1974). Today, there are two hundred radiosonde observing stations in the United States and another five hundred or more on land around the world, and numerous ascents are made from ships at sea. Modern radiosonde balloons regularly reach heights of 27,400 m (90,000 ft), and special balloons may carry radiosondes as high as 42,700 m (100,000 ft). Satellites are the most recent addition to the tool kit of meteorologists and climatologists. They give visual and infrared images of cloud cover, cloud heights, snow cover, and sea ice, and they now have ability to measure land and sea temperatures.

Computers add another dimension to instrumental climatology. Not only do they process vast quantities of data in a short time, but also they make possible the construction of simulation models approximating atmospheric and oceanic circulation during past climatic epochs, and predictive models of possible future climates. As computer models help provide a better understanding of the present global climate and how it relates to the coupling of the ocean-atmosphere system, we will be better equipped to decipher past climates, and to make reasonable predictions about future changes.

A Brief Description of Climates of the Past

Knowledge of past climates is essential to gain some insight into possible climatic futures and the effect they might have on human populations and other life forms. Although most paleoclimatologists believe that the earth's "normal" geological climate was considerably warmer and more benign than it has been during the Pleistocene epoch, evidence exists that a number of glacial periods have interrupted this normal climatic state, although they have been brief in terms of total geologic time (Brooks 1970[1949]).

Precambrian Climates before 600 Ma

Existing evidence offers a dim and rather fragmented view of that vast expanse of geologic time, the Precambrian era, comprising 86 percent of Earth's existence. It was in the Precambrian period that life first appeared. In chapter 1, we reviewed some theories concerning the nature of the early atmosphere and how it evolved into the present oxygen-rich mixture of gases. The origin of life, perhaps as early as 4,000 Ma,[2] is involved intricately with the evolution of Precambrian climates. The development of green plants and photosynthesis, maybe by 2,700 Ma, the removal of CO_2, and the enrichment of the oxygen content of the atmosphere created a climate near the end of the Precambrian era in which oxygen-breathing animal life could evolve (fig. 4-1). The stage was set for the evolutionary development of more complex plants and animals, and for the great proliferation of life in the Cambrian period of the Paleozoic era.

We can make inferences about climates of the early Precambrian era, but definitive evidence of glaciation is found only in rocks of middle and late Precambrian age (fig. 4-1). Although major glaciation possibly occurred in

[2] Because geologic time encompasses such vast numbers of years, abbreviated notations are used instead of long strings of zeros. *Ma* stands for "million years ago"; *Ya*, meaning "years ago," is used for dates more recent than 1 Ma. Also, the term *billion* has different meanings: In the U.S. and French system, a billion is 1,000 millions, or 1,000,000,000 (10^9), whereas in the British and German system a billion is 1,000,000 millions, or 1,000,000,000,000 (10^{12}). To avoid confusion, in this text we use multiples of a million. Thus an American "billion years" becomes 1,000 million years; a "billion years ago" becomes 1,000 Ma. Exponential notation also is used, so that four and one-half billion (American) is expressed as 4.5×10^9.

earlier periods, the rocks have been so greatly altered in subsequent periods (*metamorphosis*) that identification of glacial materials seems unlikely. Geologic evidence exists (some of it disputed) for at least four major ice ages in the Precambrian. The first records date from around 2,300 Ma (*Huronian*), and consist of deposits of *tillite* (solidified ice-deposited materials called till) in Canada, and possibly in South Africa and India. Evidence for the *Gnejsö* Ice Age of some 950 Ma comes from Greenland, Spitsbergen, and Norway; for the *Sturtian,* 750 Ma, from Australia, China, southwest Africa, and Scandinavia; and the *Varangian,* 660 to 680 Ma, from North America, Greenland, parts of northern Europe, the U.S.S.R., China, India, Australia, Africa, and South America (Gribbin 1978). Some geologists believe that at least some of the Varangian deposits are not really of glacial origin because paleomagnetic evidence indicates that they were formed at low latitudes, and because glacial striations on bedrock are not present (Girdler 1964).

Climates of the Lower Paleozoic Era (640–340 Ma)

Warm ice-free conditions of what C.E.P. Brooks called "normal" geological climate seems to have persisted throughout most of the Paleozoic era, but evidence exists in rocks in the Sahara Desert of a glacial epoch during the Ordovician period about 460 Ma (fig. 4-1). Contained in these rocks are remnants of U-shaped paleovalleys with ice-polished and striated floors overlain by glacial deposits (tillites), and glacial outwash materials containing frost polygons and other evidence of permafrost. Ordovician deposits of possible but questionable glacial origin also have been identified in other parts of Africa, South America, Newfoundland, and Great Britain (Gribbin 1978).

Upper Paleolithic Climates (340–230 Ma)

Of all pre-Pleistocene ice ages, glacial deposits of Permo-Carboniferous age are the best documented and most studied. These occur in all the Gondwana continents—Antarctica, Australia, Africa, and South America—and probably account for support given to the idea of continental drift by Southern Hemisphere geologists long before their Northern Hemisphere colleagues accepted the idea. As during the Ordovician glaciation in the lower Paleolithic, the Permo-Carboniferous or Gondwanan glaciation probably occurred when the configuration of Pangaea placed these areas in south polar regions (fig. 4-5).

When glaciers were present in what are now parts of the Southern Hemisphere continents, the paleoclimate of North America was certainly tropical and subtropical. The Carboniferous (Latin, coal-bearing) period, which is designated as Pennsylvanian (lower Carboniferous) and Mississippian (upper Carboniferous) in North America, was the period of luxuriant forests, the fossil remains of which form the major coal deposits in North America and Europe. Coal formation continued into the Permian, but the younger deposits of the Mesozoic and Cenozoic are mostly lignites (brown coals). The centers of extensive coal formation are in areas of Pangaea that lay in tropical and subtropical latitudes.

Climates of the Mesozoic Era (230–65 Ma)

The Triassic period saw the beginning of the breakup of Pangaea and the evolutionary proliferation of reptiles from their modest beginning in the Pennsylvanian. The reptiles were the first vertebrates to take full advantage of the increasing supply of oxygen in the air and in the oceans, brought about by the proliferation of green plants. The most famous of the reptilian orders was the dinosaurs, which dominated land, sea, and air throughout most of the Mesozoic era. Plate movement, following the initial breakup of Pangaea, transported the continents toward their present geographical positions. Mammals, extending back to at least the late Triassic, remained small but continued to evolve, replacing the dinosaurs as the dominant land-dwelling animals in the following era, the Cenozoic.

During the last part of the Mesozoic era, from about 100 Ma to 65 Ma, the global climate seems to have been considerably warmer than it is today, and there were no polar icecaps. These mild climatic conditions seem to have persisted well into the Cenozoic era.

Climates of the Tertiary Period of the Cenozoic Era (65–4 Ma)

Although the global climate is thought to have been mild during the first 10 million years of the Cenozoic era, a long cooling trend set in about 55 Ma, and it continued with only minor interruptions until the great ice age of the late Pliocene and Pleistocene epochs. By 10 Ma, extensive mountain glaciation and substantial growth of the Antarctic ice marked the beginning of the present glacial age. Although the present geologic epoch is referred to as postglacial or *Holocene,* it is probably one of the numerous interglacial stages that have separated glacial advances throughout the last 10 million years.

The Great Ice Age of the Pleistocene (4 Ma–10,000 Ya)

Continental ice sheets first appeared in the Northern Hemisphere about 3 Ma when the North Atlantic Glacial began. Although the boundary between the Pliocene and Pleistocene is set officially at 1.7 Ma, based on the first appearance of cold conditions in the Mediterranean basin, a more logical boundary would be some 3–4 Ma; by that time glaciation clearly had been initiated in the Northern Hemisphere (fig. 4-7).

During the ensuing time between initiation of the present glacial age and retreat of the last glacial ice some ten thousand years ago, there may have been seven or more glacial cycles, separated by interglacial episodes, each cycle lasting some hundred thousand years (Roberts and Lansford 1979). For many years, glaciologists identified only four major glacial episodes

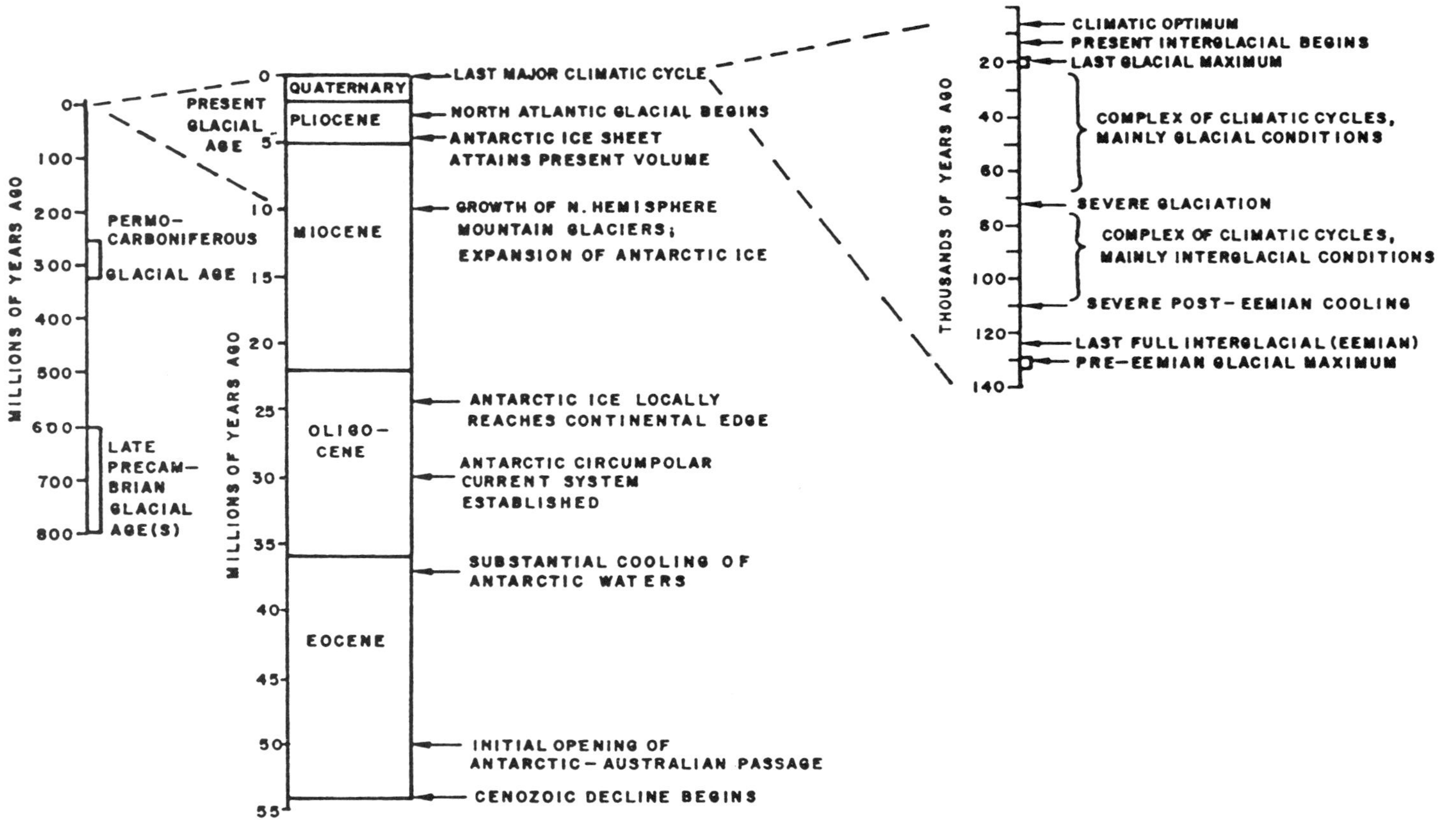

Fig. 4–7. History of glacial ages over the last 1,000 million years. Intervals when extensive polar ice sheets occurred are indicated. (From National Research Council 1975, 161.)

during the Pleistocene, named (from oldest to youngest) the Nebraskan, Kansan, Illinoian, and Wisconsinan in North America (fig. 4-8), and the Günz, Mindel, Riss, and Würm in the European Alps.

Geological evidence is preserved best in the younger Wisconsinan glacial deposits, whereas the Nebraskan deposits are discontinuous and often buried beneath younger materials. Recent evidence suggests that the Nebraskan represents several glacial stages, and it seems probable that it will be re-designated "Pre-Kansan" until the individual stages can be identified. Figure 4-9 indicates the known and inferred outer limits reached by ice advances during the four glacial stages in central North America.

The last ice advance, the Wisconsinan, reached its greatest extent some 14,000 Ya, when it covered about one-fourth of the earth's land surface. The ice front then began to recede because wasting exceeded the rate of advance. By 8,500 Ya, the continental glaciers had disappeared from Europe, leaving only mountain glaciers, and by 7,000 Ya, North American ice had shrunk to its present extent, with ice sheets remaining only in Greenland and on some neighboring islands in the Canadian Arctic. Today, glaciers cover about 10 percent of the Earth's land surface.

Postglacial Climates of the Holocene (10,000 Ya–Present)

By 10,000 Ya, most formerly glaciated areas were ice free, and postglacial time—the Holocene—was well under way. A picture of the climate of the past ten thousand years emerges from numerous palynological studies, oxygen-isotope analyses, tree-ring investigations, records of changing sea level, and critical reading of historical accounts. The Holocene has not been a period of progressive and gradual warming; rather, there is increasing evidence that several widespread warming and cooling trends have marked the period (figs. 4-10 and 4-11), associated with variations in the global pattern of precipitation.

Although climate is always changing, four climatic epochs stand out in the record and are of particular interest to climatologists—the *Altithermal* (warm), the *Iron Age* (cold and rainy), the *Secondary Climatic Optimum* (warm), and the *Little Ice Age* (cold).

The Altithermal

The earliest Holocene climatic epoch, the Altithermal (sometimes called the *Climatic Optimum*), reached its peak between 7,000 Ya and 5,000 Ya. It probably was the warmest time since the end of the Pleistocene. During the Altithermal early civilizations developed in Egypt, the Middle East, and China. Climate and vegetation belts shifted northward, with a well-developed monsoon bringing rainfall to the Sahara Desert, giving it a climate similar to present savannas to the south. Summer temperatures in Europe averaged higher than today by 2–3°C (3.6–5.4°F), and Hawaii experienced increased rainfall as the tropical rainy belt moved northward in

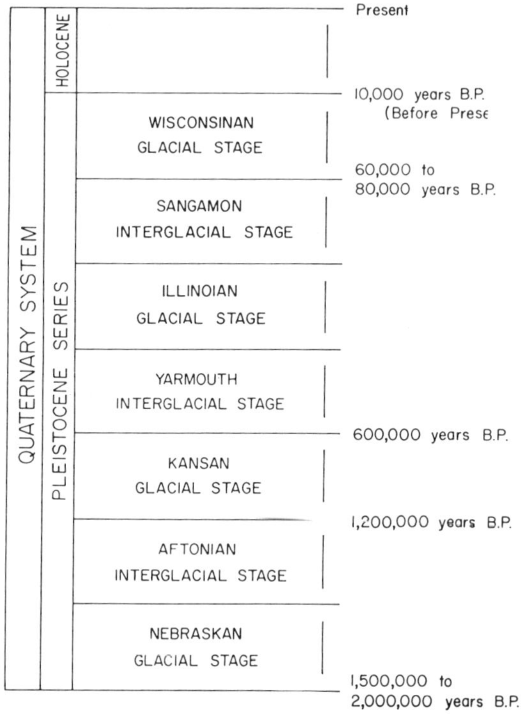

Fig. 4–8. Stratigraphic column of Quaternary system for North America. B.P. means "before present." (From Jean C. Prior, *Iowa Landforms,* Iowa Geological Survey Educational Series 3, Iowa City, 1976.)

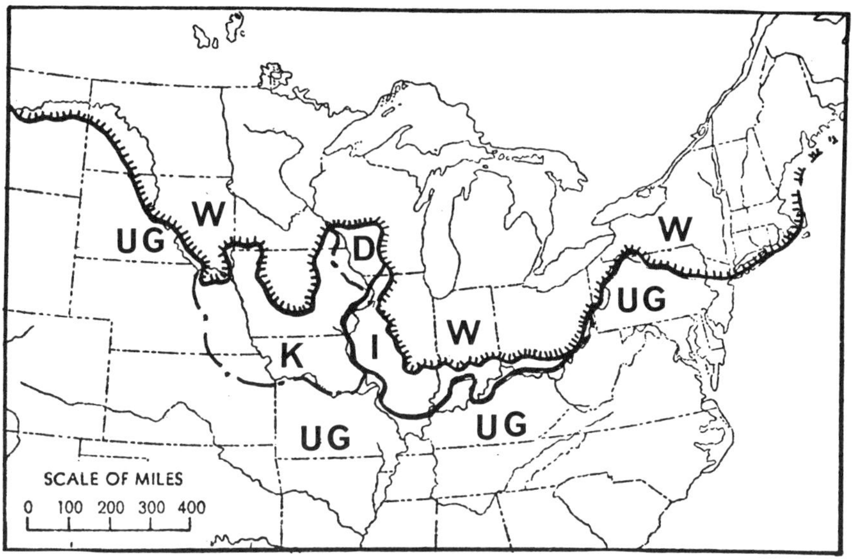

Fig. 4–9. Known and inferred outer limits of glacial stages in eastern North America. Key: W = Wisconsinan; I = Illinoian; K = Kansan; D = "Driftless Area" probably glaciated by Pre-Kansan glaciers; UG = Unglaciated. Pre-Kansan (Nebraskan?) drift borders are mostly covered by more recent ice advances. (Generalized from USGS maps.)

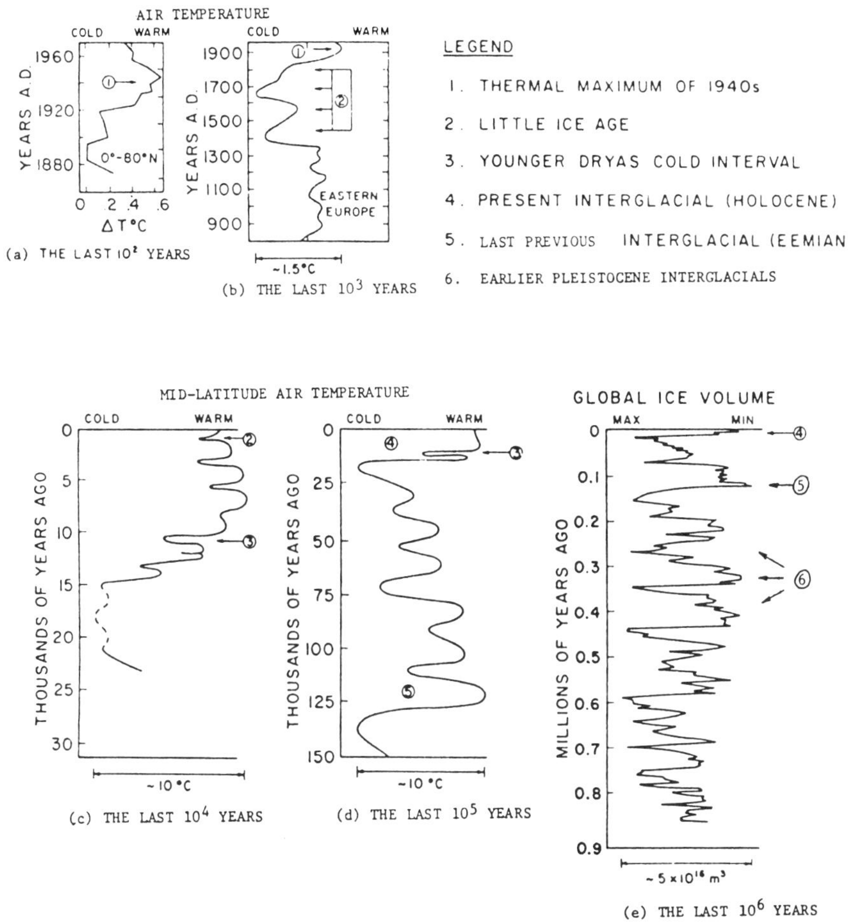

Fig. 4–10. Generalized trends in global climate. (From National Research Council 1975, 130.)

the Pacific Ocean. Simultaneous warming occurred in the Southern Hemisphere which indicates that the Altithermal was a global event (Gribbin 1978; Huntley and Prentice 1988).

The Iron Age

The Iron Age of northern Europe, between about 900 and 300 B.C. (2,900 and 2,300 Ya), was a period of cold, wet climatic conditions responsible for the regrowth of peat bogs and the spread of heaths into what had been forest lands (Trolls-Smith 1961). Beech replaced oak-elm forests on the better soils, and heath replaced these forests on the sandy outwash plains of western Denmark and northern Germany (see chapter 5). In the Mediterranean

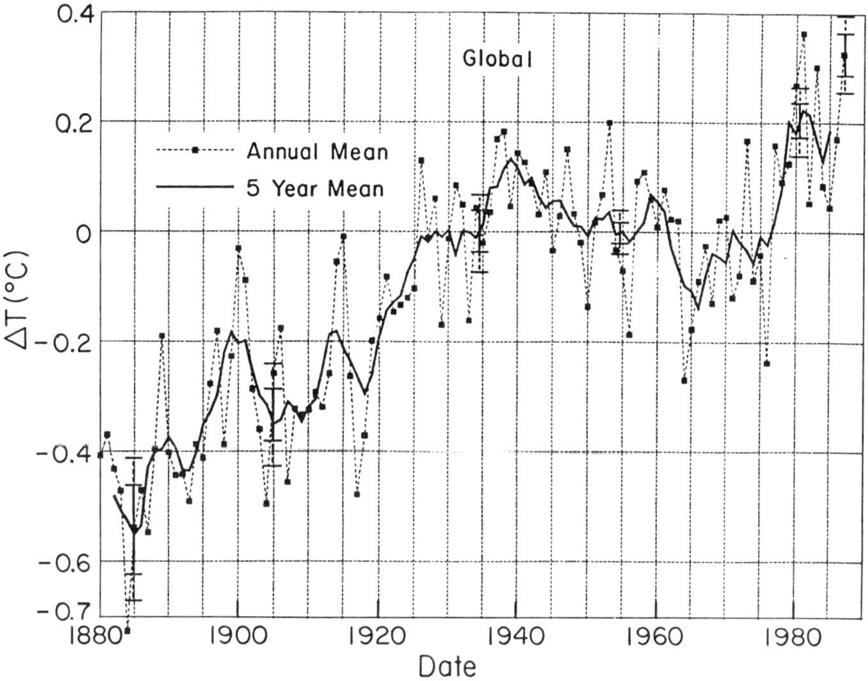

Fig. 4–11. Global surface air temperatures for the last 100 years. The notation △T* means change in temperature. (From U.S. Department of Commerce 1988. Based on Hansen and Lebedeff 1988.)

basin and in North Africa, the climate apparently became drier, and Roman records indicate that the cultivation of olives and grapes spread northward in Italy. This northward extension of warmth-loving crops seems to imply that cooler conditions may have prevailed in the Mediterranean before 100 B.C. Also evidence exists for cooler conditions in the Southern Hemisphere between 500 and 300 B.C. (Gribbin 1978).

The Secondary Climatic Optimum

This warm period occurred around A.D.1000–1200 and it may have lasted for another century in North America. Historical climatology furnishes more descriptive data than are available for previous climatic epochs. This was the era of Norse exploration and settlement in Iceland, Greenland, and Newfoundland; in central and Western Europe, evidence exists of vineyards farther north and at higher elevations than in previous times, and the tree line in mountains was higher. These changes indicate that mean summer temperatures were a little more than 1°C higher than at present.

Evidence from other parts of the world indicates that the secondary optimum was a global phenomenon; for instance, there were milder climates in Antarctica, and moister conditions in Central America, Cambodia, the Mediterranean, and the Middle East (Gribbin 1978). In North America, archeological investigations of Indian farming communities of the eastern Great Plains and adjacent areas to the east indicate a severe drought and abandonment of many farming communities. Drought conditions in this area coincided with an expansion of westerly air flow, a possible cause of the warmer and drier conditions preceding the abandonment of agricultural villages by A.D. 1400 (Bryson and Murray 1977). Excavations at the Mill Creek site in northwestern Iowa indicate that just before the abandonment of that agricultural settlement, the source of meat in the diet of the farmer-hunters shifted from deer to bison. This change in diet probably indicates disappearance of deer with the coming of drier conditions as shortgrass steppes replaced tallgrass prairies.

The Little Ice Age

Harsher conditions prevailed in Europe and elsewhere between A.D. 1430 and 1850, a period known as the Little Ice Age. Although colder and more stormy conditions predominate in the climatic record for those centuries, some periods of relatively mild climate occurred. In most areas the fifteenth and seventeenth centuries probably were the coldest, with severe winter conditions unheard of in modern times. The canals of Venice were frozen several times, with ice thick enough to support people and vehicles; the Ebro in Spain and the Tiber in Italy were frozen in some years, and during the Thirty Years' War, which lasted from 1618 to 1648, the Swedish army often crossed the frozen Baltic Sea in winter (Neuberger et al. 1973).

In the British Isles, the winter of 1683–84 was the coldest recorded since the invention of the thermometer, and probably was the coldest since early postglacial times (Gribbin 1978). One of the most vivid descriptions of that winter is found in the novel *Lorna Doone,* by R. D. Blackmore. Although written in the 1860s, the following description of the terrible winter weather in Devon in 1683–84 is historically correct:

> That night such a frost ensued as we had never dreamed of, neither read in ancient books, or histories. . . . The kettle by the fire froze, and the crock upon the hearth-cheeks; many men were killed, and cattle rigid in their head ropes. Then I heard that fearful sound, which never I had heard before, neither since have heard (except during that same winter), the sharp yet solemn sound of trees, burst open by the frost-blow. Our great walnut lost three branches, and has been dying ever since. . . . And the ancient oak at the cross was rent, and many score of ash trees. But why would I tell all this? the people who have not seen it (as I have) will only make faces, and disbelieve; till such another frost comes; which perhaps may never be.

In North America, the coldest year in historic times, at least in New England, may have been 1816—"the year without a summer." Weather records for that year at New Haven, Conn., indicate a June mean temperature 7°F below normal, and snow fell throughout most of northern New England on 5–6 June. The cold continued without respite, from the 6–10 June, with hard freezes five nights in succession at Danville, Vt. (Stommel and Stommel 1983). Western Europe also experienced an exceptionally cold summer in 1816, resulting in extensive crop failures, food shortages, and famine. Famine in Ireland in 1817, following the failure of the potato crop the preceding year, may have led to a typhus epidemic that killed sixty-five thousand people in Ireland and spread from there to all of Europe.

This cool summer of 1816 possibly was the result of the massive eruption of the volcano, Mount Tambora, on the island of Sumbawa east of Java in the southern Indonesian chain (Sigurdsson and Carey 1988). The Tambora eruption was the most explosive in the past ten thousand years, ejecting ten times as much material into the atmosphere as the more famous Krakatoa eruption in 1883, and a hundred times as much as was ejected by Mount St. Helens in 1980. Atmospheric dust from this eruption reduced direct sunlight received at the surface by an estimated 15–20 percent (Stommel and Stommel 1983). Although the volcanic dust from the Tambora eruption may have contributed to the exceptionally cool summer, the immediate cause of the prolonged extreme temperatures probably was the Rossby waves in the upper westerlies, which became locked in a pattern that steered polar air southward over New England and Western Europe throughout most of the summer of 1816.

The Past Hundred Years

The last century has seen rapid development of instrumental and theoretical meteorology, and proliferation of climatic data. Our view of the past, filtered through hundreds or thousands of years and many levels of proxy data, may leave us with a false sense that past climatic changes followed relatively simple patterns shifting from glacial to nonglacial, hot to cold, and dry to wet. The instrumental record, on the contrary, presents a detailed, if somewhat confused, pattern of climatic events requiring the aid of computers and statistical analysis to sort out trends and patterns in temperature and precipitation. As more detailed knowledge of the earth's heat balance and its effect on circulation of the atmosphere-ocean system is gained, it may be possible to formulate more realistic simulation models of conditions responsible for past, present, and future climatic changes.

The most widely accepted climatic trend during the last hundred years is a general warming indicated by global and hemisphere-averaged indices (Hansen and Lebedeff 1988). These indices indicate a worldwide warming trend from the 1880s to about 1940, and a cooling into the mid-1970s. In the late 1970s this cooling trend ended in what appears to be a return to the general warming trend persistent since the end of the seventeenth century. The 1980s have proven to be the warmest decade in the instrumental record, with 1988 so far the warmest year of the century. These overall warming and cooling trends, most pronounced in the polar regions of the Northern Hemisphere, are associated with changes in large-scale atmospheric circulation in response to changes in sea-level pressure; the role of an enhanced greenhouse effect and global warming is more difficult to substantiate.

Long-term variations in precipitation, more complex than temperature, have received less attention from climatologists. As a very general rule, when warmer conditions prevail in middle and high latitudes, drier conditions seem to prevail in some areas and increased monsoonal activity occurs in others; conversely, when high latitudes cool, monsoons tend to become more erratic or to fail altogether. The failure of the monsoon results from the displacement equatorward of subtropical highs and the intertropical convergence zone (ITC).

The warming trend from 1880 to 1940 probably was a factor in the relative dependability of the Asian-African monsoon during that period, contrasting with its greater variability in previous centuries. The recent cooling trend possibly contributed to the periodic failure of the Asiatic-African monsoon since 1940, leading to recurring disastrous drought in the 1970s and resulting in famine in the Sahel region of Africa (Bryson and Murray 1977). The effect of the present warming trend on the monsoons remains to be seen, although heavy summer rains and flooding have occurred in India, Bangladesh, and the Sudan in 1988.

Why Do Climates Change?

Processes operating in the atmosphere-hydrosphere climatic system are so complex that it is difficult to designate a single cause for climatic change. Furthermore, the concept of climatic change is somewhat confusing. Some climatologists make a distinction between *climatic variability* and *climatic change;* variability is the normal short-range fluctuation around the mean (seasonal change, cold years, warm years, droughts), whereas change involves long-range trends in the mean (a cold or warm sequence of years, centuries, or millennia). One thing is certain: climate does change. How and why it changes are debatable issues. Some possible causes of climatic change are sumarized in table 4-2.

Solar Radiation

Variation in solar radiation received by the atmosphere and the earth's surface has a direct effect on climate because it is the prime source supplying energy to run the system. An obvious example is the seasonal change resulting from the response of the internal climatic system to changes in angle of the sun, length of daylight, and, to a lesser degree, the sun-Earth distance determined by the Earth's position on its orbit. In addition to variations in solar radiation intensity resulting from the well-understood seasonal variations, changes in solar radiation intensity can result from: (1) changes in

TABLE 4-2 Some Possible Causes of Climatic Change

Possible Cause	Approximate Trends, Periodicity, or Time Scale (years) Involved
Solar Output	
Change in ultraviolet radiation	Probably random
Sunspots	11 and 22
Galactic Rotation	
Passage through interstellar dust	?
Cosmic year (1 galactic rotation)	250,000,000
Earth's Orbit and Axis	
Eccentricity of earth's orbit	100,000
Obliquity of earth's axis	42,000
Precession of equinoxes	21,000
Sun-earth distance (radius)	1
Atmospheric Composition	
Volcanic dust	Random
Dust (*aerosols*) from human activities	Increasing rapidly
Carbon dioxide from natural sources	Increasing gradually
Carbon dioxide from human sources	Increasing rapidly
Other greenhouse gases from human and animal sources (N_2O, CFCs, O_3, CH_4, etc.)	Increasing rapidly
Changes in Earth's Surface	
Land-sea configuration	>1,000,000 to 100,000,000
Topography (elevation and barriers)	>1,000,000 to 10,000,000
Urbanization, cultivation, change in vegetation (albedo, heat)	1 to 1,000s
Cryosphere (albedo, air temp., ocean temp., etc.)	
Land ice	100+
Sea ice	Seasonal +
Snow	Seasonal +

Source: Modified from Griffiths and Driscoll 1982, 267.

spectral composition of the sun's energy output; (2) variations in the Earth's orbit and inclination of the axis to the plane of the orbit; and (3) changes in the composition of the atmosphere. An additional cause might be the position of the solar system in its long orbit around the galaxy and the presence or absence of galactic dust, but there is little convincing evidence for this theory.

Variation in Solar Output

Chapter 1 discussed the possible effects of very long-range variations in the sun's radiant energy output on the Earth's early atmosphere and climate; the short-range effects are not so clear. The measured variations of solar output indicate small random changes in ultraviolet intensity associated with the eleven-year sunspot cycle. Although these changes are small, their effects may be magnified by feedback or triggering mechanisms. Attempts to correlate rainfall patterns with the sunspot cycle have not yielded statistically conclusive results, although both positive and negative correlations appear to exist in climatic records for some areas (Waite and Partington 1986).

The Milankovich Model

Considerably more evidence exists to substantiate a climatic relationship with variations in the orbit and inclination of the Earth. Three astronomical cycles form the basis of the *Milankovich Model.*[3] They are: (1) variation in eccentricity of the Earth's orbit; (2) variation in tilt of the Earth's axis from vertical to the plane of the orbit (obliquity); and (3) changes in the season of closest approach to the sun (precession of the equinoxes).

Eccentricity. Over time, the Earth's orbit varies from nearly circular to more elliptical and back to circular, a cycle requiring from ninety thousand to one hundred thousand years for completion. The difference between minimum and maximum insolation varies by as much as 30 percent between maximum and minimum eccentricity.

Obliquity. Changes in the tilt of the axis varies between 21.8° and 24.4° from vertical to the plane of the orbit, with a period of about forty-one thousand years. This tilt is what produces seasonal change on the Earth—the greater the angle, the greater the amount of solar radiation received in middle and high latitudes in summer, and the less in winter. Less obliquity means less contrast between winter and summer.

Precession. Because the Earth wobbles on its axis, the season of closest approach to the sun (*perihelion*) varies through a cycle of approximately twenty-two thousand years. The Earth is now closest to the sun during the Southern Hemisphere winter (South Pole inclined away from the sun), and farthest away (*aphelion*) during the Northern Hemisphere summer (North Pole inclined toward the sun); in some eleven thousand years, the situation will reverse. With the exception of Antarctica, the Southern Hemisphere land masses in middle and high latitudes are not extensive, and dominance of water masks the effects of greater contrasts in seasonal radiation received (see chapter 2).

Configuration of Oceans and Continents

Already we have discussed the climatic significance of plate tectonics. On a scale of tens of millions of years, significant changes take place in the shapes of ocean basins and distribution of land on the globe. Obviously, the arrangement of land and water plays a significant role in determining global climates. Many of the variations in solar radiation mentioned above are rela-

[3]Named for Milutin Milankovich, Yugoslav astronomer and meteorologist, who in the 1920s calculated the influence of changes in Earth-sun geometry on solar radiation received on Earth. Recent calculations extend the record to 120,000 Ya.

tively small in relation to total energy received from the sun; thus, it is probable that fortuitous distribution of land and water magnifies these seemingly minor fluctuations. For example, the Milankovich variables probably are not of sufficient magnitude to initiate an ice age unless there are land or shallow seas in the polar regions—a situation existing during the Pleistocene Ice Ages and still prevailing today.

Changes in Atmospheric Composition

In chapter 1 we already have discussed modifications in atmospheric composition resulting from natural evolution of the atmosphere, involving changes in its relative nitrogen, oxygen, and CO_2 content. We also treated earlier the possible role of recent volcanic eruptions in reducing the effectiveness of solar radiation and in producing cold periods. Here, we deal with modification of the atmospheric radiation balance by progressive enrichment of the content of greenhouse gases during the present century. The CO_2 content of the atmosphere has increased by possibly 20 percent since the 1880s, chiefly from burning fossil fuels; since then there has been a warming trend (figs. 4-12 and 4-13). Human activity also contributes to increases in other greenhouse gases; these gases include nitrous oxide (N_2O), chloroflurocarbons (CFCs), methane (NH_4), ozone (O_3), and others. Some estimates indicate that in the next forty years, the warming effect of the other greenhouse gases will equal or exceed that caused by CO_2 (fig. 4-14). Paralleling the increase in CO_2 and other greenhouse gases is atmospheric dust from human activity, particularly from industry and agriculture.

Increase in CO_2 and other greenhouse gases enhances the greenhouse effect, in turn increasing surface temperatures. Increased temperatures produce a positive feedback effect by causing more evaporation, and water vapor content of air increases, enhancing the greenhouse effect. Increased water vapor also increases cloud cover, which has a negative influence on temperature by reflecting about 30 percent of the sunlight back to space, while at the same time contributing to the greenhouse warming by trapping infrared radiation from the surface. Data from the Earth Radiation Budget Experiment (ERBE) for 1985 indicate that clouds have a small net cooling effect, particularly in high latitudes (Ramanathan et al. 1989)

Aerosol particles both scatter and absorb solar radiation. The part that scatters may travel downward and upward, the upward component being lost to space, so that less sunlight reaches the ground. Solar energy absorbed by particles adds heat to the upper atmosphere, decreasing the vertical lapse rate and inducing subsidence, a process that discourages precipitation. Consequently, not only does the presence of atmospheric dust cool the surface and warm the upper atmosphere, it may also produce drier climatic conditions (Bryson and Baerris 1967).

An extreme case of particle loading of the atmosphere undoubtedly would be one result of a major exchange of nuclear weapons, after which a so-

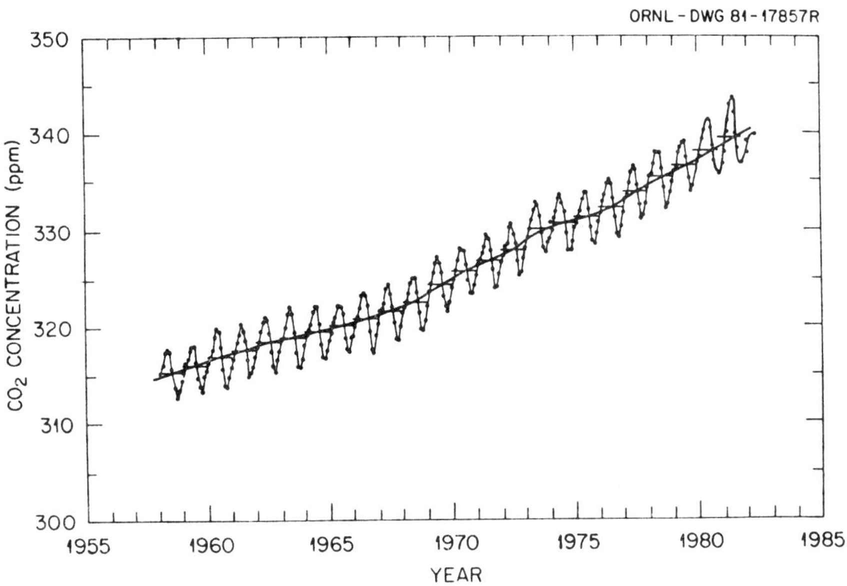

Fig. 4–12. Mean monthly concentrations of atmospheric CO² at Mauna Loa, Hawaii. The yearly oscillation is explained mainly by the annual cycle of photosynthesis and respiration of plants in the Northern Hemisphere. (Geophysical Monitoring for Climatic Change, National Oceanic and Atmospheric Administration.)

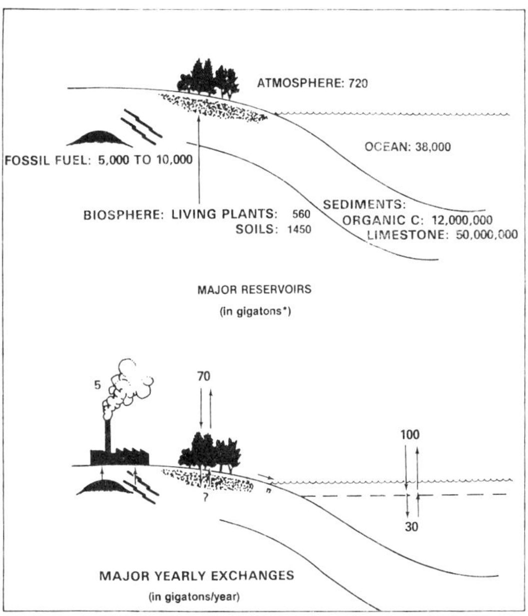

Fig. 4–13. The carbon cycle in the Earth's atmosphere, biosphere, hydrosphere, and lithosphere. All numbers are in gigatons. One gigaton equals 10^9 (1,000 million) metric tons. (From U.S. Dept. of Energy, *The Carbon Dioxide Research Plan, A Summary*, 1983.)

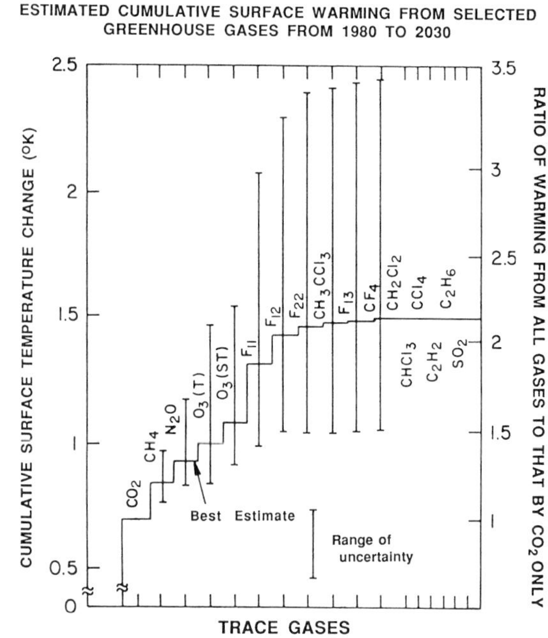

Fig. 4–14. Estimated cumulative surface warming due to increase in CO_2 and other trace gases by 2030. The "best estimate" line represents a consensus of the investigators as to the most probable effect of each gas. The vertical bars represent the range of uncertainty for trace gas amounts: the top of the bar indicates the possible but not probable maximum effect; bottom the possible but unlikely minimum effect. Note that the combined warming from other greenhouse gases will exceed that by CO_2 by 2030 (CO_2 warming is 1; CO_2 + other gases is > 2). (Modified from V. Ramanathan et al. 1985. Copyright © 1985 by the American Geophysical Union. Reproduced with permission.)

called nuclear winter might threaten life on Earth. A small-scale preview of such an event was the aforementioned "year without a summer" that followed the massive injection of dust into the atmosphere from the 1815 eruption of Mount Tambora. Within 600 km (360 mi) of the volcano, the sky remained pitch dark for one or two days, with the air first becoming hot and then cold (Stothers 1984).

The Climatic Future

The suggestion made at the beginning of this chapter that "the past is the key to the future" is valid for predicting natural climatic change, but forecasting future climates involves taking into account the increasing effects of human activity. All the possible effects summarized above probably will play some role in determining future climates, but for the immediate future, the mechanisms enhanced by human activity seem to be of paramount importance.

A Warmer or a Colder Future?

According to most climatologists and glaciologists, there is no reason to assume that the last major advance of continental glaciers was the final such event of the Pleistocene. It is logical to assume that the present may be an interglacial stage preceding another glacial event. Nonetheless, most researchers expect a warming trend for the next century, brought about by human activities, largely through increase in atmospheric CO_2 and other greenhouse gases (figs. 4-12 and 4-13). Most assumptions are for a 25 percent increase in CO_2 over the levels of the 1980s by A.D. 2000, and a doubling by A.D. 2050. It is estimated that the cumulative effect of other greenhouse gases will equal or exceed that of CO_2 by 2030 (fig. 4-14). Although the atmosphere is becoming dustier—a condition that probably induces a cooling effect—the warming seems likely to dominate. Most projections indicate that warming will be modest—from 1°C to 3°C (1.8–5.4°F) by A.D. 2000, and from 3°C to 6°C (5.4–10.8°F) by A.D. 2050, with greater warming in high latitudes. More serious concerns are the shift in atmospheric circulation and the accompanying changes in the world's precipitation pattern.

Climatologist William Kellogg (in Gribbin 1978) suggests that the Altithermal of 4,000–8,000 Ya might serve as a model for conditions that might prevail on a warmer Earth, particularly in light of global rainfall patterns (fig. 4-15). Generally, computer models seem to substantiate his suggestion. Because warmer conditions seem to favor a better developed and more reliable monsoon in Africa and Asia, and increased rainfall in most of Europe, the resulting climatic patterns would seem to be more favorable to more of the world's people than the present pattern. Also, in middle and high latitudes, a 1°C increase in summer temperature would extend the length of the average growing season by roughly ten days. Increased atmospheric CO_2 also will have a favorable impact on photosynthesis and should favor higher crop yields for some crops.

One exception to this generally favorable scenario for world food production would be drier conditions in the wheat- and corn-growing regions of North America (the Prairie Peninsula is thought to have been formed during the dry Altithermal; see chapter 6). Although the growing season in this area would be longer, frequency of drought probably would increase, not only adversely affecting crops but also causing severe water problems in the arid West of the United States. Other atmospheric pollutants, such as those producing acid rain, are increasing parallel to the increase in CO_2, and may damage vegetation and other life forms.

Warming probably will be greatest in polar regions (50° to 90° latitude), in part because less snow cover will permit a longer warming period because of reduced surface albedo. One obvious result of warmer conditions

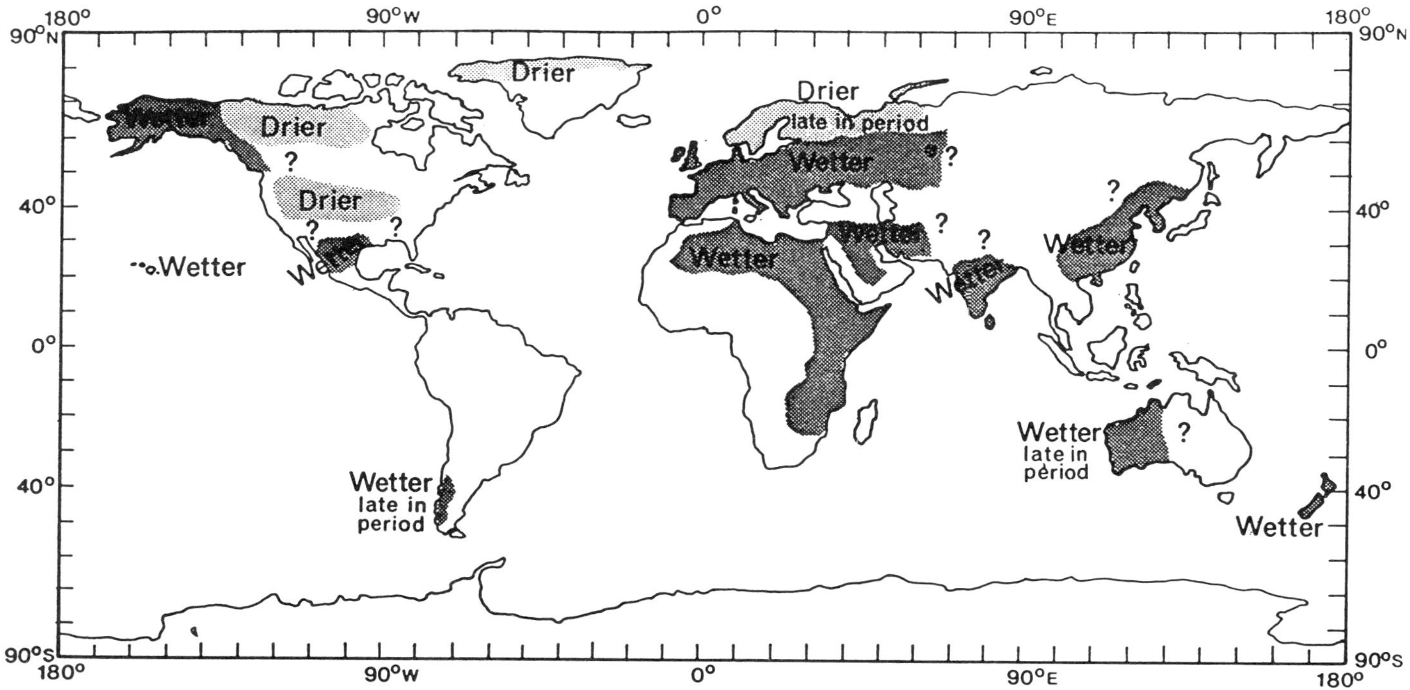

Fig. 4–15. Distribution of global precipitation during the Altithermal, a warm period 4,000 to 8,000 Ya, may indicate possible changes in a warmer future. (From William E. Kellogg in John Gribbin, ed., *Climatic Change,* copyright © Cambridge University Press, London, 1978. Reproduced with permission of the publisher.)

will be a reduction in glacial ice, and a parallel rise in sea level as the water formerly locked in ice reenters the hydrologic cycle. Sea level has risen about 150 m (492 ft) since the peak of the last glacial period around 15,000 Ya, for an average rate of 1 m per century. The maximum rate of rise immediately following the glacial period was perhaps 5 m (16 ft) per century; thus, the present rate is only 10–20 cm (4–8 in) per century, indicating that compared to the past, the present is a time of quiet sea level.

Projections for the next century indicate that warming from increasing CO_2 and other greenhouse gases will cause sea level to rise about 40 cm (15.6 in) from melting ice and another 30 cm (11.7 in) from expansion of the upper ocean water as its temperature rises, for a total of 70 cm (28 in). Recently observed thickening of the Greenland ice sheet, suggesting that higher precipitation may be characteristic of warmer climates in polar regions, may indicate that these estimates are too high (Zwally et al. 1989). Although these predictions seem modest, even a small increase would exacerbate shoreline problems caused by coastal erosion, damage by storm surges, and saltwater intrusion into groundwater aquifers.

Sea level probably will continue increasing after the next century, and the rate could reach 1–2 m per century should melting of the great Antarctic ice sheet accelerate. Warmer temperatures related to higher precipitation in polar regions, although initially causing sea level to fall, might accelerate glacier flow and cause a rise in sea level over the long run. Some concern centers on the West Antarctic ice sheet, most of which is grounded on bedrock below sea level. Should a rise in sea level and reduction in ice mass cause this ice sheet to float, it could collapse rapidly, in two hundred to five hundred years, and produce an additional worldwide rise in sea level, of 5–6 m (16–20 ft). Rises of this magnitude would cause major displacement, especially on the densely populated river deltas, and would affect most of the world's great port cities.

The Human Impact of Climatic Change

The impact of climatic change would be both favorable and unfavorable, yet outside the scientific community there has been little planning for possible disastrous consequences. There is a pressing need for cooperation between the social and physical scientific communities in evaluating the serious worldwide economic, social, and geopolitical effects of even modest climatic changes. A warming climate, at least in the short run, would seem to favor agricultural production in some parts of the world, particularly in monsoon lands of Southeast Asia and Africa. Decreased rainfall would have adverse affects on North American grain belts. The warming trend of the 1980s culminated in 1988, globally the warmest year of the instrumental record (fig. 4-11). The higher temperatures were accompanied by devastating drought in the North American grain belts and record floods in parts of monsoon Asia, a pattern paralleling Kellogg's reconstruction of rainfall patterns during the Altithermal. Perhaps the 1980s are indicative of climatic patterns to be expected in a warmer future.

If the minority of atmospheric scientists—those who see a cooling trend rather than a warming one in the climatic future—are correct, the scenario will have equal but different impacts. The most serious would be a new ice age, which might be the ultimate climatic event as the present interglacial period comes to an end. Disruption of the monsoonal circulation and failure of rainfall in some of the world's most densely populated areas is another possibility of a cooler future. We have seen the devastating results of minor changes, perhaps brought about by the relatively short cooling period in the 1960s and 1970s. On the other hand, limited cooling might have a beneficial effect on the grain producing areas of North America by increasing average precipitation, although this might restrict the northward limits of wheat in the prairie provinces of Canada.

One thing is certain: there is a pressing need to plan for the climatic future, not only on a national basis, but worldwide. Otherwise, human societies will continue the pattern of the past, merely reacting to what comes, even though what comes could be catastrophic. Climatic planning, combining the best thinking of both the social and physical scientific communities, should be given the status of military planning among nations of the world, since climate recognizes no boundaries or ideologies. Perhaps such planning might add a measure of rationality to what is presently the pseudoscience of geopolitics.

Part 2

VEGETATION

The one great law of vegetation dynamics is that the composition of vegetation tends always to constitute itself in relation to the environmental factors that influence the site occupied by the vegetation. Use by man is an environmental factor. Where use by man is intensive, that factor overshadows all others.

EDWARD H. GRAHAM, "The Re-creative Power of Plant Communities"

Chapter 5 THE NATURE AND DYNAMICS OF VEGETATION

Vegetation, more than any other feature, lends a unique character to the landscape. The palm trees and lush vegetation of the humid tropics, the stately Douglas firs of the Pacific Northwest, and the sagebrush, mesquite, and cactus of the desert Southwest—all are regional characteristics etched in the memory of a visitor. Equally, the checkerboard of corn and soybean fields in the American corn belt, the broad wheat fields of Kansas, and the neat citrus groves and vineyards of California create distinctive landscapes. Thus, both wild natural vegetation and domesticated plants contribute to the unique character of geographic regions.

Although *natural vegetation* may have a variety of meanings, we will restrict its use to that wild vegetation that is indigenous to an area. It is that which probably would re-occupy an area if *conscious* human intervention ceased to shape the vegetation cover. According to this usage, for example, the wheat fields of the Ukraine or the olive groves of Spain are not considered natural vegetation. Inadvertent or even intentional acts by humans, however, play a role in establishing and maintaining some natural landscapes. A spectacular example is fire, either deliberately or accidentally set by humans, which creates and maintains many of the world's grasslands. Although evidence exists of human intervention in establishing most grasslands, these landscapes are considered natural because in most instances fires are not set with the specific intention of maintaining the dominance of the grass.

Approaches to Plant Geography

There are two major approaches for studying plant geography—*floristic* and *ecologic*. Floristic plant geography deals primarily with the evolution, migration, and present geographic range of the many genera and species of plants in relation to their prehistoric and historic distribution. It is more concerned with taxonomic composition of vegetation cover than with dynamic environmental interrelationships. In the floristic approach rare plants receive particular emphasis.

Ecologic plant geography, on the other hand, is description, interpretation, and classification of *plant communities*. A particular class of plant community, such as tropical rainforest, consists of vastly different genera and species in different geographic areas, but the structure and ecology of plant communities in the same class are very similar. This structural resemblance is the result of *convergent* or *adaptive evolution*, which produces plants that are physiognomically similar (they look alike) but taxonomically diverse (they are not related genetically). Also, the ecologic approach deals with the suc-

Table 5-1 Taxonomic Divisions of Plants
Phyla (sing. *phylum*) (ending *-phyta*)

Schizophyta:	Bacteria and Blue-green algae
Thallophyta:	Algae, Diatoms, Fungi, Lichens
Byrophyta:	Liverworts, Mosses
Peteridophyta:	Horsetails, Club-mosses, Ferns
Spermatophyta:	Seed Plants
Classes of Spermatophyta	
Gymnospermae ("naked seed")	
Orders of Gymnospermae (living) (ending *-ales*)	
Coniferales:	Conifers
Cycadales:	Cycads
Genetales:	Ephedras
Ginkgoales:	Ginkgo (only 1 living species)
Angiospermae (flowering plants, "enclosed seed")	
Subclasses of Angiospermae[a] (ending *-eae*)	
Monocotyledoneae (single seed-leaf)	
Selected Families of Monocotyledoneae (ending *-ceae* or *-ae*)	
Gramineae:	Grasses
Cyperaceae:	Sedges
Orchidaceae:	Orchids
Palmae:	Palms
Liliaceae:	Lilies
Dicotyledoneae[b] (2 seed-leaves)	
Selected Families of Dicotyledoneae	
Compositae:	Daisy family; many herbaceous plants
Leguminosae:	Pea family; peas, beans, leguminous trees
Rosaceae:	Rose family; includes apples and pears

[a] Orders of Angiosperms are seldom used since the categories are too large to be of much use.
[b] Most trees, shrubs, and herbaceous plants are dicotyledons.

TABLE 5-2 Taxonomic Classification of the European Oak and White Spruce

	European Oak	White Spruce
Phylum:	Spermatophyta	Spermatophyta
Class:	Angiospermae	Gymnospermae
Subclass:	Dicotyledoneae	(not used)
Order:	(not used)	Coniferales
Family:	Fagaceae (beech)	Pinaceae (pine)
Genus:	*Quercus*	*Picea*
Species:	*Quercus robur*	*Picea glauca*

cession of plant communities that eventually culminates, at least theoretically, in a climax community in equilibrium with environmental factors, such as climate, land forms, and soil.

In this chapter our primary emphasis will be on the dynamics of plant communities rather than on the taxonomic classification of plants. Nonetheless, in both approaches to plant geography, it is inevitable that references will be made to taxonomic divisions of the plant kingdom. Tables 5-1 and 5-2 outline the systematic classification of plants.[1]

PLANT COMMUNITIES

Plant communities in the broadest sense are described according to their predominant life forms—for example, grasslands or forests. Grasslands can be further separated into subdivisions, such as tallgrass prairie or shortgrass steppe. Forests are classified as coniferous or broadleaf, evergreen or deciduous, or more specifically as pine forest, beech forest, or oak-hickory forest. Of course a forest almost never consists of a single species; a pine forest, for example, will contain several species of pines and a number of taxonomically unrelated trees, but pines will predominate.

Vegetation comprises a mosaic of plant communities forming patterns repeated in the landscape much as tiles form patterns on a pavement (fig. 5-1). A plant community and the environment in which it grows and interacts form an *ecosystem* (*eco-* = *oikos* = home), or a *biogeocenose* (*bio-* = life; *-geo-* = earth; *-cenose* = *koinos* = common). A shorter term for essentially the same thing is a *biome*, a community consisting of plants and animals, the soil penetrated by plant roots, and the layer of atmosphere into which the plants extend. The grassland biome of the Great Plains, for example, includes (a) grasses and flowering plants (*macroflora*); (b) the soil and its microorganisms (*microfauna* and *microflora*); (c) the larger animals, such as ro-

[1] In addition to their widely recognized common names, the scientific names of most plants mentioned in this chapter are given in parentheses. Colloquial names may vary from place to place, even within the United States, and are almost never the same in other languages; consequently, taxonomic names that have worldwide recognition are preferable for a wide audience.

Fig. 5–1. A mosaic of upland and bog plant communities in northern Minnesota. In the foreground center and left center is a 49-year-old stand of mature aspen (*Populus tremulodies* and *P. grandidentata*); in the lower right corner is a stand of 54-year-old balsam fir (*Abies balsamea*) on a steep hillside; in the lower center is a 98-year-old stand of black spruce (*Picea mariana*) growing on a perched *Sphagnum* bog. On the far side of the bog, right half, is a strip of balsam fir paralleling the bog perimeter. Beyond this is a strip of red pine–white pine–birch (*Pinus resinosa, P. strobus, Betula lutea*); right of center is a small "island" of aspen between the pine and fir strips. Upland forest in the top half of the picture has a predominantly sphagnum community growing on a mat of *Ericaceae* roots with some sedge (probably *Carex trisperma*) and cotton grass (*Eriophorum tennellum*). The dominant *Ericaceae* shrub is *Ledum groenlandicum*. The *Ericaceae*-sphagnum mat is floating near the open water. (U. S. Forest Service photo by Bluford W. Muir, June 1964; description of vegetation supplied by Elon S. Verry of the Northern Conifers Laboratory.)

dents, grazers, predators, and birds (macrofauna); and (d) all other living things that are present there.

Sometimes it is possible to explain a natural landscape in terms of the dynamic interrelationships existing between elements of that landscape. Geology, climate, vegetation, soils, and human activities all interrelate in a complex way that is not readily apparent to observers. It is important to be cautious in concluding that simple cause and effect relationships exist between these diverse environmental factors. Sometimes geographers oversimplify these relationships, particularly those between climate, vegetation, and soils. On the other hand, it is necessary to introduce a certain degree of generalization and synthesis to describe world vegetation so that observers can grasp the broad global patterns without becoming lost in complexity. Obviously, one cannot locate every tree, or even every plant species, but by grouping plants into associations and then generalizing their distribution according to a classification scheme it is possible to portray the broad picture.

Categories of Plant Communities

Researchers employ various categories to describe and map plant communities: (a) actual vegetation, (b) original vegetation, (c) natural vegetation, (d) cultural or anthropogenic vegetation, and (e) potential natural vegetation.

Actual vegetation. This is the plants that are present at the time of observation, without considering the stability of the present plant community or the nature of the former vegetation.

Original Vegetation. This includes former plant communities occupying the landscape before significant human intervention. In the nineteenth century, when surveyors made original land surveys of the United States west of the Appalachian Mountains, they recorded presettlement vegetation. Their field notes are invaluable in recreating original vegetation. Where actual records are not available, indirect evidence from soils and fossil pollen trapped in sediments may aid in reconstructing past vegetation cover for a particular time. Where human settlement has been of long duration, however, it may be difficult to determine original vegetation with any real degree of confidence.

Natural Vegetation. This is not necessarily the same as original vegetation; it is vegetation cover existing where human impact has been minimal. It is in balance with environmental factors, including human activities not consciously directed toward eradicating or changing vegetation cover. Obviously, clearing land for crops or pastures, or replacing native trees with exotic ones would destroy natural vegetation; but the human role often is unclear. Is a grassland that is maintained by fire, often of human origin and deliberately set, natural or cultural? Plant geographers and ecologists disagree on this and similar questions.

Cultural or Anthropogenic Vegetation. The question of natural vs. cultural vegetation can be partly resolved by classifying human-induced and human-altered vegetation into two categories: *cultural crops* and *semi-natural vegetation.* Cultural crops include weeds that prosper where land is disturbed (fig. 5-2). Ragweed (*Ambrosia* spp.) is such a weed, a camp follower since the dawn of agriculture.[2] In fact, the discovery of an abundance of ragweed pollen in an ancient alluvial stratum may signify to an archaeologist that

[2] *Ambrosia* is the scientific name for the *genus* (pl. *genera*) of plants to which all ragweeds belong. The abbreviation *spp.* means any or all species of *Ambrosia.* To identify a particular species of ragweed, such as the giant ragweed, the species name is added to the genus: thus the giant ragweed is *Ambrosia trifida.*

Fig. 5–2. Saint Paul in southern France, as viewed from La Colombe d'or—a cultural landscape strikingly modified by centuries of human occupation. Man-made terraces reduce slope and control erosion on cultivated land. Steeper slopes are occupied by woodland or scrub vegetation which can be classified as semi-natural (USDA, Soil Conservation Service photo by C. W. Lowdermilk, Dec. 4, 1938.)

humans were on the scene with their primitive agriculture and attendant weeds when that stratum was deposited. (A weed has been described as "a plant growing where it is not wanted.") Semi-natural vegetation includes natural pastures, some tropical savannas created and maintained by fire of human origin, and vegetation occupying land that was formerly cultivated or overgrazed and abandoned to a less luxuriant vegetation cover than was present before human interference. Large areas of the Mediterranean basin covered with scrub forests of *chapparral* or *maquis* would fall into this category. The human role in determining such vegetation landscapes, however, is still controversial.

Potential Natural Vegetation. This is the vegetation community that eventually would occupy land following a series of successive plant communities, called *successions.* A term sometimes used is *climatic climax* vegetation, implying an equilibrium with the environment so long as no drastic climatic change occurs. Present climatic conditions are assumed in theorizing about potential natural vegetation; thus, original vegetation may not have been the same, particularly if there have been changes in climate or irreversible changes in the landscape. Take, for example, the heaths of Northwestern Europe which, developing through a combination of agricultural and pastoral activities and climatic change, replaced original forest.

Classification of Plant Communities

As with classifying climates, vegetation classification involves several approaches. In many instances, the difference between classifications is more apparent than real; some classifications merely use different terms to identify the same thing, while others actually differ in principle. Commonly encountered classification systems often involve one or both of the following criteria: (1) *floristic criteria*, such as "oak-hickory forests," which use the dominant taxa or groups of taxa; and (2) *physiognomic criteria* that recognize dominant life forms, as in "sclerophyllous woodlands" (hard-leaved), combinations of life forms, as in "savanna" (trees and grass), and seasonal loss of leaves, as in "summer-green" or "deciduous summer forest."

Physiognomic criteria generally are used to map vegetation on a global or continental scale, while floristic criteria become important in more restricted areas; thus the "broadleaf deciduous forests" of middle latitudes, but the "oak-hickory forests" of the Ohio Valley. Most classifications using life forms are based on the concept of climatic climax vegetation, but generally they include subdivisions, particularly savannas and some grasslands, which may owe their existence to factors other than climate.

A comparison of climatic and vegetation maps shows that similar climate does not necessarily mean similar life forms. Areas with comparable climates, such as the humid subtropical (*Cfa*) of China and the southeastern United States, have different plant formations. The same is true of the west coast marine climates of Western Europe and the Pacific Northwest of North America. This seeming contradiction can be explained in part by the complex geologic and climatic history of the earth, from continental "drifting" during the past 200 million years to more recent continental glaciation. Certain plant species may not be present in an area, no matter how favorable the climate might be for them. On the other hand, if introduced into the area they often prosper; a case in point is the successful introduction of eucalyptus from Australia into California and the Mediterranean region of Europe. A number of broadleaf evergreens from China are successful as ornamental trees in mild winter sections of the United States.

Chapter 6 describes distribution of world vegetation according to widely recognized *formation types*, determined by life forms making up climatic climax plant communities. A formation type subdivides into *formations*; these in turn subdivide into *associations*. An example of this hierarchy of climatic climax communities appears in figure 5-3.

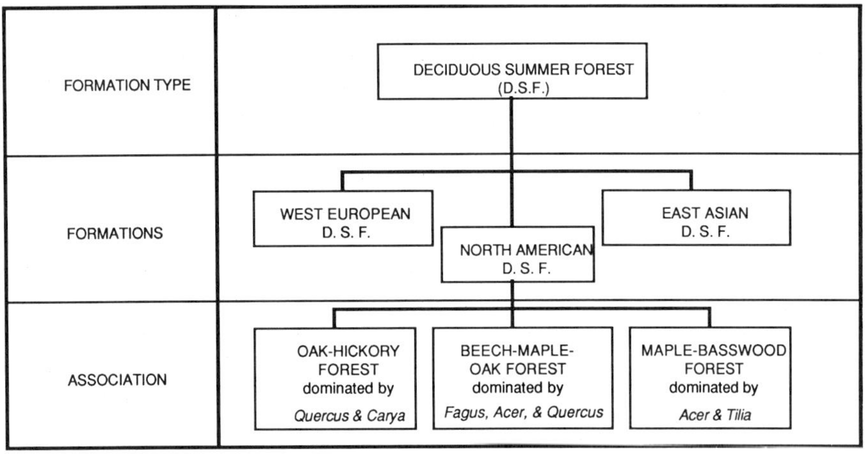

Fig. 5–3. The hierarchy of climatic climax plant communities as represented by the Deciduous Summer Forest Formation Type and its subdivisions. Only the North American Formation has been subdivided into three of its several associations; the West European and East Asian Formations can likewise be subdivided. (Modified from Eyre, 1963.)

World vegetation maps indicate large expanses of territory covered by a single formation type; however, almost always large areas exist within these boundaries that are not climatic climax communities. Obviously, cultivated land and improved or planted pastures fall into this category. In addition, large areas of wild vegetation are not in equilibrium with the physical environment and are not climatic climax associations.

How is a climatic climax association established? Sequences of plant communities will invade areas of bare soil or rock, such as mud flats, sand dunes, recent volcanic ash and lava, and rock or mineral soil uncovered by erosion or landslides. Pioneer plant communities give way to subsequent communities, each replacing the preceding one, until a climatic climax is reached. These temporary plant communities, under which new soil develops, are designated *seral communities*. The pioneer plant invaders constitute the *primary sere* or *prisere*; where colonization takes place on land, the community is a *xerosere*; in fresh water, as in a shallow lake, it is a *hydrosere*; if the water is salty, as in coastal marshes, it is a *halosere*. Figure 5-4 illustrates a succession of seral communities in a northern bog similar to the one shown in figure 5-1.

Environmental influences, such as frequent flooding, repeated burning, or continued erosion on steep slopes, may arrest the progression toward a climatic climax plant community. If these disrupting events are irreversible,

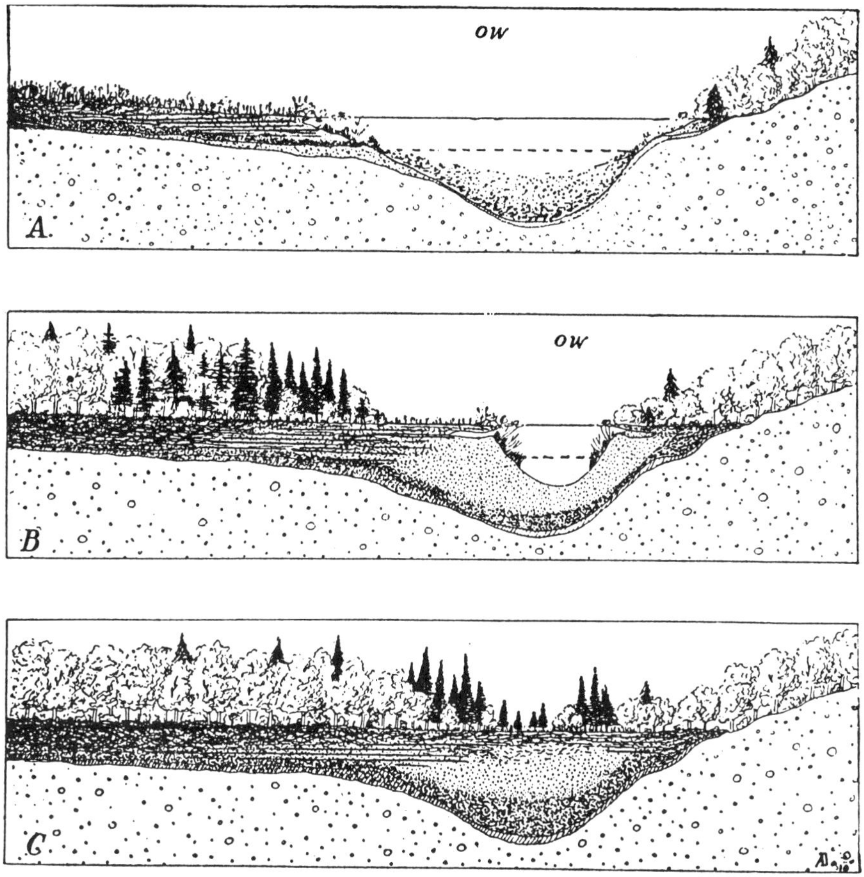

Fig. 5–4. A succession of seral plant communities in a northern lake. (A) The aquatic vegetation is accumulating in open water (OW), and marsh vegetation has encroached on the shallow area of the lake to the left. (B) Plant debris has filled a large part of the lake and marsh vegetation is closing in on the reduced area of open water. (C) Swamp forest has advanced over the bog that was formed by the closing of the lake by marsh vegetation, producing a forested bog similar to the one shown in figure 5-1. (From USDA, *Soils and Man,* Yearbook of Agriculture, 1939, 1,129.)

the succession will be more or less permanently arrested and will be *sub-climax*. Plant communities in a state of essentially permanent subclimax through human activities are designated as a *plagioclimax* or *disclimax*. Examples of plagioclimax plant communities include heaths and permanent grasslands of Western Europe, and at least some grasslands and savannas apparently maintained by fire of human origin and by grazing. Other plant

Fig. 5–5. An example of plagioclimax vegetation cover of sagebrush (*Artemisia*) in an area of Nevada that was formerly dominated by shortgrass steppe. The sagebrush is maintained through the constant pressure from grazing livestock and the absence of periodic burning. (USDA, Soil Conservation Service photo by Rod Bradford, n.d.)

communities, such as mesquite and sagebrush (fig. 5-5), invade grasslands because of overgrazing and protection from fire. Controlled burning can remove the invaders and re-establish dominant grasses (fig. 5-6). Plagioclimax essentially is synonymous with seminatural anthropogenic vegetation.

Climate and Natural Vegetation

Lack of a precise match between boundaries on climatic and vegetation maps does not necessarily prove a lack of relationship between broad distribution patterns of vegetation and climate. More likely, it demonstrates that factors for climatic classification overlook many complex climatic elements affecting plant growth. Climatic classifications generally use macroclimatic data obtained at standard meteorological stations comprising readily available statistics. Plants, on the other hand, more closely adjust to *microclimates*—climate near the ground. A closer look at the multitude of climatic

Fig. 5–6. An area in Ravalli County, Mont., formerly in heavy sagebrush (*Artemisia*), three years after controlled burning. Fire has eradicated the sagebrush and helped return the area its former dominant grassland vegetation. (USDA, Soil Conservation Service photo by C. E. Hitch, 20 October 1963.)

elements related to plant growth, including light, temperature, moisture, and wind, will demonstrate some of the complex interrelationships.[3]

Sunlight and Photosynthesis

Photosynthesis is the process by which plants convert solar energy into chemical energy. It is the basis for the food chain for life on Earth. This life-sustaining process also releases oxygen to the Earth's atmosphere while at the same time absorbing CO_2. The essential chemical reaction that takes place in photosynthesis is:

$$CO_2 + H_2O + \text{solar energy} \rightarrow CH_2O + O_2 - \text{energy}$$

[3]*Macroclimate* is the climate represented by data obtained at standard meteorological stations from instruments in a standard instrument shelter usually at least 1.5 m (5 ft) aboveground. *Microclimate* is identified by observations made below the level of a standard instrument shelter, or in special situations, such as the north or south side of a hill, in a forest or open grassland, in a city or open country.

The process of photosynthesis requires as a catalyst chlorophyll, the green pigment in plants, and is much more complicated than this equation implies. The simple carbohydrate, CH_2O, is the building block for much more complex compounds, including sugar, starch, and cellulose. All carbohydrates consist of carbon combined with hydrogen and oxygen in the same ratio as in water, and a general formula is $(CH_2O)_n$; for example, sucrose, or common sugar, is $C_{12}H_{24}O_{12}$. Plants, primarily through their green leaves, are capable of using the sun's energy to convert CO_2 and water into carbohydrates and to release oxygen as a byproduct. Carbon dioxide derives from the atmosphere where plant respiration, animal respiration, decaying organic matter, and the burning of fossil fuels replenish it. Burning of petroleum and coal releases solar energy received on the Earth millions of years ago and converted to organic compounds by photosynthesis. Plants obtain water for this process from soil moisture through their root systems.

Plants also respire up to half the organic matter produced by photosynthesis in the form of CO_2, water, and heat of combustion:

$$CH_2O + O_2 \rightarrow CO_2 + \text{heat}$$

Note that this reaction is the reverse of photosynthesis. Although plant respiration produces CO_2, photosynthesis is a much more vigorous process, so that green leaves of plants increase the proportion of oxygen in the air at the expense of CO_2. Consequently, Earth's oxygen-rich atmosphere depends on green plants.

Plants grow because photosynthesis produces more organic material than they lose through respiration. This gain in dry matter by plants is called *net photosynthesis* and is usually measured in units of dry organic matter produced in a given area on the earth's surface, in grams per square meter per day (gm/m^2/day). If an adequate CO_2 and water supply exists, the theoretical maximum production of organic matter is *potential net photosynthesis* (Chang 1970).

What determines the rate of photosynthesis? Primarily, it is light intensity when temperature is above a certain threshold—for most plants outside polar regions, some temperature above 0°C (32°F). On a graph, this rate increases to a certain point almost as a straight line, then levels off, even though light intensity continues increasing. This point on the graph is the *light saturation point*, at which a leaf is light saturated (fig. 5-7). Light saturation levels differ for different plant species: in general, levels are higher for tropical plants than for those of middle and high latitudes, and in a forest setting they are higher for tall tree species than for undergrowth. A plant with a high light saturation level will grow best in full sunlight and is called a *heliophyte* (light-loving); conversely, a plant with a low light saturation level will grow best in shade and is called a *sciophyte* (shade-loving).

Plant respiration rates increase as the temperature increases, reaching a maximum rate at an optimum temperature level. The optimum temperature

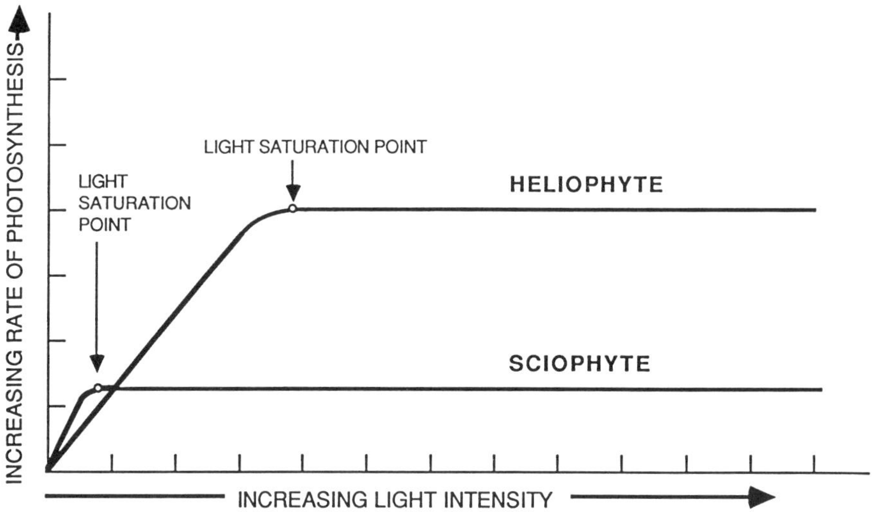

Fig. 5–7. Light saturation curves for a hypothetical sun-loving *heliophyte* and a shade-loving *sciophyte*. There is considerable variation in shape and magnitude of the light saturation curves among different plant species. Note that increases in light intensity beyond the light saturation point do not increase the rate of photosynthesis.

for maximum respiration differs from species to species. Day and night temperatures have opposite effects on net photosynthesis because although respiration continues at night, photosynthesis ceases. High night temperatures, therefore, increase respiration loss, which decreases net photosynthesis. Conversely, high day temperatures, up to about 30°C (86°F), may increase the rate of photosynthesis and thereby increase the rate of organic matter production.

Photosynthesis involves water, but a plant uses only about 1 percent of the volume entering through its root system. A shortage of water, however, may cause plant dehydration and cessation of photosynthesis. When plants wilt, the *stomata* or leaf openings close; thus, no CO_2 intake is possible. Faced with a water shortage, plant species will vary greatly in their ability to photosynthesize.

Photoperiodism

During a year all locations on Earth receive an equal number of hours of daylight; thus, there is little variation in the quantity and quality of sunlight from latitude to latitude. It is apparent that potential net photosynthesis alone cannot account for the varied distribution of vegetation types. On the other hand, the length of periods of daylight and darkness varies greatly, from almost constant twelve hours of daylight and twelve hours of darkness at the equator to long periods of continuous light or darkness in the polar regions. Duration of daylight is the *photoperiod*, and the response of plants to length of night and day is called *photoperiodism*. Actually, the length of the dark period is more critical to plants than length of light, but the term *photoperiodism* dates from a time when this was not yet understood (Chang 1968).

Plants are adjusted genetically to the photoperiod of the region where they evolved, and this effectively may restrict some species to a particular latitude range every bit as much as temperature. Plants are classified as *long-day plants*, flowering only when daylight periods are more than fourteen hours; *short-day plants*, flowering only with less than fourteen hours of daylight; *day-neutral plants*, flowering under any period of light; and *intermediate-day plants*, flowering with twelve to fourteen hours of daylight, but not with shorter or longer periods (Oliver 1973).

Temperature and Plant Distribution

Relationships between plant distribution and temperature are more obvious than are relationships between distribution and length of light period. Tropical plants cannot survive in the cold winters of middle and high latitudes, just as plants adapted to cool climates may wither and die in the tropics. Nevertheless, the temperature range in which biological activity occurs is relatively narrow, with little plant growth below 0°C (32°F) or above 50°C (122°F). For most plant species, the range of tolerable temperature is much less than this. All plants have upper and lower temperature limits above or below which growth ceases. Between these limits is an optimum temperature that assures the most rapid growth for a particular species. These three critical temperature values are *cardinal temperatures*; they differ from one plant species to another, and even within the same species from one stage of development to another.

Some plants respond favorably to the complex rhythmic variations of a number of environmental factors brought into play by the shift from night to day. Among these diurnal variations in addition to photoperiodism, temperature has the most obvious effect on plant growth. The term *thermoperiodism* describes the response of plants to temperature changes occurring between night and day. Without sufficient contrasts between night and day temperatures, certain plants will not flower and produce seed; seeds of some plants will not germinate; in others, proper growth may depend on optimum night-day temperature range. Thus, the small diurnal range of the lowland tropics will have undesirable effects on nontropical plants that have evolved in areas with greater diurnal temperature ranges.

Seasonal change in temperature, or lack of it, may determine the geographic range of certain plants. Those having a dormant season during winter require a period of cold temperature for stimulating growth and terminating the rest period. For example, peach trees require a dormant period with 400 or more hours of temperatures at or below 7°C (44.6°F) for proper later development of leaves, flowers, and fruit. Peaches cannot be produced profitably in the lowland tropics because of this lack of winter stimulation (Daubenmire 1959, 188).

Moisture Requirements and Plant Distribution

Availability of moisture is another environmental factor having a profound effect on the distribution of vegetation. Water constitutes a major portion of plant tissues, it is a major reagent in photosynthesis and other chemical processes involved in plant metabolism, it carries plant nutrients from the soil into the plant through the root system and disperses them throughout the plant, and it is essential in maintaining *turgidity* (the state of rigidity of plant cells that prevents their wilting). A large share of water moving through a plant is transpired from the leaves and has a cooling effect that may prevent overheating of leaves if they receive direct sunlight.

Plants fall into three broad classes according to the moisture conditions of their habitat: *hydrophytes*, *xerophytes*, and *mesophytes*. These terms derive from the Greek roots *hydro-* (wet), *xero-* (dry), and *meso-* (intermediate); *-phyte* means plant.

Hydrophytes. This classification includes aquatic plants growing in water or in soil so saturated as to be uninhabitable by other plants. Included in this classification are the water lilies (*Nymphaea*), cattail (*Typha*), and bald cypress (*Taxodium distichum*) (fig. 5-8). *Phreatophytes*, such as willow (*Salix*), are transitional between hydrophytes and mesophytes and are grouped under the latter. Some writers use the term *hygrophyte* (*hygros* = moist) as synonymous with hydrophyte, but our choice in this text is the more restrictive definition of hygrophyte as a subdivision of mesophytes.

Xerophytes. Plants in this category can withstand periods of severe moisture deficiency destructive to mesophytes and hydrophytes. Although the means by which plants can escape or endure drought is extremely varied, there are three general types of xerophytes: *ephemeral annuals*, *succulents*, and *nonsucculent perennials*.

1. *Ephemeral annuals* can complete their life cycle in a very short time, taking advantage of infrequent rains to sprout, grow, flower, and produce seed. As seeds they remain dormant until rain again moistens the soil, and the short cycle recurs. Although these plants can withstand extremely dry atmospheric conditions, they cannot tolerate dry soil. Because of this characteristic they are sometimes considered not to be true xerophytes.

2. *Succulents* have the ability to store water in the cellular structure during brief rainy periods and to use it during drought. For survival, the transpiration rate must be low during dry periods; this is assured by the fact that succulents have a relatively high weight and volume compared to their surface area. Succulence may occur in roots, leaves, or stems. A good example

Fig. 5–8. Bald cypress (*Taxodium distichum*) growing in shallow water of Reelfoot Lake in northwestern Tennessee. This tree is representative of plants belonging to a group referred to as *emergent anchored hydrophytes*. The bald cypress is a valuable lumber tree in the South Atlantic and Gulf Coastal Plains, and in the Mississippi Valley north to southern Illinois. (Photo by W. E. Akin.)

of a leaf succulent is the common jade plant (*Crassula argentea*), a popular decorative house plant. Stem succulence is well developed in the cacti, the best known of the desert plants (fig. 5-9). Cacti have an additional adaptation in their shallow root system, which permits them to obtain water from the occasional light shower that does not penetrate more than a few inches below the soil surface.

3. *Nonsucculent perennials* are a large group of plants with many diverse features enabling them to survive drought. Some plants survive through their ability to send roots downward faster than the soil can dry out. Others

Fig. 5–9. The giant saguaro cactus (*Carnegiea gigantea*) growing at the foot of the Superstition Mountains in Arizona. These xerophytes belong to a group called *stem succulents*. They store water in the tissues of their massive stems, which have relatively small surface areas compared to their volumes. Also, leaves are modified into spines, providing for a minimum loss of water by transpiration. (USDA, Soil Conservation Service photo by B. Brixner, n.d.)

have root systems of great lateral extent in relation to that part of the plant exposed aboveground; this contributes to the wide spacing of desert shrubs (fig. 5-9). Some desert trees and shrubs send roots to great depths to keep contact with perennially moist subsoil. Mesquite (*Prosopis*) roots can penetrate depths greater than 20 m (65 ft) below the surface, and alfalfa (*Medicago sativa*) roots reach to a depth of 40 m (120 ft).

Other perennials have internal structures permitting them to survive desiccation of their cells; still others have the ability to reduce transpiration to a very low level during wilting. Such structurally adapted plants are *xeromorphic*; and some botanists consider these to be the only true xerophytes, eliminating from that category most other dry land plants that store water or are especially efficient in obtaining water from soil.

Included with the nonsucculent perennials are the *sclerophylls* (hard-leaved plants), a group of evergreens whose leaves are covered by a thin shiny layer of noncellular material called *cutin*, or by a waxy coating that retards loss of moisture during drought. Surprisingly, with adequate water xerophytic plants do not have lower transpiration rates than do less drought-resistant plants (Chang 1968).

Mesophytes. Most other plants are mesophytes; they cannot tolerate wet soil or survive in water for any extended period of time. On the other hand, neither can they tolerate prolonged dryness of the soil. Mesophytes grow where soil moisture is plentiful, but their roots will not penetrate into the saturated zone below the zone of aeration. Sometimes plants that have no specific adaptation for withstanding drought are called hygrophytes, but most botanists reserve *hygrophyte* for mesophytes that are extremely sensitive to dry air and grow only in habitats where relative humidity is always high. Many plants found in rainforests or cloud forests are hygrophytes.

Wind as a Factor in Plant Growth and Distribution

Wind has an important effect on many plant communities; it is most apparent along seacoasts, on open plains, and in high mountains. Wind effects may be favorable or unfavorable to plants. Unfavorable effects include desiccation caused by strong winds that increase transpiration rates; such winds can also cause mechanical damage and alter the shape and size of plants (fig.

Fig. 5–10. An example of wind training in a farm windbreak near the North Sea coast of Denmark. The prevailing westerly winds have given the trees a characteristic sloping profile with stunted trees toward the west. The arrow indicates the most frequent wind direction. (Photo by W. E. Akin, August 1962.)

Fig. 5–11. "Elfin forest" at timberline in Arapaho National Forest, Colorado. The trees are bristlecone pine (*Pinus aristata*) which have been dwarfed and misshapen by wind. These bristlecones are among the oldest of living trees. (U.S. Forest Service photo by H. L. Shantz, 4 September 1941.)

5-10). Near the tree line in high mountains, persistent strong winds produce a dwarf forest of gnarled, misshapen trees that bear little resemblance to members of the same species in their normal growth habitat. Such plant communities are called *elfin forests* (fig. 5-11). In tropical mountains, the upper limit of the forest often is marked by a windswept, dwarf forest of gnarled and crooked trees covered by mosses. In some areas of tropical South America local people refer to this elfin woodland as *Ceja de la Montaña*, "eyebrow of the forest."

Wind has favorable effects as well. It is the most efficient agent for dispersing pollen and seeds of land plants. The floating fluff of dandelions and cottonwood trees and the whirling winged fruit of the maples attest to the success of this method of seed dispersal. The tumbleweed (*Salsola kali* or

Russian thistle) of western folklore is an interesting example of seed dispersal by wind. When mature, the entire aboveground part of the plant breaks off, and the wind rolls it along as its seeds scatter over vast areas. The Russian thistle is only one of a number of similar tumbling weeds.

Vegetation and Soil—The Edaphic Factor

Soil and vegetation essentially form a single unit. Root systems often occupy a volume of soil greater than the space occupied by the plant aboveground. From the measurement of the root system of a single rye plant grown for four months in one cubic foot of soil, agronomists were able to make estimates of the magnitude of the root system. Roots numbered about 13.8 million, with a total length of about 620 km (385 mi), and a surface area of 237 m^2 (2,550 ft^2). They estimated 14 billion root hairs with a total length of 10,621 km (6,600 mi), and a surface area of some 401 m^2 (4,320 ft^2) (USDA 1957, 46).

Figure 5-1 shows the intimate relationship between plant communities and soils as reflected in a pattern of upland and swamp forest in northern Minnesota that is determined by moisture and drainage conditions of the underlying soil. In addition to moisture, other soil factors, including texture, structure, and chemistry, affect plant growth.

Soil chemistry depends on elements and compounds required for plant growth, and also on those elements that may inhibit growth or kill certain plants. Soil texture determines the ease with which roots can penetrate and the capacity of soil to store water and make it available to plants.

Plant Nutrients from the Soil

The sixteen elements known to be required for plant growth are listed in table 5-3. Except for carbon, oxygen, and hydrogen, which are obtained from water and CO_2, these plant nutrients come from the soil. Other soil elements, such as iodine, sodium, and cobalt, are found in plants but have not been proven necessary for plant growth. Some of these, however, are essential for growth and survival of animals.

Both organic matter and inorganic minerals within the soil supply plant nutrients. Soil organic material of plant and animal origin is called *humus*; when present in sufficient quantity, it gives a dark color to soils. Humus also is an important source of nitrogen, phosphorus, and sulfur. A continuous turnover of these elements occurs among soils, plants, animals, and sometimes the atmosphere, giving rise to nutrient cycles, including carbon, nitrogen, phosphorus, and sulfur cycles.

Since the *carbon cycle* involves all living organisms, including humans, it is sometimes designated the "cycle of life." The primary source of this life-supporting element in soils is the decay of organic matter derived from plants and animals growing on and in soil, and from animal wastes. The oxidation of carbon is a major source of energy required by the flora and

TABLE 5-3 Important Plant Nutrient Elements in Soil

Element	Symbol	Principal Ions
Aluminum	Al	Al^{+++}
Boron*	B	$B_4O_7^{--}$
*Calcium**	Ca	Ca^{++}
*Carbon**[a]	C	CO_3^{--}
Chlorine*[b]	Cl	Cl^-
Cobalt*	Co	Co^{++}
Copper*	Cu	Cu^{++}
*Hydrogen**[a]	H	H^+, OH^-
Iron*	Fe	Fe^{++}, Fe^{+++}
*Magnesium**	Mg	Mg^{++}
Manganese	Mn	Mn^{++}, MnO_4^-
Molybdenum*	Mo	MoO_4^{--}
*Nitrogen**	N	NH_4^+, NO_2^-, NO_3^-
*Oxygen**[a]	O	(combined)
*Phosphorous**	P	$H_2PO_4^-$, HPO_4^{--}
*Potassium**	K	K^+
Silicon	Si	Si^{++++}
Sodium*[b]	Na	Na^+
*Sulfur**	S	SO_4^{--}
Zinc*	Zn	Zn^{++}

[a] Carbon, hydrogen, and oxygen obtained mostly from air and water.
[b] Formerly thought not essential to plants; now known to be essential in minute quantities. Sodium may partially substitute for potassium.
Note: Starred elements are important plant nutrients; italicized elements are macronutrients.

fauna within the soil. A large quantity of CO_2 comes from this biological activity, and it escapes to the atmosphere where plants may again use it through photosynthesis (Buckman and Brady 1969).

In the *nitrogen cycle* organic compounds containing nitrogen must be converted to inorganic compounds before nitrogen is available to plants. Soil bacteria perform this conversion, called *nitrification.* Nitrogen from the atmosphere also reaches the soil by *nitrogen fixation,* a process involving a number of soil bacteria, including those living on roots of legumes.

The inorganic or mineral part of the soil, derived from the disintegration of rocks through the process of chemical weathering, supplies most remaining mineral nutrients. Potassium, in terms of quantity used, is one of the three major plant nutrients; its source is the weathering of several soil minerals, among them are *orthoclase* and *microcline* feldspars and *muscovite* and *biotite* micas. The following chemical reaction releases soluble potassium compounds from the weathering of orthoclase feldspar:

$$2KAlS_3O_8 + H_2CO_3 + H_2O \rightarrow K_2CO_3 + Al_2(OH)_2Si_4O_{10} + 2SiO_2$$

orthoclase feldspar	carbonic acid	water	potassium carbonate (soluble)	clay mineral (kaolinite)	silica

Another important nutrient source released by this reaction is potassium carbonate.

Weathering of biotite mica [$K(Mg_1Fe)_3AlSi_3O_{10}(OH)_2$] produces potassium bicarbonate ($KHCO_3$), a particularly important source of potassium in tropical soils where bedrock weathers rapidly. Weathering biotite and the mineral dolomite—a calcium magnesium carbonate, $CaMg(CO_3)_2$—release magnesium, which is particularly important for green plants, since one atom occurrs in each molecule of chlorophyll.

Generally, soils contain more of these minerals than plants can use, but much of this potential nutrition, combined in mineral or organic compounds, is not immediately usable by plants; an example is the unweathered orthoclase feldspar mentioned above. Weathering of minerals and mineralization of organic materials release soluble forms of the nutrients. Soil moisture containing dissolved nutrients is the *soil solution*. These soil nutrients are in the form of electrically charged atoms or molecules called *ions*. Positively charged ions are *cations* (Ca^{++}, K^+, Mg^{++}, NH_4^+); others with negative charges are *anions* (NO_3^{--}, SO_4^{--}, $H_2PO_4^{--}$, BO_3^{--}). Some cations held out of solution on the surface of clay crystals and on humus colloids are available to plant roots through *ion exchange*. In this process root hairs release hydrogen ions (H^+) to the soil solution or to the clay or humus colloids in exchange for nutrient cations which they then absorb—a positive ion for a positive ion. The root releases two hydrogen ions for each calcium ion absorbed, and one hydrogen ion for each potassium ion (Figure 5-12). Nutrient anions, such as nitrates, sulfates, phosphates, borates, and molybdates, are not available to plants from clay and humus but must be taken directly from the soil solution. Again, when the plant root takes in nutrient anions, it releases OH^- or HCO_3^- ions in exchange.

Trace minerals are minerals required by plants in minute quantities. While the role of some of them in plant metabolism is unclear, others certainly are of great importance to plant and animal life. In parts of southern Australia, because of a deficiency in cobalt in pasture grasses, sheep and cattle lose weight, become anemic, and even die. The addition of cobalt salts to pastures corrects the problem; the yearly amount of cobalt required per hectare is only 0.69 grams (0.01 oz/acre). Cobalt is essential for synthesizing vitamin B_{12}, and this cobalt-containing vitamin is a necessity for all animals. Copper and molybdenum deficiencies also may adversely affect plant and animal health (Thompson and Troeh 1978, 320–41).

Because of diversity in the mineral content of rocks that are the source of soils, some nutrient soil minerals may be deficient, while others may occur in excessive concentrations. Variability in mineral content of soils may profoundly affect vegetation in some plant communities. Residual soils, developed in place by weathering of certain types of bedrock, often support a thin discontinuous vegetation cover because of low availability of certain essential elements, or because of overabundance of elements that may be toxic in

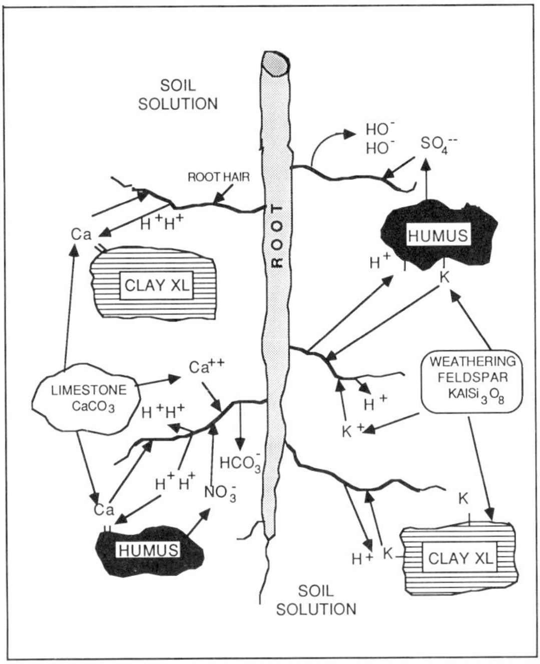

Fig. 5–12. A diagrammatic illustration of how root hairs obtain nutrients from the soil solution and from exchangeable cations on clay and humus particles. Limestone fragments give up calcium ions (Ca^{++}), and the weathering of orthoclase feldspar produces potassium ions (K^{+}). Both the Ca and K ions may remain in the soil solution or be held out of solution on the surface of clay or humus particles. Anions of sulfate (SO_4^{--}) and nitrate (NO_3^{-}) are produced by decomposition of humus and occur as available ions in the soil solution. Root hairs release H^{+} in exchange for Ca^{++}, K^{+}, and other nutrient cations, and they release OH^{-} or HCO_3^{-} ions in exchange for SO_4^{-}, NO_3^{-}, and other nutrient anions. Under actual conditions the root hairs would be in direct contact with the humus and clay particles. Several different kinds of cations may be attached to the same clay or humus particle.

high concentrations. Such areas, often called "barrens," are clues for locating ores of such metals as nickel, cobalt, and zinc. Some botanists maintain that some tropical savannas were not a product of climate and fire; rather, growth of forests was prevented by mineral deficiencies in soils. Conversely, certain mineral surpluses may be just as effective in eliminating sensitive plants from an area. For example, excess salt (NaCl) in some arid regions produces saline soils and promotes growth of salt-tolerant plants called *halophytes* (from *halite*, the mineralogical name for common salt).

Vegetation and Soil pH

Soil acidity or alkalinity also may determine the type of vegetation prospering in an area. Climate plays an important role: most soils of humid areas are acid, while those of drier areas are alkaline. What is important to plants is the acid or basic reaction of the soil solution. This reaction is measured by the concentration of hydrogen ions (H^+) and hydroxol ions (OH^-) in the soil solution. The pH scale from 0 through 14 indicates the concentration of hydrogen ions; the pH of soils ranges from about 3 to 10. The pH scale is based on the negative log of the H^+ concentration in a solution. A neutral solution has a pH value of 7, an abbreviation of 10^{-7}, meaning that there is a hydrogen ion concentration of 0.0000001 gm of H per liter of solution (1×10^{-7}). A solution with a pH of 6 would have 0.000001 gm of H per liter of solution (1×10^{-6}), or ten times as much active H^+ as a pH 7 solution. As the pH number decreases by one, acidity increases tenfold, so that pH 5 is ten times more acid than pH 6, and pH 4 is a hundred times more acid than pH 6. Conversely, the higher the pH value, the higher the OH^- concentration and the lower the soil acidity (fig. 5-13).

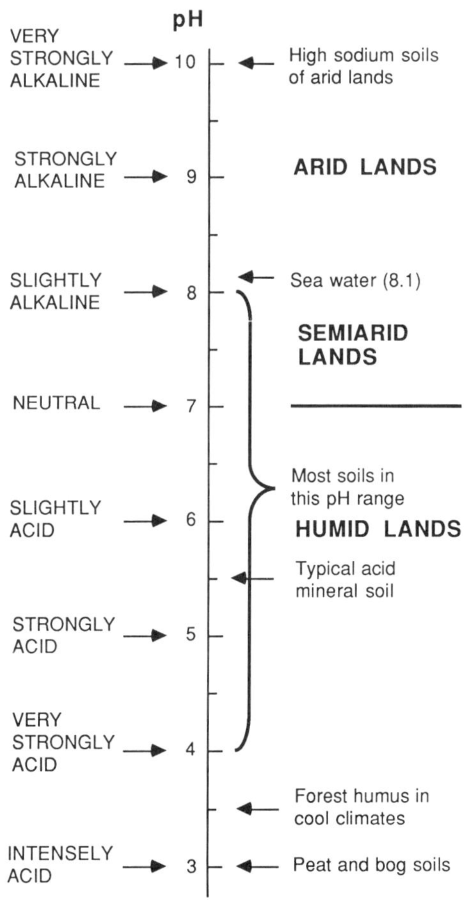

Fig. 5–13. The pH values for soils range from intensely acid to very strongly alkaline. Most soils fall within the range pH 4 to pH 8.

Acids form in soils from the decomposition of light-colored igneous rocks, from waste products of plants and soil organisms, and from decomposition of some types of plant litter. A source of growing concern is *acid rain.* The detrimental effects of acid rain on aquatic life in hundreds of lakes in North America and Scandinavia have been well documented, but the effects on terrestrial environments are more complex and less clear. Acid deposition on foliage adversely affects many plants, as does increased acidity of the soil solution. Additional damage to vegetation by high ozone concentrations in air pollutants adds complications to the situation.

An abundance of hydrolyzing salts produces basic or alkaline soil reaction. In humid areas, calcium carbonate lime ($CaCO_3$) is most important in this process, whereas in arid regions both calcium and sodium carbonate (Na_2CO_3) contribute. These basic salts derive from decomposition of limestone or dark-colored igneous rocks; soluble in water, they are readily removed in solution. Where rainfall is heavy, soils leach rapidly; consequently, humid soils tend to be acid, dry soils alkaline.

Figure 5-14 shows the so-called "lime line" in the United States that divides the humid East from the dry West. East of this line, soils mostly are neutral to acid in reaction; to the west, they are neutral to alkaline, except in forested mountains and along the humid parts of the West Coast. In spite of

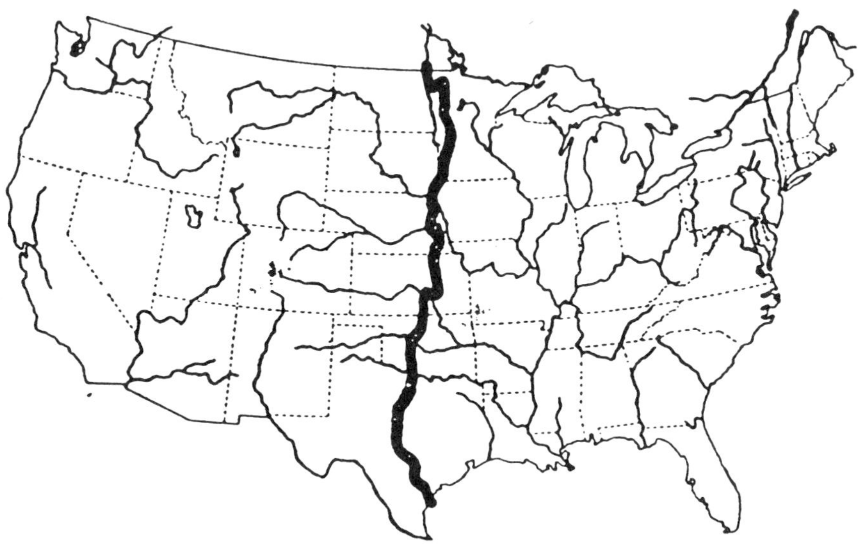

Fig. 5–14. The "lime line" in the United States. The line approximates the pH value of 7. East of the line most soils range from neutral to acid in reaction. West of the line soils are generally neutral to alkaline with the exception of forested mountains and humid west coast areas. There are local occurrences of moderately alkaline soils in the eastern United States where parent materials happen to be highly calcareous. (USDA, *Soil,* Yearbook of Agriculture, 1957, 187.)

humid climate in eastern states, there are scattered areas of naturally calcareous soils, high in $CaCO_3$ and neutral to moderately alkaline.

Soil pH values may vary greatly according to depth from the surface. Surface soils generally are more acid than subsoils because of stronger leaching action by percolating rain water and a greater abundance of acid-forming organic matter in upper layers. Bases leached from upper layers may be deposited in subsoil. In arid regions this action often produces a strongly cemented layer of $CaCO_3$ called a *caliche* (see chapter 7).

All plants require at least small amounts of calcium for their metabolic processes, but these requirements vary greatly among species, with those of cool, moist areas requiring very little and those of dry areas requiring considerably more. Calcium availability is correlated closely with soil pH; generally, it is readily available in the pH range above 6 and almost completely unavailable below 3.8. Only plants that can endure extreme calcium deficiency survive under these extremely low pH conditions. Free lime usually occurs in soils with pH above 8.3 where it will effervesce when treated with hydrochloric acid.

Availability of plant nutrients, with the exception of calcium, correlates closely with soil pH in a relatively complicated way. Around pH 6.5 all minerals are sufficiently soluble to satisfy most plant needs, but deviation in either direction makes certain nutrients less available. For example, levels of pH below 6 and above pH 8 impair nitrification, and at extreme pH levels certain nutrients may reach such high concentrations in the soil solutions that they are toxic to some plants (Thompson and Troeh 1978). Under natural conditions, plant communities adjust by species composition to the pH of their soil habitat. As the plant community progresses toward a climax, the pH may shift toward more favorable levels for the community. Farmers often apply lime to raise the pH level of their soil.

Soil pH is the most important chemical characteristic related to plant ecology. Some plant communities alter soil pH by cycling minerals from the subsoil to the upper horizons. For example, some grasses absorb calcium from the calcium-rich subsoil, where leaching concentrated it, and it accumulates in plant tissues. The calcium returns to surface layers of the soil upon death and decomposition of the plant.

The Role of Fire in Plant Communities

Since land plants first appeared on the Earth, fire has played a major role in vegetation dynamics. Fossil evidence of charred vegetable matter suggests that periodically lightning ignited fires long before humans came on the scene. Only very wet regions, or very dry or cold regions having very sparse vegetation, manage to escape large scale burning. With the use and control of fire by primitive peoples, burning became more frequent and widespread.

Today, the process sometimes reverses, and protection from fire may alter vegetation. Prairie fires once were common in the grasslands of North America, but cultivated fields and dense road networks make effective fire barriers, and fire is no longer a major ecological force in those areas. Consequently, there has been an invasion of trees and shrubs into some former grasslands, particularly mesquite (*Prosopis*) in dry southwestern areas of the United States, sagebrush (*Artemisia*) in areas farther north, and oak (*Quercus*) along the more humid eastern margins of the Great Plains. The practice of controlled burning in some areas improves grasslands (fig. 5-5 and 5-6).

In areas where fires frequently occur, some plants evolved special adaptive features for survival in the face of repeated burning. Such plants are *pyrophytes*, and fire-resistant associations are *pyrophytic*; they require occasional burning to maintain their characteristic floral composition. Pyrophytes have a number of survival adaptations, including (1) seeds that require fire for germination; (2) seeds that germinate well in mineral soils after fire has burned off the accumulation of plant litter; (3) ability to produce sprouts from roots after the upper part of the plant has been destroyed by fire (burning may even stimulate sprouting); (4) physiologic features, such as thick, insulating bark that protects mature plants from heat; and (5) plants of the palm family, fire resistant because they are not dependent on vulnerable cambium tissue beneath the bark for circulation of water and nutrients.

Human Impact on Plant Communities: The Case of the Danish Heath[4]

Although the role of humans in plant ecology still is being debated, in many areas evidence abounds of human impact on some plant communities. An outstanding example is the development of large areas of heath on the peninsula of Jylland (Jutland) in western Denmark during prehistoric and historic times, and modern efforts to convert this heath into more productive agricultural land and forests. The Danish heath offers an excellent case study of the human role in changing the character of vegetation, slowly and incidentally over thousands of years, rapidly and deliberately during the last century.

Origin of the Heath

A heath is an area covered with a complex of low-growing plants and shrubs dominated by the common heather or ling (*Calluna vulgaris*) and members of the heath family (*Ericaceae*). Figure 5-15 shows a heath in western Denmark. This vegetation cover, in combination with a damp marine climate, has caused extreme *podzolization*[5] of the sandy soils, giving rise to some of the poorest soils in Europe in their natural state. Through archaeological

[4]This section is taken, with minor changes, from W. E. Akin, "The Reclamation of the Danish Heath," *The Journal of Soil and Water Conservation* 18 (No. 3, May–June 1963): 97–107. Reprinted with permission of the publisher.

[5]*Podzolization* is a soil-forming process occurring in areas with a moist, cool climate under vegetation, such as coniferous forests and heaths, that produces highly acid litter.

Fig. 5–15. A remnant of preserved heath in Jylland in western Denmark. The dominant plant is common heather (*Calluna vulgaris*). (Photo by W. E. Akin, 1962.)

evidence and analysis of pollen trapped in peat bogs, botanists have reconstructed the probable sequence of vegetation in Jylland, providing some insight into the role of humans in changing the plant and soil environment that existed at the end of the last glacial period twelve thousand years ago.

Geological events played a major role in differentiating western Jylland from other parts of Denmark. Only this area escaped the last ice advance, although it was covered by earlier glaciers (fig. 5-16). As the last ice sheet melted, great quantities of gravel and sandy outwash spread over large areas of the western part of the peninsula. This coarse-textured outwash served as the parent material for the poor, sandy soils on which the heath reached its greatest development.

This development did not depend solely on geological events. Pollen studies show that the heath plain was occupied by a sequence of forests, the last predominantly oak (*Quercus robur*). This oak forest contained numerous open areas of grass and heather. Not only did oak dominate the forest of western Jylland, but also it was the principal forest tree in all of Denmark from about 5,000 B.C. until the beginning of the Iron Age, about 600 B.C. The beech (*Fagus sylvatica*) thrived only in isolated stands, and not until the Iron Age did it begin to spread throughout the heavier soils of the country to become the major tree of historic times. Remnants of the old oak forest survive today in Denmark, one of the best examples being near Viborg at

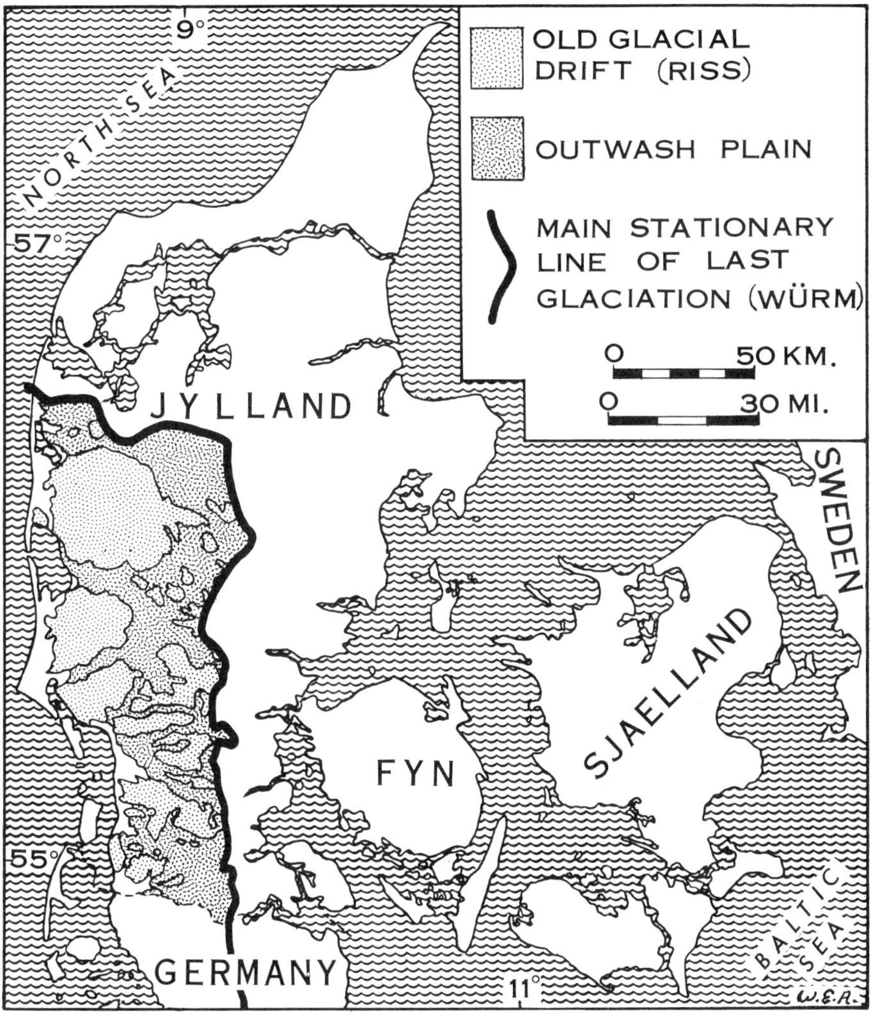

Fig. 5–16. The boundary of the last glacial advance into Denmark—the "Main Stationary Line." West and south of the line, vast quantities of sand and gravel spread out from the margins of the melting glacier, forming a vast outwash plain. Higher areas of older glacial materials form "hill islands" (*bakke*). (Based on A. Schou, *Atlas of Denmark,* vol. 2, Copenhagen: C. A. Reitzels, 1949.)

Hald Ege (fig. 5-17) Danish place names often give hints about the former vegetation cover; for example, *lyng* (heather), *eg* (oak), *skove* (forest), and *hede* (heath). These terms, along with other vegetation designations, are used as prefixes or suffixes in many Danish place names (a sample is listed in table 5-4).

Fig. 5–17. A remnant of the oak forest that probably covered most of western Jylland some 5,000 years ago, before destructive agricultural practices and a cooler and damper climate prevented the forest from re-establishing itself after the land was abandoned, and encouraged the spread of the heath. (Photo by W. E. Akin, 1962.)

Much evidence exists of the importance of the human role in the development of the heath both here and in other parts of northern Europe. Traces of agriculture first appeared in the late Stone Age (*Neolithic*) about five thousand years ago. The light, sandy soils of Jylland were well suited to primitive methods of prehistoric farmers. The attack on the forests began with the introduction of primitive *swidden* (slash and burn) agriculture. Farmers cleared small fields and reduced larger tracts by burning to improve pastures. By the beginning of the Bronze Age, some thirty-five hundred years ago, much of western Jylland probably consisted of open oak forests with many grass-covered clearings, producing a park-like landscape. The

TABLE 5-4 Elements of Danish Place Names that Provide Clues About Former Vegetative Cover

Term (Meaning): Sample place names containing term
Eg, Ege,[1] Egen[2] (Oak): Hald Ege, Egebærg, Egebæk, Egedal, Egemark, Egeskov
Bøg, Bøge, Bøgen (Beech): Bøgeskov, Bøgebakke, Bøgehøj
Birk, Birke (Birch): Birkmose, Birkhøj
Lyng (Heather): Lyngby, Lyngsaa, Lyngsø
Hede (Heath): Hedeby, Hedegaard, Almindhede, Bannerlundshede
-rup (Clearing in forest): Allerup, Durup, Geestrup, Stenderup, Hjelmerup, Tommerup, Torup, Ostrup
Other terms common in place names: gaard or gård (farm); sø (lake); å or aa (river); mose (bog, moor); bjerg or bakke (hill); bæk (brook); dal (valley); mark (field); by (town); høj (hill, mound); skov (forest)

[1]The suffix -e creates the plural form: Eg = oak, Ege = oaks.
[2]The suffix -en means "the": Egen = The oak.

Neolithic and Bronze Age apparently had average temperatures higher than those of today, the so-called "climatic optimum" or *Altithermal* having been reached around five thousand years ago (see chapter 4). Evidence exists of considerable concentrations of population during the Bronze Age in these areas presently covered by the heath. Numerous Bronze Age burial mounds dot the landscape.

During the Iron Age, beginning around 600–500 B.C., cooler and damper climatic conditions prevailed, and heather and associated heath plants spread at the expense of forest. When farmers abandoned cultivated land, it became heath because forest was unable to regenerate under the poorer climatic conditions. Undoubtedly, burning in order to maintain open pastures further contributed to the demise of the woodland. The extension of the heath, with the help of human agricultural and pastoral activities, probably continued into relatively recent historic times.

By the eighteenth century, the great heath of Jylland was inhabited only along water courses where farms could have a supply of hay from low meadows and where sheep and cattle could graze on the poor heath pastures. Today, old farms still are seen along the water courses where former heath lands and flood plains meet, contrasting with new farms on the reclaimed lands. Between water courses, large areas of heath lay virtually uninhabited. In 1800, about 11,000 km^2 (4,200 mi^2) of heath and heath-covered bogs and dunes existed in Denmark. Most of this area was in Jylland, where about 48 percent of the peninsula was heath (fig. 5-18).

Heath Reclamation in the Nineteenth and Twentieth Centuries

Although early efforts to reclaim the heath date from the mid-eighteenth century, the greatest impetus came with the establishment of the Danish Heath Society (*Det Danske Hede Selskab*) in 1866. The success of its endeavor

Fig. 5–18. The extent of heath areas in Jylland in 1800 (left) and in 1950. There is a clear relationship between the boundary of the last glacial advance and the distribution of heath (Modified from Nielsen 1953, 394, 395)

becomes apparent when one considers that heath, dune, and bog lands occupied 25 percent of the total land area in all of Denmark in 1850, but only 8 percent a hundred years later. (Figure 5-18).

Although early emphasis was on land for agriculture, forestry played an increasing role in reclamation; today the remaining small areas of reclaimed heath are almost all in trees. Also, some previously reclaimed agricultural land is now planted in forest because returns from agriculture are disappointing economically. Figure 5-19 shows the extent of forests in Jylland in 1800, compared to 1950.

The sequence of vegetation on the Danish landscape since the last glacial period replicates that in many other areas of Europe. In areas such as the Mediterranean basin, human impact has been a powerful ecological force for

Fig. 5–19. Forested areas in Jylland in 1800 (left) and 1950. Forests were scarce in western Jylland in 1800, compared with the extensive planted forests in the same area in 1950. (Modified from Nielson 1953, 394, 395)

a much longer time than in Denmark. In the Western Hemisphere, and in large areas in the tropics, the results of human activity on vegetation are less well documented and more difficult to reconstruct. Some plant ecologists, however, hold that natural vegetation, in the sense of vegetation unaltered by human activity, probably is a rare thing, limited to inaccessible areas such as tundra.

The Human Role in Retrospect

Four major categories sum up the role of humans in altering the nature of vegetation: (1) the use of fire, (2) agriculture, (3) exploitation of forests for lumber and fuel, and (4) introduction of exotic plant and animal species.

Use of Fire. Fire probably was the earliest human tool capable of causing major changes in vegetation, either through accidental burning or intentional fires to drive game, improve browse for game or livestock, or clear forests for slash and burn agriculture.

Agriculture. Many forests and grasslands were turned into "cultural steppes" by clearing, plowing and planting crops, or using land for pastures. Cultivation further alters the nature of soil, which in turn helps determine future vegetation that may reoccupy land after its abandonment. In extreme cases, such as the Danish heath, alteration of natural vegetation is complete, and original vegetation does not return until humans intervene once more.

Exploitation of Forests for Lumber and Fuel. Since prehistoric times, forests have served as a source of wood for fuel and building materials. Even today possibly two-thirds of the world's people use wood for cooking, often in the form of charcoal. In ancient times great forests of cedar (*Cedrus libani*) covered the middle slopes of the mountains of Lebanon and other parts of the Mediterranean basin. Today, only a few protected stands of these famous cedars of Lebanon remain (fig. 5-20). At present, the area resembles a desert

Fig. 5–20. "He shall grow like a cedar in Lebanon" (Psalms 92:12). Cedars still grow in Lebanon although their numbers are greatly diminished since the days of the Psalmist, some 3,000 years ago. Magnificent forests of *Cedrus Libani* once covered this landscape before they were cut down to supply timbers for ancient Mediterranean civilizations. The monastery walls that surround this grove, one of the four remaining, protect it from grazing and browsing goats; probably goats were a factor in the failure of the forest to regenerate after cutting. (USDA, Soil Conservation Service photo by C. W. Lowdermilk, September 1939.)

landscape with bare, eroded hills and sparse vegetation. Yet, it receives on average about 1,000 mm (40 in) of precipitation annually, an amount and seasonal distribution similar to parts of California where luxuriant forests grow today.

Why did this great forest disappear? Records from nearly five thousand years ago indicate that Egyptians depended on the eastern Mediterranean for their timber. The Old Testament account of Solomon's construction of the temple in Jerusalem (966–958 B.C.) speaks of "fourscore thousand hewers in the mountains" to obtain the prized cedars of Lebanon. The Phoenicians, living in what is today Lebanon, were the greatest seafarers and merchants of the ancient world, and they used the cedars for shipbuilding. Their commercial fleets, and those of Greece and Rome, annually carried millions of tons of cargo of lumber, wheat, olive oil, and wine throughout the Mediterranean Sea basin. Availability of excellent timbers for shipbuilding undoubtedly aided the Phoenicians in dominating Mediterranean shipping for many centuries, but it also led to the depletion of this once magnificent forest.

Mining and smelting of metals also took a heavy toll of forests of the ancient world. Forests supplied wood for charcoal to fuel furnaces and to reduce metallic ores long before coke from coal replaced charcoal in relatively recent times. Coal and coke did not replace wood until the last half of the eighteenth century in England, and wood remained an important fuel until about 1850. Also, vast areas of forests were devastated to supply timber for mine props, carts, sluices, pipes, and other equipment, in addition to the demand for fuel. Charcoal is still used today to make steel in small mills, particularly in Brazil.

Introduction of Exotic Plants and Animals. New plant species may be brought into an area, either intentionally or by design. For example, guinea grass (*Panicum maximum*) probably arrived in the Caribbean from Africa in the bedding of slave ships, and it is now a major pasture and wild grass in the region. The stately eucalyptus is a native of Australia, but it is an important intentional import into many areas having a Mediterranean climate, including Spain and California. According to the U.S. Forest Service, more than a thousand species of foreign trees grow in the United States, especially in subtropical areas from Florida to California.

Plant importations are not always desirable and they may become weeds. Two plants introduced into the United States during the 1930s to aid in soil and wildlife conservation are kudzu (*Pueraria thunbergiana*) and multiflora rose (*Rosa multiflora*). Kudzu, a native of Japan, showed promise in erosion control and as a pasture legume, but today it is a serious problem in the southeastern United States, where it has invaded forests and roadsides and actually strangles large trees. Multiflora rose plantings were encouraged in the Midwest as windbreaks and for wildlife shelter. Today, the multiflora rose hedges have spread until they interfere with mechanized agriculture; they are difficult and expensive to remove.

Insects and insect-borne diseases may have profound effects on vegetation. In the 1930s, the Dutch elm beetle was introduced into the United States in a load of lumber, and today a fungus spread by this insect and the American bark beetle has virtually destroyed the American elm (*Ulmus americana*). Domesticated animals—particularly browsers, such as goats and camels, which eat twigs, leaves, and bark as well as grass—are especially destructive of natural vegetation. They are responsible in large part for the barren nature of much of North Africa and the Middle East.

In this chapter we stressed the dynamics of vegetation and the complex interrelationships between plant communities and their environment. We attempted to go beyond simple climate-vegetation correlations. For ease of presentation and comprehension, we described separately the environmental factors influencing vegetation, but these processes take place simultaneously, and each may influence or be influenced by the others. An exhaustive treatment of plant-environment relationships, many of which are still not clearly understood, was not the aim of this chapter. Rather, its purpose was to illustrate the complex nature of those relationships in determining the vegetation of Earth. Chapter 6 focuses on the great global plant formations that covered the land in the past, and on their remnants still occupying large areas.

In the spring when abundant rains have bathed the rugged slopes, an almost uniform green suffuses them; in early summer they are brilliant with loco weed and other flowers; but when summer has advanced, when the rains have ceased and the blistering winds and scorching sun have robbed the southwesterly prairie slopes of their moisture, the sheltered groves and the exposed prairie surfaces stand out in sharp contrast, visible for many miles and setting out with striking effect and unmistakable precision the varied features of this singular topography.

BOHUMIL SHIMEK, *Geology of Harrison and Monona Counties* [Iowa] (1910)

Chapter 6

GLOBAL PATTERNS OF VEGETATION

Plant communities show even more variability than do climates; thus, their world distribution pattern is even more complex. One approach to simplify this complex pattern is to group naturally occurring associations of plants into the *climatic climax formation types* described in chapter 5. As with any such attempt, naturally there is much uncertainty and disagreement among plant geographers, particularly about areas where humans are important ecological agents.

Life forms, not species, determine plant formations. Most plant communities within a formation type are similar in structure, size, and general appearance, but in different parts of the world, the actual plant species that compose them vary greatly. For example, in tropical rainforests, plant species of the Old World differ from those of the New World, but life forms in both formations are very similar.

Generally, plant formations and associations adjacent to each other do not have sharp boundaries, as a vegetation map might suggest; rather, in zones of mixed life forms, one type gradually gives way to another. For example, sometimes a transition zone of mixed broadleaf and needle-leaf trees separates a deciduous broadleaf forest from a coniferous one. Plant ecologists call such transitions *zones of ecotone*, or just *ecotones*. Exceptions abound; one example is the rather sharp boundaries between some grasslands and forests.

The Vegetation Map

Plant geographers do not always agree on the criteria for mapping vegetation and thus patterns on maps of the same area sometimes differ. This ap-

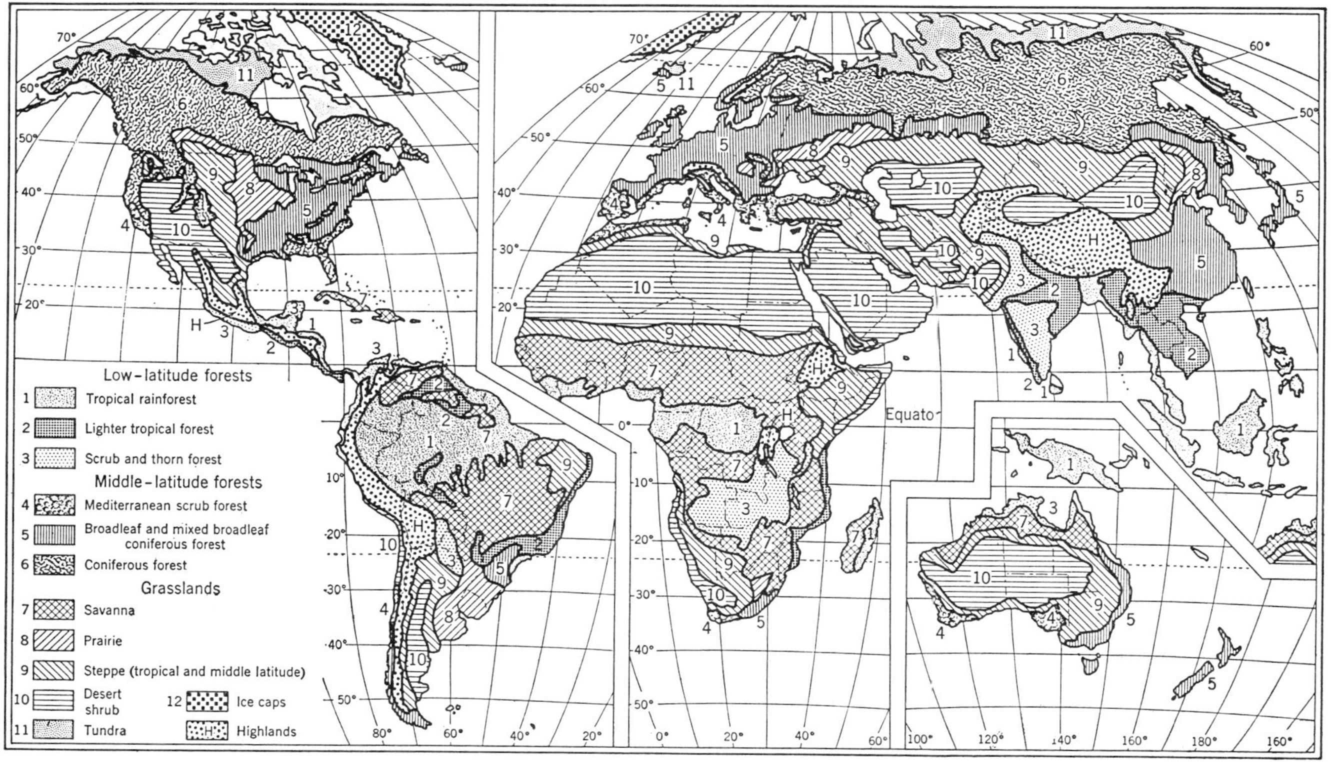

Fig. 6–1. World distribution of the principal vegetative formations. Patterns are simplified, and more than one formation type may be included within one classification; for example, large areas of tropical deciduous woodlands occur in number 3, shrub and thorn forest, and number 9, steppe, consists mostly of pure grasslands in temperate North America and Eurasia, but is composed of grass and shrubs in the tropics. (Reproduced with permission from Arthur N. Strahler, *Physical Geography,* © New York: John Wiley & Sons, 1951, 346.)

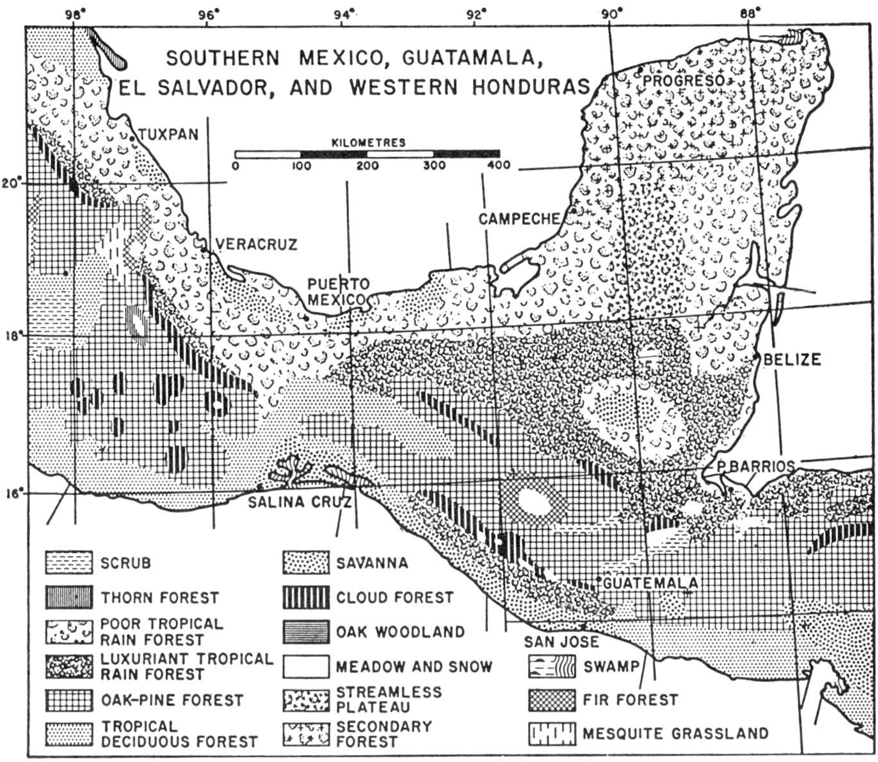

Fig. 6–2. Vegetation formations of southern Mexico, Guatemala, Belize, El Salvador, and part of Honduras. Two types of swamp are recognized, freshwater and mangrove. Compare this map with the same area on the world vegetation map; note the greater detail possible at this larger scale and the degree of generalization necessary on the small scale world map. (Reprinted by permission from V. E. Shelford, 1963, 397.)

parent disagreement is not surprising since *natural vegetation* is open to numerous and sometimes conflicting interpretations. All these maps have some validity, but their differences can be confusing unless the criteria are made clear. All vegetation maps, especially those of worldwide scope, are only approximations—sometimes very rough ones at that—and not to be taken as absolute truth.

In figure 6-1 world vegetation has been grouped into eleven broad formation types for treatment in this text. The degree of generalization necessary for indicating vegetation on a small-scale world map can be seen by comparing figure 6-1 with the larger scaled map of Mexico and Central America (fig. 6-2), or with the vegetation map of the United States (fig. 6-3). Some of

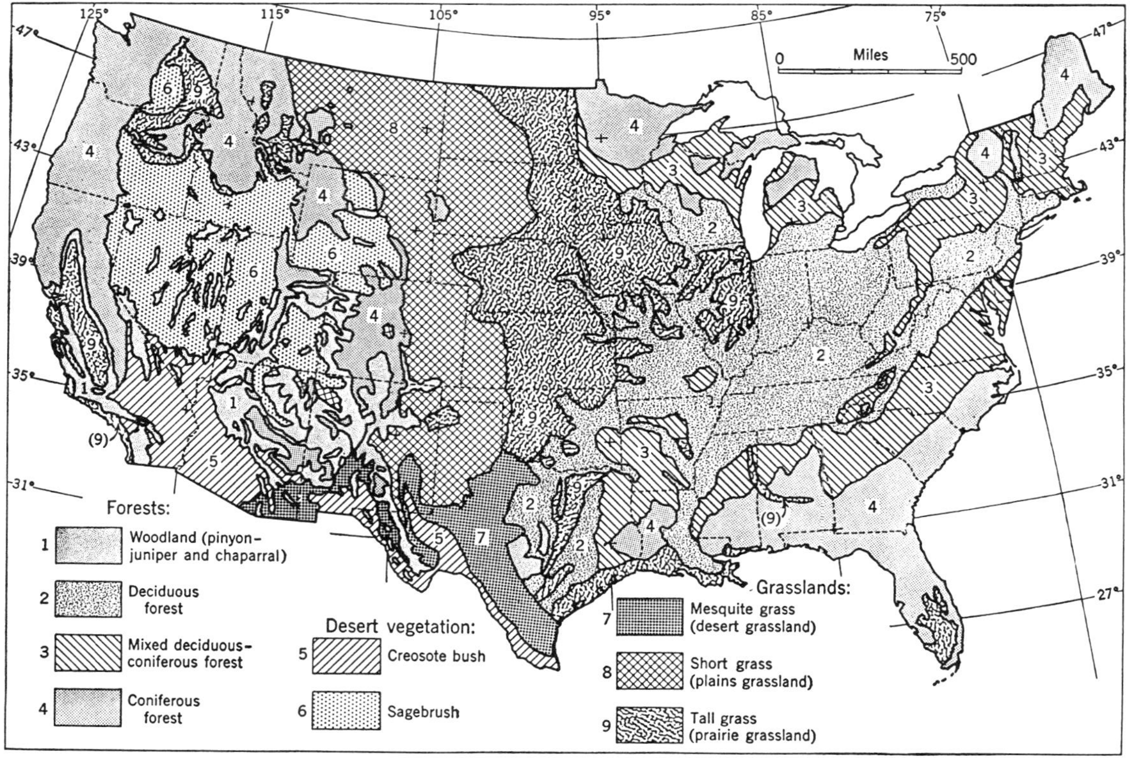

Fig. 6–3. Map of vegetation in the conterminous United States. The unpatterned areas are tallgrass, midgrass, and shortgrass prairies in the Great Plains and eastern states; mesquite (*Prosopis*) and grass in central and southeastern Texas, southern New Mexico, and southeastern Arizona; creosote bush (*Larrea*) in southwestern Arizona and southeastern California; sagebrush (*Artemisia*) in the Great Basin areas of Utah, Nevada, Idaho and Oregon. (After H. L. Shantz, USDA Yearbook of Agriculture, 1938 [Reproduced with permission from A. N. Strahler, *Physical Geography*, New York: John Wiley & Sons, 1951: 404.])

the formations described in this chapter are not delineated on the world map; for example, broadleaf and mixed broadleaf-coniferous forests are shown as a single pattern. Also, the areas designated scrub and thorn forests include the tropical deciduous woodlands of southern Africa and other regions.

Vegetation of Tropical Regions

Let us begin with the tropics. This part of the Earth is little understood by most people who live elsewhere. Because they have only vague and often erroneous ideas about it, stereotypes define it for them. The stranger to the tropical world envisions it as a disease ridden area of intolerable heat and steaming, impenetrable jungles, filled with dangerous animals and people. Motion pictures, television, and novels often perpetuate this false image of that vaguely conceived region between the tropics of Cancer and Capricorn.

The ubiquitous term *jungle* sums up most popular ideas about tropical vegetation. The term originated in India where it meant areas of wild uncultivated land with dense undergrowth; when used as a synonym for tropical rainforest, as in "the jungles of the Amazon," the usage bears little resemblance to the original meaning. Consequently, the term *jungle* is of little value to plant geographers. In fact, tropical vegetation varies from the great rainforests to desert scrub, and it includes many varieties of woodlands and savannas as well.

In the tropics, the most important environmental variables affecting vegetation are the amount and seasonality of precipitation, and edaphic factors such as soil, slope, and drainage. Temperatures vary little from season to season or from place to place except as modified by elevation. Only at very high elevations does freezing occur, and consequently the temperature-determined growing season is continuous. Variations in vegetation within the tropics generally reflect amount and seasonality of precipitation and elevation above sea level—a correlation, however, that is far from perfect.

Tropical Rainforest

This most magnificent of biomes, the tropical rainforest, girds the Earth's equatorial region in a discontinuous belt coinciding roughly with the wet tropical climates where mean annual rainfall exceeds 2,000 mm (79.4 in). Biogeographers often use the Brazilian term *selva* as synonymous with tropical rainforest. The tropical luxuriance of the rainforest has impressed travelers since Columbus first described those of Hispaniola in 1492, and many early observers assumed the soil of the rainforest must be "marvelously fertile." As we shall see, for the most part, this assumption was erroneous.

The tropical rainforest formation type subdivides into three great formations occupying large areas of land in the equatorial zone: (1) South American; (2) African; and (3) Indo-Malayan. Seasonal forests and savannas separate several outliers of tropical rainforest from the main centers. Rainforest occupies an estimated 400 million ha (1.5 million mi^2), with the most extensive surviving area located in the Amazon region of South America. The impact of human activity has destroyed large areas, some having been obliterated. Nevertheless, recognizable tropical rainforest still exists throughout much of its former range.

Correlations Between Rainforests and Climate

Climatic and vegetation maps show a relatively good correlation between the heart of the tropical rainforest formations and the wet tropical (*Af*) climate. Some areas of monsoon tropical (*Am*) and wet-dry tropical (*Aw*) climates also have tropical rainforests, whereas others have seasonal forests and savannas, a pattern reflecting in part the length and severity of the dry season (fig. 6-4). On the other hand, boundaries of vegetation formations and climates are not so well matched. In South America, tropical rainforest is considerably more extensive, with rainforest occupying large areas of

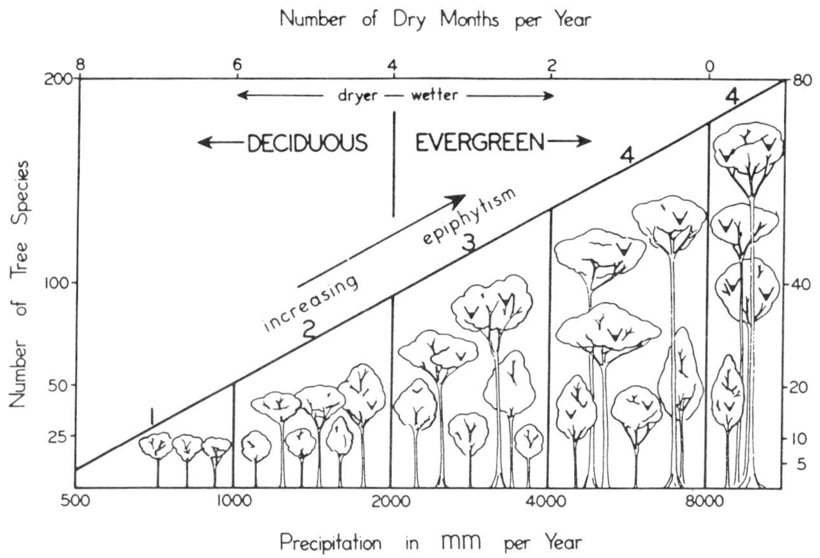

Fig. 6–4. A relationship exists between precipitation and the complexity of the forest biome in tropical lowlands. In areas with more than 2,000 mm (79.4 in) of average annual precipitation, broadleaf evergreen rainforests predominate; in areas with lesser amounts, semideciduous and deciduous tropical forests are the rule. With increasing precipitation there is also an increase in the number of tree species and numbers of epiphytes. (From National Research Council 1982, 30. Reproduced with permission of the National Academy Press.)

wet-dry tropical (*Aw*) climate. There is a similar lack of correlation between climate and vegetation in Africa, where some areas with *Af* climate have savannas, and some have rainforests.

Life Forms and Structure in Tropical Rainforests

Although the tropical rainforest biome is the richest on earth in terms of the numbers of species of plants growing there, life forms of the forest show a remarkable similarity throughout the formation type. Broadleaf evergreen trees dominate, and trees of widely variant families and species have leaves that are remarkably similar in size, shape, and texture. Dark green, leathery leaves of moderate size are the rule. They often have a pronounced elongation of the terminal end, called a "drip tip." Contrast this similarity in shape with the great diversity of leaf shapes in a typical broadleaf forest in the United States, in spite of the fact that there are many more species in the tropical rainforest.

Rainforest trees exhibit other common characteristics; most have very thin bark, possibly because there is no need for insulation from extreme temperatures, and the trees need little protection from water loss. Many trees and shrubs bear flowers and fruit directly on their limbs and trunks, a phenomenon called *cauliflory*. Many of the largest species develop "plank buttresses" extending outward from the base of the trunk (fig. 6-5), while others have "stilt" roots growing out from the trunk. Stilt roots are particularly well developed in trees, such as the mangrove (*Rhizophora mangle*), that grow in flooded areas, but they occur on high ground as well. Plant morphologists still debate the survival value of such features, but it seems likely that these buttresses compensate for shallow root systems.

Life forms in the rainforest evolved over millions of years; during this long time span, these plants have had to make continual morphological adaptations for survival in the competition for nutrients and sunlight. Tall, slender, straight-trunked trees dominate, but many other life forms compete successfully. In a typical rainforest, the crowns of trees tend to be stratified into three or four layers, although to an observer on the forest floor, this separation may not be obvious. These layers are referred to, from the top downward, as the A, B, and C layers. Below these is the D layer, consisting of a sparse growth of shrubs, young trees, small palms, and tall herbaceous plants and ferns. The plants of the D layer average about 1 m (3 ft) in height.

Statistical studies made in a number of tropical rainforests and reported by P. W. Richards (1966), indicate that the average height of the top of the A layer seldom exceeds about 45–55 m (160–180 ft) and may be considerably less at some locations. Early accounts often exaggerated the size of tropical trees, although some individual trees may reach more than 60 m (200 ft), and a rare one may exceed 92 m (300 ft). Nevertheless, rainforest trees seem

Fig. 6–5. Tropical rainforest on Malaita, Solomon Islands. 'Umbuni, of the Kwaio people of the area, carries the ubiquitous bush knife, universal tool of the rainforest. Note the buttressed roots of the tree in the foreground. (Photo by W. E. Akin, 1980.)

never to reach the massive proportions of California redwoods (*Sequoia sempervirens* and *Sequoiadendron gigantium*) or the gum trees (*Eucalyptus*) of Australia. Trees of the B layer average between 20 and 27 m (65 and 90 ft), whereas those of the C layer average about 8–15 m (26–50 ft).

Taller trees, generally speaking, are more light demanding (*heliophilous*) than shorter trees and low vegetation of the forest floor. Many herbs are native to the floor of tropical rainforests; some, including *Aspidistra* (*A.* spp.) and African violets (*Saintpaulia* spp.), have become popular house plants well adapted to indoor conditions in the middle latitudes where warm temperatures and lack of direct sunlight duplicate conditions in their native habitat.

While receiving only limited diffused sunlight, tree seedlings of rainforest species are able to survive in a semiarrested state of growth until space becomes available. Those of tall trees must bide their time in the half-gloom of the forest interior until the death or destruction of a mature tree offers an opportunity for sunlight to penetrate to the ground and initiate furious competition for the vacated space. Since young trees of light-demanding species respond more quickly to light stimulus than do more shade-tolerant species, they can take advantage of an opening and occupy it before other species overtop them and shade them out.

Other plant forms have special adaptations for gaining their place in the sun. Climbing vines, the *lianas*, use trees for mechanical support to reach the sunlight. Some lianas may attain lengths of more than 60 m (200 ft). *Epiphytes* live on limbs and twigs of trees in upper levels of the forest; they represent many families and species, including orchids, cacti, ferns, and bromeliads (fig. 6-4). Although their position high in trees assures favorable light conditions, the reservoir of soil moisture is unavailable to them. Depending directly on rainfall for water, most epiphytes have the ability to store water in their tissues in the manner of xerophytes. Many have succulent leaves; some have aerial roots that absorb water during showers; and the bromeliads collect water in funnel-like bases of their leaves where the plant tissues absorb it directly.

Another group of rainforest plants is the "stranglers"; they begin as epiphytes but later send roots down to the soil, eventually engulf the supporting tree with their branching root systems, and often kill it. When the dead supporting tree rots, the strangler becomes a hollow but independent tree, often of great dimensions. Most familiar and plentiful of the stranglers in the tropical rainforest are numerous species of strangling figs (*Ficus* spp.).

Unquestionably, the outstanding characteristic of the tropical rainforest, when compared to other formation types, is its floristic richness and diversity. Even in limited areas, it is remarkably rich in the number of families and species represented. In middle latitude forests, it is most unusual to find more than a dozen or so species of trees in one area, and in the boreal forests the number of species may be only four or five. In contrast, the tropical rainforest often has more than 120 species per hectare (50 species per acre). The Amazon rainforest has at least twenty-five hundred species of tall trees, about a thousand in the state of Pará alone. Similarly, the Malay Peninsula is home to about the same number of tree species as the Amazon forest (Eyre 1963). It is likely that these figures are an underestimation. Many species have never been catalogued and, because of accelerated forest destruction, some may never be listed.

The Plant Nutrient Cycle in a Rainforest

Soils and soil-forming processes are the subjects of chapter 7; in this chapter we consider only the seemingly paradoxical situation of luxuriant forests growing on soils highly leached of nutrients and low in humus—soils that often prove to be very poor and unproductive when cleared for agriculture. Most ecologists conclude that the undisturbed rainforest is in a continuous state of equilibrium, with the nutrient supply locked within a closed system. Plant nutrients are released continually from decaying vegetation and are absorbed immediately by plant roots and then recycled in the *biomass*[1] before they can be leached from the soil.

In the past, bedrock was much closer to the surface than it is today. Weathering of this bedrock through the ages produced the present supply of nutrients available in the biomass of some rainforests. Now, the weathered materials lie so deep that nutrients released from freshly decomposing bedrock are no longer within the reach of the shallow-rooted forest plants. This fact may explain why some rainforests will not regenerate after large areas have been cleared for agriculture or pastures.

Clearing rainforests for agriculture breaks the nutrient cycle, and the nutrients formerly trapped in the cycle are released through burning of the debris and by rapid decay of the sparse soil organic matter (humus). Also, the exposure of the surface to direct sunlight encourages rapid growth of weeds. It is not surprising that crop yields, which may be quite good in the first year after clearing, soon decline, so that farmers must abandon these plots and clear new ones. Under this system of farming, called *shifting cultivation, swidden agriculture,* or *slash-and-burn agriculture*, disruptions are small. The forest may be able to regenerate, but only if the fallow period is prolonged (fig. 6-6). Large-scale clearing for cultivation and pasture, however, can be disastrous, not only for the pioneer agriculturists, but for the survival of the forest itself. Many plant and animal species may be lost forever, the biota of Earth may be impoverished, and the genetic pool on which future biological evolution depends will be much restricted.

Tropical Seasonal Forests

In areas where the dry season is more prolonged and where there is a season of critical moisture deficiency for some plant species, the tropical rainforest gives way to a very complex series of vegetation types collectively called *tropical seasonal forests*. These forests are not only less luxuriant in

[1] *Biomass* is the sum total of the plant and animal organisms living in the biome. It can be measured in a variety of ways, including living weight (fresh biomass), dry weight, organic matter weight, weight of carbon, or energy equivalents.

Fig. 6–6. Slash and burn agriculture on Malaita, Solomon Islands. Crops are mostly taro and sweet potatoes. (Photo by W. E. Akin, 1980.)

structure and size, but also, as mean annual rainfall decreases, they contain fewer and fewer species of trees. The rainforest formation type is relatively easy to recognize and map because of the unique features characteristic of all the formations within it. Not so seasonal forests, and the savannas and thorn forests with which they often intermingle. Interpretation and mapping of the distribution of this vegetation classification varies considerably among plant geographers. One map will show savannas in an area where other maps will show thorn forests or seasonal forests; all depends on the subjective decision of the map maker.

Semideciduous Seasonal Forests

Toward margins of the tropical rainforest where the less rainy season approaches a true dry season, some trees in the A layer become deciduous, but the percentage is small and the typical rainforest structure prevails. Where the dry season is more pronounced and mean annual precipitation falls below 2,000 mm (80 in), the predominantly evergreen forest gives way to a distinct formation type, the *semideciduous seasonal forest* (also called *semi-*

evergreen). This semideciduous forest forms a true ecotone between the evergreen rainforest and the tropical deciduous forests in a climatic zone where mean annual rainfall is between 1,500 and 2,000 mm (60 and 80 in) (National Research Council 1982). This formation type includes American, African, and Indo-Malaysian formations that are similar in life forms and structure but vastly different in species.

S. R. Eyre (1963) describes the structure of an undisturbed semideciduous seasonal forest in Trinidad that seems to be representative of the formation type. This forest has two stories; an upper one reaching a height of about 21 m (70 ft) and containing between 20 and 30 percent regularly deciduous trees, and a lower one that is predominantly evergreen. Most of the evergreen species in the upper layer, however, are "facultatively deciduous"—that is, they can lose their leaves in an abnormally dry season but remain evergreen in all seasons during a normal year. As with the neighboring rainforest, the semideciduous forest is rich in lianas. However, epiphytes are much less common here, since they are ill equipped physiologically for withstanding drought (fig. 6-4).

In tropical America, this formation occurs from southern Mexico to southeastern Brazil, in areas with 750–1,250 mm (30–50 in) of rainfall. In Africa, no large areas of semideciduous seasonal forests remain, but evidence suggests that in the past, savanna areas bordering directly on the West African rainforest may have supported such a formation. The present moist savannas appear to be plagioclimax formations, as a direct result of clearing for agriculture; human pressure and fire prevent forest regeneration.

In the Indo-Malaysian region, this formation type often is the "moist monsoon forest," while the deciduous seasonal forests are "dry monsoon forests." The famous teak forests (*Tectona grandis*) of Burma and other Southeast Asian areas belong, in part, to this formation—the "moist teak forest" in local terms. Annual rainfall in some areas of semideciduous seasonal forests in the Indo-Malaysian area is much greater than in the American. The dry antimonsoon season, however, may be drier and more prolonged than its American counterpart, an indication that the severity of the dry season and not the total rainfall is the critical factor (Eyre 1963).

Deciduous Seasonal Forests

In areas where the dry season is even more prolonged than in the semideciduous seasonal forests, and where average annual precipitation may be as low as 500 mm (20 in), vegetation is better adapted for drought survival. Here one encounters deciduous seasonal forest formations. As in the semideciduous forest, two stories are present: an upper one consisting almost entirely of deciduous forms, and a lower one in which evergreens may predominate. Trees often are gnarled, crooked, and branched near the ground, rather than straight trunked and well formed as in the semideciduous and evergreen forests. Dominant trees in deciduous forest often have umbrella-shaped crowns, whereas those of the semideciduous forest generally are conical or rounded in shape. Lianas are scarce and epiphytes are missing in most deciduous communities. Trees in both layers are mostly small-leafed—another adaptation to drought.

In general, tropical deciduous seasonal forests are confined to areas in the American, African, and Asiatic tropics where a dry season extends to at least five consecutive months, each month averaging less than 100 mm (4 in), with two of the dry months averaging less than 25 mm (1 in). Edaphic[2] deciduous forests, however, may occur throughout the range of semideciduous forests, and some grow even within the range of tropical rainforests, where sandy soils or shallow limestone bedrock may cause excessive loss of soil moisture through rapid infiltration and percolation of rainfall.

The African deciduous seasonal forest formation differs from those of tropical America and Asia, and it is deserving of our special attention. A mosaic of open deciduous forest, intermingled with savannas, it stretches from Angola eastward across Zambia and southern Zaire to western Tanzania. Included are the Katanga Province of Zaire to the north, and a strip in Namibia, Botswana, and Zimbabwe to the south. It is an open woodland dominated by flat-topped, spreading trees, and because of its open nature, grass is an important constituent. Variously mapped as "savanna," "wooded savanna," "deciduous woodland," "grassy forest," and "Miombo forest," it is unlike most other tropical deciduous formations in that there is but a single story, and this story seldom exceeds 18 m (60 ft) in height.

Tropical Savannas

The term *savanna* describes a variety of plant communities in which grasses (Graminae) and grass-like sedges (Cyperaceae) prevail, but in which trees also are present. We touched on some ambiguity involved in the use of the term in the section on tropical seasonal forests; *savanna* applies to associations ranging from pure grasslands to woodlands with a continuous cover of grass under an almost closed canopy (fig. 6-7). Heinreich Walter (1973) suggests that *savanna* applies only to homogeneous grasslands where woody plants are more or less evenly distributed, and recommends that *grassland* be used where there is an absence of woody plants. He holds that *steppe* should be reserved for semiarid grasslands of the middle latitudes.

Along savanna water courses, where soil moisture is at high levels even in the dry season, belts of trees called *galeria forests* occupy the valleys. The composition of the galeria forest is similar to that of the rainforest or semideciduous forest, the moist soil environment being extended into the cli-

[2] *Edaphic* refers to soil-related factors, such as slope, drainage, texture, parent materials, etc., especially as related to vegetation.

Fig. 6–7. A grassy woodland near Port Moresby, Papua New Guinea. The trees are broadleaf evergreens typical of formations in northern Australia. (Photo by W. E. Akin, 1980.)

matically drier savannas. The fact that fire is less likely to move downslope into the moister environment of valleys may be another factor contributing to the existence of galeria forests.

South of the Sahara, Africa is the land of savannas par excellence. African savannas are grouped into three major categories: (1) high grass–low tree savanna, (2) acacia–tall grass savanna, and (3) acacia–desert grass savanna. Considerable variation exists in the amount of tree cover in each category, and types grade imperceptibly one into the other and into bordering seasonal forests.

Limited to Africa, high grass–low tree savannas are the most luxuriant kind. In these savannas, elephant grasses (*Pennisetum* spp.) dominates, so that they are commonly called "elephant grass savanna." The pennisetums and associated grasses are tussock-forming bunch grasses (they do not form a sod or turf). Here, most trees are deciduous, seldom exceeding 12 m (40 ft) in height, although some evergreen species common to neighboring forests may be taller. In West Africa and the Congo (Zaire) Basin, high grass–low tree savannas occur adjacent to evergreen forests, with the boundary between forest and savanna often abrupt.

The acacia–tall grass savannas resemble other tropical savannas, with a continuous cover of tussock grasses, 1–1.5 m (3–5 ft) tall, between widely scattered trees. In Africa, deciduous acacias are the dominant tree forms; they range in height from 3 to 15 m (10–50 ft), giving the area a park-like appearance. This association extends in a continuous belt from the west coast of Africa, in Gambia and Senegal, eastward to Sudan along the northern margin of the high grass–low tree savanna. The belt continues southward through large areas of Kenya, Tanzania, Mozambique, Zambia, Zimbabwe, and Botswana as far as the margins of the Kalahari. It is also the chief vegetation type in western Madagascar. This is the "big game" country of Africa made famous by motion pictures and travelogues.

In territory with less rainfall and longer dry season, the more luxuriant grass is replaced by a discontinuous cover of drought-resistant desert grasses and small thorny trees and shrubs—the acacia–desert grass savanna, or *Sahel*, of West Africa. Northward, this association gives way to the semidesert scrub of southern Saharan borderlands. Again it appears in East Africa, in southeastern Ethiopia and northern Kenya, and in southern Africa it is the vegetation of much of the Kalahari.

Savannas abound in the Indo-Malayan region, particularly in peninsular India and Australia. South America has two extensive areas, the *Llanos* of the Orinoco basin of Venezuela and Colombia, and the *campos* of Brazil, located mainly in Mato Grosso and Goiás and parts of neighboring states. In comparison with the African acacia–tall grass savannas, the Llanos and campos are similar in structure, but different in taxa. Brazilians recognize three types of campos: *campo cerrado* (literally a "closed grassland"), *campo sujo* (dirty grassland), and *campo limpo* (clean grassland). The campos of interior Brazil grade imperceptibly from a semideciduous woodland with an almost closed canopy (*cerradão*), through *campo cerrado* where trees are plentiful but do not form a closed canopy, and into *campo sujo,* grassland with widely scattered trees and shrubs (James 1959). *Campo limpo* is more characteristic of middle latitude grasslands of the subtropical southern states of Brazil; nevertheless, some areas of pure tropical grasslands do exist in southern Mato Grosso.

Large areas of South American savanna seasonally are flooded because of their low topographic position. An example is the *pantanals* of Brazil adjacent to the Rio Paraguay near the town of Corumbá in western Mato Grosso. Near the Rio Orinoco and its tributaries, extensive areas of the Llanos flood annually. On higher parts of the Llanos, general flooding does not occur, but low areas collect water and fine sediments in which only palms (*Copernicia tectorum*) will grow in association with grasses—the picturesque *palmares* of the Orinoco plain.

The Origins of Savannas

Although many texts and atlases still refer to "tropical savanna climate," savannas occur under a relatively wide range of climatic conditions, ranging from permanently wet (*Af*) to semiarid (*BSh*). Most extensive areas of savanna have climates with a well-marked dry season. But the causal effects of climate may be indirect, in that a dry season can set the stage for fire.

Fire is an important ecological agent in savannas, acknowledged even by those who do not accept the *pyrogenic* (fire-caused) origin of savannas (fig. 6-8). Some researchers give credit to fire for the origin and maintenance of all upland savannas, excluding only edaphic savannas where poor drainage or frequent flooding precludes the development of forests (but these are likely to be grasslands rather than savannas, since trees are entirely excluded). In Africa, the use of fire by humans probably contributed to the wide expanse of savannas and the restricted range of rainforest and semideciduous forest. In some locations, an abrupt transition from rainforest to high grass–low tree savanna gives some credence to this theory.

Burning the savannas seems to be an ancient practice. Hano of Carthage

Fig. 6–8. Burning savanna near Port Moresby, Papua New Guinea, during the dry season, July 1980. (Photo by W. E. Akin).

reported seeing great fires inland as he sailed down the west coast of Africa around 480 B.C. Archaeological evidence indicates that Africa may be humankind's most ancient home, and that humans there have controlled fire since the upper Pleistocene epoch, some one hundred thousand years ago (Butzer 1964). It is significant that in Africa large areas of savanna have climates similar to areas on other continents that support forests.

Annual firing of the savannas not only occurs in Africa, but also seems to be almost a universal practice. One of the best descriptions of savanna fires and their effects is by Peter Fleming in his book *Brazilian Adventure* (1934), in which he recounts a trip into northeastern Mato Grosso between the Rio Xingú and Rio Araguía during the dry season of 1933:

> There is always a fire burning somewhere. I suppose that the Indians burn the jungle as keepers in Scotland burn the heather. I know it seemed to me that about a tenth of the Interior of Brazil was perpetually in flames. . . .
>
> Much more campo. The prospect before us was encouraging. Open country, dead flat and recently burnt, stretched as far as the eye could see. . . . Unburnt campo is heavy stuff to walk through because the grass is long and thick and grows in tufts of which the base makes a fairly solid protuberance on the surface of the ground. But when it has been newly burnt, as this had, walking is almost as easy as on a road. . . .
>
> The swart little trees had a blasted look, though in reality the fire had hardly harmed them. [Quoted with permission of Jonathan Cape Ltd., London, in behalf of the Estate of Peter Fleming.]

Observations like those of Fleming, and results of experiments in which savannas have been protected from fire for years, lend weight to the theory that most, if not all, upland savannas are plagioclimax associations in which woodland development has been arrested and the dominance or codominance of grasses is maintained by fire. Where there is no burning, trees invariably become dominant, in some cases completely shading out the grass.

Thorn Forests

Where the dry season is even more severe, tropical deciduous forests and savannas give way to low-growing associations of deciduous and evergreen trees and shrubs; many of these species bear thorns. The evergreen forms are quite distinct from those of tropical rainforests in that they are highly xerophytic, with small hard leaves and other xeromorphic adaptations. Many of the woody plants are leguminous, including many species of mimosa (*Mimosa* spp.) and acacia. The American formation, in particular, is rich in cacti (*Cereus* spp.), often of the giant variety. In African and Asian formations, on the other hand, similar cactus-like life forms appear, comprising primarily various species of *Euphorbia*. Grasses are almost absent in many thorn forests, while "grassy woodland" often prevails in others.

In the Western Hemisphere, the thorn forest formation type occurs in the tropical and subtropical areas where severe seasonal drought prevails;

namely, from coastal Mexico southward to Argentina. Particularly extensive thorn forests occupy dry northern coasts of Colombia and Venezuela, southwestern coastal areas of Ecuador, vast *caatingas* (light forests) of northeastern Brazil, and the Gran Chaco of western Paraguay, eastern Bolivia, and northern Argentina. Other thorn forest communities grow within the confines of semideciduous and deciduous forests where excessive drainage on sands and porous limestones produces droughty soil conditions.

The Brazilian *caatinga* is one of the most extensive areas of thorn forest in the Americas. Here, low thorny woodland often is overtopped by giant columnar cactus and palms, including a very unusual plant, the barrigudo or Brazilian bottle tree (*Carnillesia arborea*). Its massive bottle-shaped trunk may reach a diameter of 4.5 m (15 f) and acts as a water reservoir. A counterpart to the Brazilian form is the Australian bottle tree (*Sterculia* spp.), also associated with thorn forests.

Grass is rare over large areas of the *caatinga*, and *bromeliads* dominate the ground cover[3], in spite of the fact that *caatinga* is more open than most thorn forests, a characteristic seemingly favorable to grass. The very pervious nature of the sandy soils could be an explanation for the scarcity of grass. The upper soil dries out rapidly even after heavy rains. Here, trees and shrubs have roots capable of extracting water from deep within the soil, while the shallow-rooted grasses are at a disadvantage. The people of northeast Brazil experience many climatic catastrophes, alternately suffering from disastrous floods and devastating droughts. Although average annual rainfall in parts of the *caatinga* exceeds 1,270 mm (50 in), the occasional severe drought period may prevent invasion of the area by the semideciduous or deciduous forests that average precipitation figures would seem to favor.

Other continents also have rather extensive thorn forests. India, Burma, and Thailand have a long history of human alteration of the environment; thus, although thorn forests abide there, mapping their original distribution is difficult. In general structure and life forms, remnants of the original thorn forests of Southeast Asia resemble their American counterpart. In Africa, there are many species characteristic of the formation, but they occur in savannas and do not form a true forest. The major area of African thorn forest is the "bush woodland" of central Tanzania. Although other thorn forests do exist, in most areas of Africa where they are an important element, they do not form a true forest but rather are part of thorn tree–tall grass or desert grass savannas. In south-central Australia, the sclerophyllous *mallee* of the south coast gives way toward the interior to thorn forests in isolated patches up to 9–12 m (30–40 ft) in height. On vegetation maps, this formation generally is thorn forest, although its classification as forest is questionable.

[3] *Bromeliads* are members of a large group of plants, the *Bromeliaceae*, of which the pineapple (*Ananas sativus*) is representative.

Semidesert and Desert Vegetation of the Dry Tropics and Subtropics

Plant geographers use the term *semidesert* for areas with a relatively continuous cover of plants, reserving *desert* for areas with little or no vegetation, except for isolated plants or groups. Most areas of semidesert scrub actually are within the boundaries of climatic deserts (*BWh*) or steppes (*BSh*) of the Köppen climatic system.

Only in the great climatic deserts of the world does one find extensive true deserts—those regions having little or no plant cover. Large areas of true barren desert in the tropics and subtropics are found only in the central Sahara, Arabia, central Asia, and the west coasts of South America and southern Africa. Smaller areas without significant plant cover exist in other arid regions, but they are too small and scattered for mapping on a global scale. Many of the so-called "plant deserts" are edaphic; plants will not grow in the dune landscapes of the sandy *erg* deserts, nor on gravel plains of the *reg*, nor on the bare rock of the *hamada* (these are native Saharan terms used to describe desert landscapes).

Of course, plants that compose the semidesert and desert communities are highly xerophytic; they demonstrate the various adaptations to drought conditions described in chapter 5. Most are shallow rooted, and they have the ability to intercept the rare showers that may wet only the upper few inches of soil. Although lacking in depth, root systems compensate by lateral extension—a characteristic that explains, in part, the wide spacing between desert plants (fig. 6-9).

The familiar semidesert scrub of the American Southwest is a subtropical extension of tropical formations. Numerous species of cactus, particularly giant *saguaros* (*Carnegiea gigantea*), lend beauty and majesty to the desert landscape. Other succulents important to this formation are the century plant (*Agave americana*) and other agaves, the yuccas (*Yucca* spp.), and many broadleaf xerophytic shrubs, such as creosote bush (*Laarea mexicana*). Phreatophytes, such as mesquite (*Prosopis* spp.) and acacia (*Acacia* spp.), prosper in alluvial valleys because their deep roots can tap the relatively shallow water table.

There are similar semidesert plant communities on other continents. As an example, in Africa and Asia cactus-like euphorbias replace the cacti, and in the Australian formation succulent life forms are scarce. Instead, there are other xeromorphic adaptations, such as thick, corky bark, small leaves, and low transpiration rates.

Vegetation of Tropical Highlands and Mountains

The effect of elevation on temperature determines climate and vegetation zones in tropical mountains. In high tropical mountains, temperatures range all the way from tropical near sea level to polar above the snow line. Tem-

Fig. 6–9. Semidesert vegetation near Tucson, Ariz. Former grassland has been invaded by shrub forms and cacti. The larger trees are mesquite (*Prosopis* spp.), and the cactus in the foreground is probably Engelmann prickly pear (*Opuntia engelmannii*). In the right foreground is the appropriately named "teddy bear" cholla (*Opuntia bigelovii*), and near the center of the photo are two tree chollas (*O. arbuscula*). Most of the low-growing bushes are burroweed (*Haplopappus tenuisectus*). (U.S. Forest Service photo by S. C. Martin, March 1964.)

peratures do not tell the whole story; in the highland tropics seasonal variations of temperature, length of daylight, and angle of the sun change little from month to month, while in high and middle latitudes, these climatic factors vary greatly between winter and summer. Obviously, plants sensitive to seasonal changes in these and other climatic variables discussed in chapter 5 may not prosper in the tropics, even though average annual temperatures may be similar to their natural habitat.

Where mountains rise above the lowland tropical rainforest, there is a succession of plant formations from near sea level to the snow line. As elevation increases, the composition and structure of the rainforest changes until it gives way to a two-storied forest 18–21 m (50–70 ft) in height, referred to as "submontane" or "transitional" forest. This ecotone forest contains many of the species from the rainforest below; also, it may include species from subtropical areas lying poleward of the tropics. For example, many oaks and pines abound in the tropical mountains of Mexico, resembling those of mixed forests in the southeastern United States (fig. 6-2). Naturally, because more light reaches the forest floor, ground vegetation is more plentiful than in the lowland rainforest.

The true montane forest extends above the submontane ecotone forest. It is similar, structurally, throughout the tropics—a relatively low forest, whose gnarled and twisted trees seldom exceed 9–12 m (30–40 ft) in height and form a single tree stratum. Because of the great profusion of epiphytic mosses and lichens in the montane forest, the term *mossy forest* is appropriate—the *ceja de la montaña*. This richness in epiphytes reflects the permanently wet climate of this zone—aptly called "cloud forest" by some. Owing to variations in exposure of slopes to prevailing winds and other topographic influences, the elevation of the maximum precipitation and cloudiness zone varies considerably from place to place, but on most major mountain ranges it occurs somewhere between 1,800 and 3,000 m (6,000 and 11,000 ft). Toward the treeline, the mossy forest becomes more stunted and misshapen, forming an "elfin woodland" as described in chapter 5. On isolated mountainous islands in the tropical oceans, these formations occur at considerably lower levels.

A variety of plant communities grows above the treeline in tropical mountains. In the Andes of South America, a grassy steppe with herbaceous plants, locally called *páramos*, covers large areas of the moister highlands. Desert grass and xerophytic shrubs, locally called *puna*, dominate the dry highlands of western and southern Peru and most of the Bolivian Altiplano. *Puna* is a relatively unusual plant community in tropical mountains because higher mountain slopes, even in the Sahara, usually are covered by shrub or, sometimes, woodland communities. The unique climate of the west coast of South America, with its cold Humboldt Current and stable air conditions, prevents upslope movement of air from the Pacific Ocean; thus, it creates arid conditions that may account for this vegetation anomaly. Vertical zonation of vegetation in the Andes of central Peru is shown in figure 6-10. Tropical mountains in other parts of the world show similar zonation, but the altitude of the zones may vary, and the composition of the vegetation will differ both in species and dominant life forms.

Vegetation of Middle and High Latitudes

Poleward of the tropics, the seasonal range of temperature increases and true winter occurs; here, seasonal variations in length of night and day become factors in plant adaptation. On the borders of the tropics, in areas climatically subtropical, winters are mild and frost may be rare or absent, but even rare frosts eliminate many frost-sensitive tropical plants. At the same time, seasonal rhythm is sufficient to permit plant invasion from areas on the poleward side.

Farther poleward, adaptations to cold winters and increasingly short

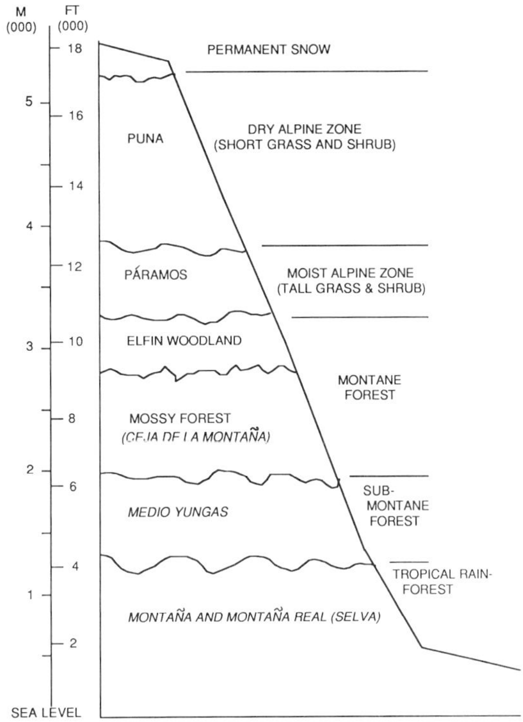

Fig. 6–10. Vertical zonation of vegetation in the Andes Mountains of Peru. (Based on James 1959 and Eyre 1963.)

summers have profound effects on vegetation formations. Yet, along some west coasts and on islands such as New Zealand, moderation of climate by marine influences may extend low latitude characteristics to much higher latitudes—for example, broadleaf evergreen trees. Particularly in the middle latitudes, quite different vegetation formations sometimes occupy different geographic areas with very similar climates. Consider the pine forests of the southeastern United States. They differ greatly in life forms and structure from broadleaved evergreen forests in south China and the grassy pampas of Argentina; nevertheless, all three formations occur in areas of humid subtropical (*Caf*) climate.

Broadleaf Evergreen and Mixed Forests of the Humid Mild-Winter Climates (*Caf, Cbf*)

In many areas where winters are mild and rainfall is substantial, broadleaf evergreen forests, often with a mixture of conifers, constitute important vegetation formations. At some time in the distant past, an extensive area of broadleaf evergreen forest apparently covered much of southern China, the southern tip of the Korean Peninsula, and southern Japan. Because of the great number of species of the laurel family (*Lauraceae*) represented in this formation, it is called "laurel forest."

In China it is difficult to reconstruct the original nature of this forest since it has been largely eradicated or greatly modified as a result of the extremely long period of relatively dense human settlement . In Japan, however, remnants of this once extensive formation remain, to give some idea of its former luxuriance. In addition to laurels, there is an abundance of native species of evergreen oaks and magnolias, along with a few conifers. In the interior of southwestern China and in northern Japan, broadleaf deciduous species become more common, and a mixed forest replaces evergreen formation. Fossil evidence suggests that these forests were once very similar floristically to those of both Europe and North America before continental glaciation during the Pleistocene epoch. The great ice sheets did not extend as far southward in Asia as they did in the remainder of the Northern Hemisphere; consequently, they had less effect on the subsequent makeup of vegetation formations.

Outside southern China, the only other extensive Northern Hemisphere area having a humid subtropical climate is the southeastern United States. Here, the broadleaf evergreen forest survives in a few areas along the Gulf of Mexico and in southern Florida. Nevertheless, isolated stands of evergreen oak and magnolia and associated broadleaf evergreens cover a fairly wide area from east Texas to coastal South Carolina, indicating that the climate does not limit broadleaf evergreen formations. The southern pine forest formation probably is not a genuine climatic climax formation, but rather reflects an edaphic adaptation to the youthful, sandy, and poorly drained outer coastal plain.

Pleistocene glaciation is another factor that contributed to the relatively minor significance of broadleaf evergreen forests in eastern North America. Here, continental glaciers advanced southward as far as the Ohio and Missouri rivers, farther equatorward than in any other part of the world. Colder climates ahead of the advancing ice may have eliminated many of the subtropical evergreens, while at the same time their former range became a refuge for the broadleaf deciduous species of the north. The broadleaf evergreen forest in turn was squeezed against the Gulf of Mexico with no contiguous land refuge available equatorward. However, when milder climates returned at the end of the Pleistocene, plants from the more tropical regions could not migrate into the zone because of the Gulf of Mexico on the south and drier climates to the west.

In the southeastern United States, most of the broadleaf evergreen formation occupies slightly elevated areas above the poorly drained coastal plain. These elevations are *hammocks,* a local variation of *hummock*, meaning high area. Here one finds various evergreen species of oak, holly (*Ilex* spp.), magnolia, and other trees and palms (fig. 6-11), and some orchids and other epiphytes, particularly the well-known Spanish moss (*Tillandsia usneoides,* really not a moss but a bromeliad). Over much of its range, this discontinuous formation constitutes an edaphically controlled mosaic with upland

Fig. 6–11. Cabbage palmetto (*Sabal palmetto*) in wet savanna grassland in Florida. In background is a hammock with typical broadleaf evergreen forest. (U. S. Forest Service photo, n.d.)

pine forests, and swamp forest dominated by bald cypress (*Taxodium distichum*) and pond cypress (*T. ascendens*).

In the Southern Hemisphere, evergreen broadleaf and mixed forests occur in New Zealand, southeastern Australia, southern Chile, and in the eastern part of Cape Province of South Africa, mostly in areas with cool summers and mild winters, corresponding to the marine climate (*Cbf*) of Köppen. An exception is an area of mixed forests in southern Brazil, famous for the Paraná pine (*Araucaria braziliensis*), which is in a humid subtropical (*Caf*) climatic area. With few exceptions, areas with marine climates in Europe are home to deciduous summer forests, whereas in North America, an almost pure coniferous formation dominates the marine west coast. Again, isolation from the more dramatic climatic and geologic changes of the Ice Age played a role in the survival of this broadleaf evergreen formation type in the Southern Hemisphere.

In southwestern South America, the Mediterranean sclerophyllous scrub of middle Chile gives way to the dense, luxuriant evergreen mixed forest at about 30°S. Before European settlement, these forests probably reached their maximum development between the present towns of Valdivia and Puerto Montt. Large areas of forest still remain, and forest-related industries are important. The araucaria pine (*Araucaria imbricata*) and several species of evergreen beech (*Notofagus* spp.) dominate. Large stands of Chilean cedar (*Libocedrus chilensis*) once existed but lumbering has greatly reduced the range of this valuable tree. Large stands of Chilean pine (*Podocarpus chilena*), mixed with broadleaf evergreen trees, grow in higher and drier areas away from the coast.

On the North Island of New Zealand, European settlements have removed most of the mixed forests in which once grew the famous kauri pine (*Agathis australis*). Farther south the kauri disappears, but the mixed forest continues south along the west coast of South Island. Here are pines belonging to the genera *Podocarpus* and *Dacrydium*, in association with evergreen broadleaf trees representing a number of families, including laurels and Leguminosae. Farther south, the forest becomes almost purely broadleaf evergreen, with many epiphytes and lianas adding to its superficial resemblance to tropical rainforest.

Mediterranean Mixed Forest

In chapter 5 we examined the impact of thousands of years of agriculture and grazing on the character of the vegetation of the Mediterranean basin. Equally important was the destruction of forests in ancient times to obtain charcoal for smelting copper and iron ores, and to procure timbers for building ships of war and commerce. Magnificent forests once covered the hills and mountains of the more humid Mediterranean lands. Paleobotanical studies indicate that this formation was an evergreen mixed forest of coniferous and broadleaf trees. Scattered throughout the sclerophyllous scrub, locally called *maquis* or *chaparral*, which occupies the uncultivated slopes today, are occasional trees, such as evergreen oaks and pines. Typical is the cork oak (*Quercus suber*) of the western Mediterranean, and the Aleppo pine (*Pinus halepensis*) of the eastern basin. It is probable that the numerous species of shrubs now dominating the Mediterranean sclerophyllous scrub were elements in the undergrowth of the original mixed forest.

Some climatologists theorize that the radically different vegetation and soils one encounters in Mediterranean lands today reflect a climatic change since the time of ancient civilizations. The destruction of the forest, however, could account for depletion of soils and alteration of microclimates sufficient to change soil types from the brown forest soils of the past to the *terra rossa* (red soil) of today. It is probably significant that both oaks and pines native to Mediterranean lands have exceptionally long and efficient root systems that would aid survival during dry Mediterranean summers. The mixed forest seems to be the true climatic climax formation of the Mediterranean basin where rainfall is adequate, while the sclerophyllous scrub is climax only in drier or more windy areas.

The Deciduous Summer Forest

Except for a small area in Chilean Patagonia, we find the deciduous summer forest formation type only in the Northern Hemisphere. Three great formations occur in eastern North America, Europe, and eastern Asia, respectively. As the ice advanced during Pleistocene glaciation in Europe, the Alps blocked southward retreat of plants and many species were eliminated. In North America, the way was open southward to the Gulf of Mexico, and only the broadleaf evergreen forest had no subtropical refuge. Consequently, the North American formation is much richer in species than its European counterpart.

Throughout most of the North American formation, the oaks (*Quercus* spp.) and hickories (*Carya* spp.) dominate (fig. 6-12), forming the "oak-hickory forest." In the eastern part of this formation, the sugar maple (*Acer saccharum*), basswood (*Tilia americana*), and American beech (*Fagus grandifolia*) share codominance with the oaks and hickories. Formerly the American chestnut (*Castanea dentata*) was important, but it has been almost eliminated by the chestnut blight. Today, another important forest tree, the American elm (*Ulmus americana*), seems destined for the same fate because of Dutch elm disease.

In Europe, the deciduous summer forest occupies climatic areas ranging from marine (*Cbf*) through humid continental (*Dbf* and *Daf*), and it is remarkably similar to its American counterpart, albeit poorer in species. Here again, oaks are an important element in the flora, along with beech and other genera represented in the North American and Asian formations. Today, a visitor to Western Europe may come away with the impression that most forests are coniferous—an impression that is correct because for eco-

Fig. 6–12. Oak-hickory forest in Ohio. Large trees are mostly red oak *(Quercus borealis)* and white oak (*Q. alba*). The seedling in the right foreground is hickory (*Carya* spp.). (U.S. Forest Service photo by R. K. Winters, May 1952.)

nomic reasons modern planted forests are chiefly conifers. The East Asian formation exists mostly in north China, the Korean Peninsula, and southeastern Siberia. It is similar in composition to the other Northern Hemisphere deciduous summer formations, except for the absence of hickories (*Carya* spp.).

An interesting variation of the deciduous summer forest formation occurs in the continental interiors where it forms a narrow band between the grasslands to the south and the boreal forests to the north. In North America this narrow belt extends from western Minnesota, across southern Manitoba and southern Saskatchewan, and into southeastern Alberta. Two species of *Populus*, the balsam poplar (*P. balsamifera*) and the aspen (*P. tremuloides*) dominate this forest-grassland transition. In central Asia a similar belt extends eastward from the Ural Mountains to the headwaters of the Yenisey River. As in North America, two species dominate, the European aspen (*Populus tremula*) and the birch (*Betula verrucosa*).

Mixed Broadleaf Deciduous and Coniferous Forests

On both the poleward and equatorward sides of the deciduous summer forest formation type, broad ecotone forests exist where coniferous and broadleaf deciduous trees are codominant; thus, they are called "mixed ecotone forest." Most likely, these forests are not climatic climax associations, but represent a zone of struggle for dominance, or they may be plagioclimax associations. In most cases, conifers and broadleaf trees are not mixed randomly in the forest. Rather, each life form shows a preference for a specific edaphic condition. The photograph of the mosaic of communities in the mixed forest of northern Minnesota (fig. 5-1) illustrates this edaphic adaptation. Of course, there is some admixture of deciduous trees throughout much of the boreal coniferous forest, and in some areas conifers are found in association with broadleaf deciduous forests.

In North America, a mixed forest association originally covered much of the area from Minnesota eastward across Wisconsin, Michigan, northern Pennsylvania, New York, and northern New England. In Canada, this mixed forest covered southern Ontario east of Lake Superior, southern Quebec, New Brunswick, and Nova Scotia.

In the North American mixed forest ecotone, most vegetation adjustments are edaphic; in sandy soils, conifers dominate to the exclusion of most deciduous trees. Where trees occupy heavier and more fertile soils, deciduous species dominate. The fact that pine is excluded from more fertile soils is no indicator that it will not grow there, only that it cannot compete with broadleaf deciduous trees. Having lower nutrient requirements, the pine has an advantage over the deciduous species on less fertile sandy soils. In some areas, slope orientation determines the distribution of species; on north-facing slopes, conifers tend to dominate, while deciduous trees are more numerous on south-facing hillsides.

Along the southern margin of the deciduous summer forest in the United States, a mixed forest is transitional to the southern pine forest of the coastal plain (fig. 6-13). Located from New Jersey south to Alabama, it occupies most of the southern Appalachians and the Piedmont and part of the northern coastal plain; it continues westward across northern Mississippi and Louisiana, into east Texas and southern Arkansas. As with the southern pine forest, the pine of the mixed forest may be seral or subclimax, with the deciduous summer forest being the true climatic climax.

Similar mixed forest transitions occur in Europe and Asia, but few examples survive there, since most were long ago destroyed to make way for agricultural land for expanding populations. The Western Eurasian mixed forests had their greatest extent in European Russia, including the Baltic states; they continue westward into southern Scandinavia, and south to the deciduous forest in the Alpine region. The East Asian formation is limited to northern and eastern Manchuria, northeastern Korea, and northern Kyushu

Fig. 6–13. Southern mixed forest of pine, oak, and hickory in the Clark National Forest of southeastern Missouri. (U.S. Forest Service photo by W. H. Shafer, June 1937.)

and southern Honshu in Japan. As with their counterparts in North America and Europe, the Asian mixed forests largely have disappeared except in rough terrain or in limited stands because, here also, increasing population continues to exert pressure for more and more agricultural land.

Coniferous Forests

Only in the Northern Hemisphere are there extensive areas where coniferous (needle-leaf) forests constitute the climatic climax formation type. Coniferous trees are important in the Southern Hemisphere, but they are largely associated with mixed forest formations, not ecotones; these are true climax associations. Northern Hemisphere conifers differ taxinomically from those native to the Southern Hemisphere; they are generally represented by species of pine (*Pinus*), spruce (*Picea*), and fir (*Abies*). Other genera occur and may even dominate in some areas; these include the deciduous needle-leaf larches (*Larix* spp.) and trees of the Pacific forests of western North America, with such dominants as Douglas fir (*Pseudotsuga taxifolia*) and redwood (*Sequoia sempervirens*).

The Boreal Forest

Where winters are very cold and summers short, needle-leaf trees are particularly well adapted for survival. When soil freezes, a physiological "drought" exists, and the small area of the needle-shaped leaves retards loss of moisture. Then when temperatures rise in the spring to levels permitting photosynthesis, evergreen trees are able to take immediate advantage of it, while deciduous trees first must grow a new set of leaves. As a result, evergreens grow faster and shade out more slowly growing deciduous species. This advantage is limited; in areas such as eastern Siberia, where temperatures are so low that any tree still in leaf in winter may not survive, deciduous species, such as the needle-leaf larch, broadleaf deciduous birch (*Betula* spp.), and aspen (*Populus* spp.), have survival advantage.

The boreal forest falls into two categories: needle-leaf evergreen forests of northern North America and western Eurasia (fig. 6-14), and deciduous needle-leaf forests of eastern Siberia where larches dominate. Occupying a broad east-west belt across subarctic regions from Alaska and Canada in North America, and from Scandinavia across European Russia and Siberia in Eurasia, this formation type comes close to corresponding in location with subarctic (*Dc* and *Dd*) climates of Köppen. Geographers sometimes apply the Russian term *taiga* to all boreal forests of the Northern Hemisphere, but the term probably should be reserved for the forest-tundra ecotone, since this more restricted meaning follows the Russian usage (fig. 6-15). The deciduous larch forest of Siberia corresponds to the *Dd* or extremely severe winter subdivision of the subarctic of Köppen. Permanently frozen ground (*permafrost*) is present in large areas of the poleward sections of the boreal forest, and it is almost continuous in northeastern Siberia.

The poleward limit of the boreal forest is not marked by a well-defined treeline. Its poleward boundary is not the northern limit of trees, for trees grow in isolated dwarfed stands in protected locations far out on the tundra. Toward its northern limit, the forest formation takes on a park-like character—the true *taiga*—with widely spaced trees or groups of trees and a ground cover of typical tundra plants. In river valleys at slightly lower elevations, tongues of true boreal forest interfinger northward into this transition zone.

Unlike forest formations farther south, most of the boreal forest remains intact, primarily because climate and soils limit agriculture. Even in Sweden and Finland, where settling of these forest lands has been national policy, the density of population remains light. In the Soviet Union, however, settlement and agriculture have pushed into these areas with some degree of success. The forest resources are considerable, and wood products, including wood pulp, paper, and lumber, are important to all countries sharing the boreal forest. Although trees are small compared with those in forests to the

Fig. 6–14. Boreal forest in northern Minnesota represented here by black spruce (*Picea mariana*), a species common in low areas and bogs from the northeastern United States, across Canada to Alaska. White spruce (*P. glauca*) and balsam fir (*Abies balsamea*) are dominant in higher areas are, sometimes lending the name "spruce-fir forest" to the boreal formation. (U.S. Forest Service photo by E. S. Shipp, 1929.)

Fig. 6–15. Forest-tundra ecotone at 700 m (2,300 ft) elevation in southern Alaska in the Chugach National Forest. Trees are small black spruce (*Picea mariana*), shrub birch (*Betula spp.),* and willow (*Salix spp.*), with a ground cover of grass and tundra plants. (U.S. Forest Service photo by R. S. Kellog, August 1909.)

south, lumber-sized trees do grow along southern margins of the boreal forest.

Subalpine and Montane Forests

At higher elevations, both in North America and Europe, the range of conifers extends southward from the boreal forests. In North America, the north-south alignment of mountain ranges makes this projection relatively continuous along the Rocky Mountains and the Cascades–Sierra Nevada ranges. In Europe and Asia, where mountains trend east-west, lowland formations isolate the montane forests from the boreal forests. In general, *subalpine* formations, occupying upper reaches of the timbered zone, contain species clearly related to the boreal forests. The *montane* formation, on the other hand, is at lower elevations and contains species indigenous to more temperate latitudes. These montane forests of conifers are found far equatorward in the highlands of Mexico and Central America (fig. 6-2). Their climatic range is vast; there are extremely humid areas supporting trees such as Sitka spruce (*Picea stichensis*), and subhumid areas with precipitation as low as 46 cm (18 in) where the ponderosa pine (*Pinus ponderosa*) thrives.

Pacific Coast Forests of North America

The magnificent coniferous forests of the west coast of North America deserve special mention. They differ floristically and structurally from the deciduous summer forests of Europe with similar marine climates (fig. 6-16). Separated by high mountains and dry climates from deciduous summer forests in eastern North America, this coniferous association west of the Cordilleras survived. In their coastal isolation they were protected from competition with broadleaf deciduous species; also, the moderating climatic effect of the adjacent Pacific Ocean protected them from the most severe consequences of glaciation.

Sitka spruce (*Picea stichensis*) dominates this formation in southern Alaska and northern British Columbia; here, summers are cool and winters are not severe for the latitude, and rainfall exceeds 2,000 mm (80 in). Farther south in British Columbia and Washington, the western cedar (*Thuja plicata*) and western hemlock (*Tsuga heterophylla*) become codominant. Along the coast ranges, from British Columbia south through Washington and Oregon, and on the western slopes of the Cascades, one finds large pure stands of Douglas fir (*Pseudotsuga taxifolia*). Here, many trees reach heights of 60–90 m (200–300 ft). Farther south, in northern California, giant redwoods (*Sequoia sempervirens*) commonly exceed heights of 90 m (300 ft), with particularly tall trees topping 100 m (350 ft) (fig. 6-16). South of these, and inland at about 1,800 m (6,000 ft) elevation in the Sierra Nevada, are the giant sequoias (*Sequoiadendron giganteum*). Although not so tall as the coastal redwoods, these towering giants, the most massive of all living things, often exceed 7.6 m (25 ft) in diameter.

Fig. 6–16. Old growth redwoods (*Sequoia sempervirens*) in Del Norte County in northern California. Redwood lumber is in great demand for house construction, fencing, and so on, because of its durability when exposed to weather. Mature stands are rapidly being depleted except in protected areas. (U.S. Forest Service photo by D. F. Roy, n.d.)

Fig. 6–17. Lumber from loblolly pine (*Pinus taeda*) on the coastal plain of North Carolina. The stand of pines in the background is typical of the southern pine forest of the United States. Note the open nature of the forest and grass ground cover. (USDA, Soil Conservation Service photo by J.W. Busch, n.d.)

The Southern Pine Forests of the United States

From the Carolinas south and westward to east Texas lies the southern pine forest, often called "pine barrens." The range of this formation falls within the humid subtropical climatic zone (*Caf*) of Köppen. One can find a northern outlier of this formation in the pine barrens of southeastern New Jersey.

Pitch pine (*Pinus rigida*) is dominant in the northern part; farther south, loblolly pine (*P. taeda*), slash pine (*P. caribaea*), and longleaf pine (*P. australis*) are more abundant. Forest industries, including lumber, pulp and paper manufacturing, and naval stores are important (fig. 6-17). The naval stores industry depends on extraction of turpentine, rosin (resin), pine tar, pine oil, and rosin oil. Trees are tapped directly for their sap, and pulp mills supply materials as by-products.

The long growing season of the humid subtropical climate provides particularly favorable conditions for tree farming. Here, trees grow rapidly and reach harvest size in less than half the time required in northern coniferous forests, and a location near the great eastern markets gives the southern pine forests an economic advantage.

Middle Latitude Grasslands

Middle latitude grassland formations are *prairies* in North America and *steppes* in Europe and Asia. Unless the term is modified, prairie applies to the tall grasslands of more humid areas—the "tallgrass prairies." In the United States and Canada, drier grasslands are "midgrass prairies" and "shortgrass prairies." All natural grasslands in eastern Europe and Asia are steppes; they comprise, however, the same life forms as in North American formations (fig. 6–18).

Grasses and many associated *forbs* (herbs other than grasses) dominate temperate grasslands; unlike most tropical grasslands or savannas, trees and shrubs are of minor importance except along stream valleys and in depressions. Major grassland formations occur in the United States and the Prairie Provinces of Canada, the steppes of eastern Europe and Central Asia, the

Fig. 6–18. Big bluestem (*Andropogon gerardi*), one of the major grasses found in the prairies of the central United States. (USDA, Soil Conservation Service photo by E. T. Jacobson, September 1967.)

pampas of Argentina and adjacent areas of South America, the *grassveld* of southern Africa, and the Canterbury Plain of southeastern New Zealand. Most grasslands in southern Australia are considered savannas because trees grow in considerable numbers there.

The origin of grasslands in middle latitudes, like that of tropical savannas, is the subject of considerable controversy. Where semiarid climates prevail, the climatic climax vegetation may well be shortgrass and midgrass steppes, such as those on the Great Plains of North America. Even here, however, fire possibly was an important ecological factor. In more humid climates, as in the tallgrass prairies of the western corn belt and the humid pampas of Argentina, there seems to be little climatic reason for absence of trees. Evidence from studies of pollen trapped in peat bogs indicates that the humid prairies are relics of a drier past (Wright 1970). Once established, fire and other ecological factors maintained the prairie formation, thus preventing or retarding the invasion by trees as moister climatic conditions returned.

Because of excellent soils and climate, cultivated land has almost entirely replaced the humid prairie formation in the North American landscape. This is particularly true in the American corn belt where the original plant communities survive only in a few prairie preserves or along railroad rights of way. Large areas of natural tallgrass prairie remain only in the Sand Hills of Nebraska and the Flint Hills of Kansas. But, even there, livestock grazing prevents the grasses from reaching their former luxuriance.

Middle Latitude Scrublands and Woodlands

Where rainfall is limited, or where dry summers occur, large areas of the middle latitudes support a cover of low-growing woody plants or small trees. Where shrubs predominate, sometimes forming a dense cover, the term *scrubland* applies; where low trees form a relatively open canopy, often over grass and herbaceous ground cover, it is called *woodland.* An exception to this generalization is in Australia, where trees of great size form a relatively open forest, resembling a woodland in many respects. There are two broadly defined subdivisions of the scrubland-woodland complex: (1) the sclerophyllous formations, found most frequently in subtropical regions with a well-defined summer drought, the dry-summer subtropical or Mediterranean (*Cs*) of Köppen; and (2) semiarid scrublands and woodlands of continental interiors, corresponding to Köppen's low latitude and mid-latitude steppe and desert climates (*BSh, BWh, BSk, BWk*).

The Sclerophyllous Formations

Sclerophyllous (hard-leaved) formations take their name from the small, hard, leathery leaves characteristic of many plants in the formation (also called "broad sclerophyll"). The close association of this type of vegetation with the dry-summer subtropical or Mediterranean climate has led many geographers to the erroneous assumption that all climates have equally well-defined associated vegetation formations. Even in areas with a Mediterranean climate (*Cs*), the sclerophyll formations range from low-growing scrub to evergreen woodlands, and forests may be present where annual rainfall totals are relatively high. It is not surprising then that vegetation formations vary considerably within areas classified as *Cs*, because the mean annual precipitation within them ranges from less than 400 mm (15 in) to more than 760 mm (50 in). Table 2-5, listing average annual precipitation and mean warm-month temperature for selected *Cs* stations, illustrates the magnitude of this variation. Although all of these stations have a very dry summer, the mean temperature of the warmest month varies considerably from station to station.

Sclerophyll formations are dominant in five well-defined areas of the world: (1) the Mediterranean Basin and adjacent areas of southern Europe and southwestern Asia, (2) California, (3) middle Chile, (4) the Cape Province of South Africa, and (5) southwestern and southeastern Australia. The five areas have common characteristics other than similar dominant life forms—similar latitude, similar location on the western margins of the continents, and similar dry-summer subtropical climates, except for the part of southeastern Australia that has a humid subtropical climate. We described the present vegetation of the Mediterranean basin in the section on Mediterranean mixed forests; here, we describe the other four sclerophyll formations.

California. Sclerophyll scrub covers the lower slopes of the mountains in southern California (fig. 6-19); it seldom exceeds 3 m (10 ft) in height (fig. 6-20). This formation is called *coastal chaparral*, to distinguish it from the semiarid formation of interior basins and ranges, which is also called chaparral. Most dominant species in this formation are evergreen except for a deciduous dwarf oak (*Quercus dumosa*). Scattered throughout the chaparral, particularly on northward-facing slopes, taller trees or patches of woodland occasionally grow; these usually are evergreen oaks and piñon (*Pinus cembroides*).

As in the case of the Mediterranean *maquis*, today the extent of chaparral is much greater than its original range. Coastal chaparral seems to have been climax only in drier parts of its present range where mean annual rainfall is between 250 and 500 mm (10–20 in). Today, it grows as far north as Oregon in areas where annual rainfall averages more than 1,270 mm (50 in). A minor sclerophyll element grows in the undergrowth in the redwood forests and in some Douglas fir forests in southern Oregon—an obvious parallel to the original position of sclerophyll in mixed forests of the Mediterranean basin (Shelford 1963). Fire contributed to the spread and maintenance of the chaparral formation. Even today, fires are a major hazard to suburban residents of southern California. Destructive brush fires are almost an annual occurrence in the Los Angeles area at the end of the dry summer and before the rains of winter; then the chaparral is tinder-dry and highly flammable.

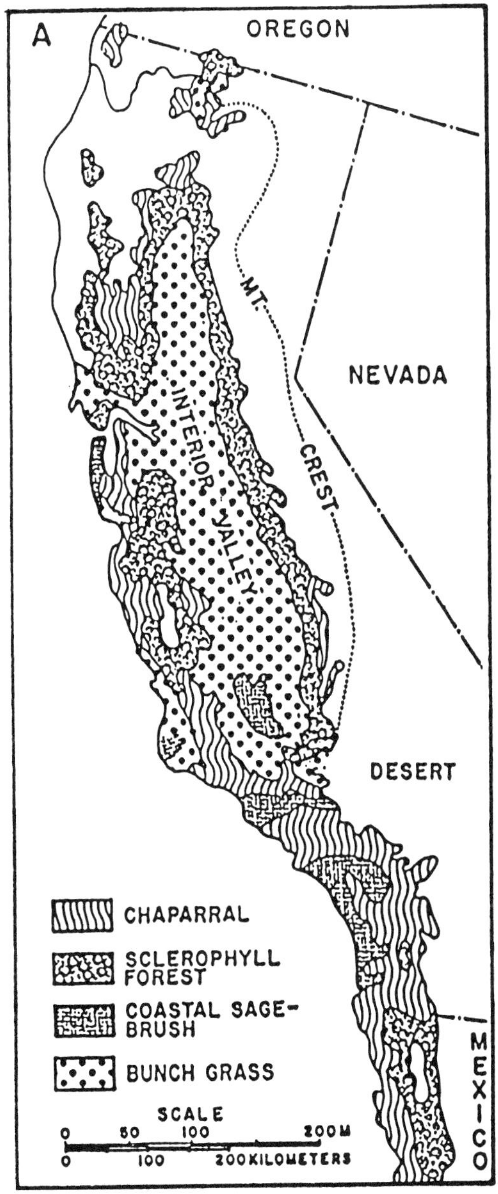

Fig. 6–19. The vegetation of "Mediterranean" California. (Reprinted with permission from Shelford 1963, 240.)

Where climate is a little cooler and moister, oak woodland may replace chaparral. Trees range in height from 6 to 23 m (20–75 ft), and usually they are widely spaced with a grassy and herbaceous substratum giving a savanna-like aspect to the woodland. Sclerophyll woodland, especially, is well developed on the foothills of the Sierra Nevada and on the inland side of the

Fig. 6–20. Chaparral-covered slopes of the San Gabriel Mountains, Calif. The light-colored area in the center of the scene is a relict grass area that has not been invaded by the spreading chaparral. (U.S. Forest Service photo by A. E. Wieslander, July 1929.)

northern coast range (fig. 6-21). It constitutes a true broad sclerophyll forest, which gradually gives way to coniferous forests toward the north as precipitation increases. Some areas of California have species of plants that are truly unique; for example, the Monterey cypress (*Cupressus macrocarpa*), native only to the Monterey Peninsula.

Middle Chile. The Chilean sclerophyll formation is remarkably similar to the *maquis* and chaparral of the Northern Hemisphere, although it has its own association of plants taxonomically distinct from those in other sclerophyll formations. In Chile, the formation is called *matorral* (thicket or bush), and it contains many species of the rose family (Rosaceae). On the higher slopes, the *matorral* gives way to a sclerophyll woodland and here, although not related, many of the trees strongly resemble evergreen oaks of California. The *matorral* is dominant from Concepción in the south, where precipitation averages 1,320 mm (52 in), to La Serena in the north, where the average is only 100 mm (4 in).

Chile is somewhat cooler than most of California and the Mediterranean basin, with warm-month temperature at both Concepción and La Serena averaging only 18°C (65°F). Because of cooler temperatures, greater effectiveness of precipitation helps explain the northward extent of the sclerophyll formation into areas of very low rainfall, but the southern extent into very humid areas is surprising.

Fig. 6–21. Open oak woodland on the foothills of the Sierra Nevada, Sequoia National Forest, Calif. (U.S. Forest Service photo by A. Gaskill, October 1903.)

Cape Province. The scrub vegetation of Cape Province, South Africa, seems to have been preceded in much of its range by rather luxuriant forests. Some areas, however, probably are climatic climax sclerophyll communities. In particular, the Cape Province flora is rich in perennials and deficient in herbaceous annuals, in spite of the relatively open nature of the scrub.

Australia. Over large areas of southwestern and southeastern Australia, a sclerophyllous scrub called *mallee* covers the landscape and bears a striking resemblance to the *maquis* of Europe, the chaparral of California, the *matorral* of Chile, and the scrub of Cape Province. Although life forms are similar, the species are very different, as the *mallee* is dominated by members of the genus *Eucalyptus.* Where rainfall is higher, the *mallee* gives way to magnificent sclerophyll forests of gum trees (*Eucalyptus* spp.) and other large eucalypts, which rival the California redwoods in size. In spite of the immense size of the trees, their leaf structure permits considerable light to penetrate to the forest floor, and here flourishes a rich understory of grass, shrubs, and herbaceous plants. Because gum forests are significantly different from other sclerophyllous formations that we have considered so far, the term *sclero-*

phyll forest might seem more appropriate; on the other hand, because of the open character of the Australian formation, in many ways it resembles a woodland.

Semiarid Scrublands and Woodlands

In middle latitudes, cool deserts and drier steppes of continental interiors support a marvelous variety of scrub and woodland formations. Typical are those of the Great Basin Province of the western United States. In warmer deserts of the southwestern United States and northern Mexico, the so-called Sonoran formation is more closely related to the tropical desert scrub formations. Other cool desert and steppe formations are found in Central Asia, from the Caspian Sea to the Gobi, and in Patagonia in southern Argentina.

In the United States, this formation, designated the upper Sonoran zone, often is called "sagebrush desert" because of the dominance of sagebrush (*Artemisia tridentata*) and to distinguish it from the lower Sonoran formation in which creosote bush (*Larrea divaricata*) dominates.

Early descriptions of the Great Basin indicate that grasses were more widespread in the sagebrush community before heavy grazing pressure eliminated large areas of native varieties. Sagebrush seems to have been a climatic climax dominant only in areas with less than 250 mm (10 in) of annual precipitation, but it has successfully invaded more humid grasslands of the western Great Plains and the Columbia Plateau. Overgrazing and reduced frequency of burning seem to have been major causative factors in its dominance at the expense of the native grasses (fig. 5-4).

A semiarid woodland formation occurs at higher elevations, between the subalpine forests of the higher country and the sagebrush of the basins. In some places, juniper (*Juniperus osterosperma* and *J. scopulorum*) dominate the woodland; in others, two species of nut pine, the single-leaf piñon (*Pinus monophylla*) and the two-leafed piñon (*P. edulis*), share dominance, forming the piñon-juniper woodlands of the Colorado Plateau and neighboring uplands. The two piñons sometimes are considered varieties of the same species (*P. cembroides*). Piñon nuts are an important staple of the Indians of the Great Basin and Colorado Plateau country.

South of the Great Basin and Colorado Plateau, and above the creosote bush desert community, scrub vegetation resembles coastal chaparral of California, with scrub evergreen oaks forming an important element. In some mountainous areas, from western Texas to Arizona and into northern Mexico, evergreen oaks, such as the Emory oak (*Quercus emoryi*, fig. 6-21), dominate areas of open oak woodlands.

Arctic and Alpine Tundra

Tundra is a term applied to the vegetation of the treeless area fringing the Arctic Ocean, from Alaska across northern Canada to Greenland, and from northern Scandinavia across the Soviet Union to the Bering Straits. Less extensive areas of tundra are found in the ice-free areas of Antarctica. Its extent approximates that of the tundra climate (*ET*) of Köppen. Frost is common, even in summer, and permafrost is almost universal except in coastal marine locations.

Tundra flora has relatively few species compared to formations of more southerly latitudes, with grasses, sedges, flowering herbs, low shrubs, lichens, and mosses accounting for most vegetation cover. Toward the southern margins of the Northern Hemisphere tundra, shrubs are common; they include evergreen heaths (Ericaceae) and thickets of birch, willow, and alder along streams. In July and August, flowering plants make a brilliant show of color, many of them familiar species of woodland and meadowland communities of middle latitudes. Prominent among these flowers are buttercups (*Ranunculus* spp.), marigolds (*Caltha* spp.), primroses (*Primula* spp.), and gentians (*Gentiana* spp.). Particularly conspicuous and widespread among tundra plants is the "reindeer moss" (*Caldonia* spp.), actually a lichen, which grows several inches tall and forms a gray, spongy mat over the ground. Northward from the boreal forest–tundra boundary, herbaceous plants and shrubs become fewer, and mosses and lichens make up an increasing proportion of ground cover.

Plant communities resembling arctic tundra are found above the treeline in many mountains, including the Rockies and Alps. This alpine tundra may contain many genera and even some species found in the arctic tundra, but there are many species from plant associations at lower elevations. For the most part, permafrost is absent in alpine tundra, mosses and lichens are less prominent, and flowering plants more conspicuous than in the Arctic. Heaths and moors, such as those in Denmark and highland Britain (described in chapter 5), resemble tundra, although they are largely plagioclimax formations resulting from human intervention.

Part 3

SOILS

Be it deep or shallow, red or black, sand or clay, the soil is the link between the rock core of the earth and the living things on its surface. It is the foothold for the plants we grow. Therein lies the main reason for our interest in soil.

ROY W. SIMONSON, *Soil: 1957 Yearbook of Agriculture (USDA)*

Chapter 7

THE NATURE AND ORIGIN OF SOILS

Soils represent the third great global pattern. This pattern corresponds, in some degree, to those of climate and vegetation. Again, one must not push the correlation too far. Climate is a major factor in the dynamics of soil formation, as it is in the growth of vegetation, but other environmental agents also may be equally or, in some cases, even more important than climate.

What is Soil?

The word *soil* means different things to different people; to an engineer it is any unconsolidated earth material lying above solid bedrock; to the layperson it is often synonymous with dirt; but to the soil scientist (*pedologist*), soil is a collection of natural bodies on the Earth's surface, containing living matter, which supports or is capable of supporting plants. Its upper boundary is the air or shallow water, and its lateral limits are marked by barren areas of rock, ice, salt, shifting sands, or deep water. Its lower limit is more difficult to define, but with few exceptions it is the depth to which the roots of common native perennial plants penetrate. Thus, soil tends to be shallow in deserts and tundra but deep in the humid tropics (Soil Survey Staff 1975).

In the past, soil was seen as the natural medium for growth of land plants. If classified at all, generally it was in terms of quality, color, or texture. Modern schemes for classifying soils developed out of revolutionary soil concepts formulated about 1870 by Russian pedologists, the most outstanding of whom was K. D. Dokuchaiev (see chapter 8). Essentially, he considered soils to be independent natural bodies, each with a unique form resulting from a combination of climate, vegetation, parent rock materials, landscape relief, and time. A certain combination of these genetic factors produces a unique soil characterized by its *profile* (Soil Survey Staff 1960).

The Soil Profile

A soil profile consists of the several horizons generally present because of soil-forming (genetic) processes; these are called *genetic horizons* to distinguish them from other horizons, such as those inherited from stratified parent material. Soils may or may not have readily discernible horizons, but their presence or absence of horizons in a soil is an important criterion in its

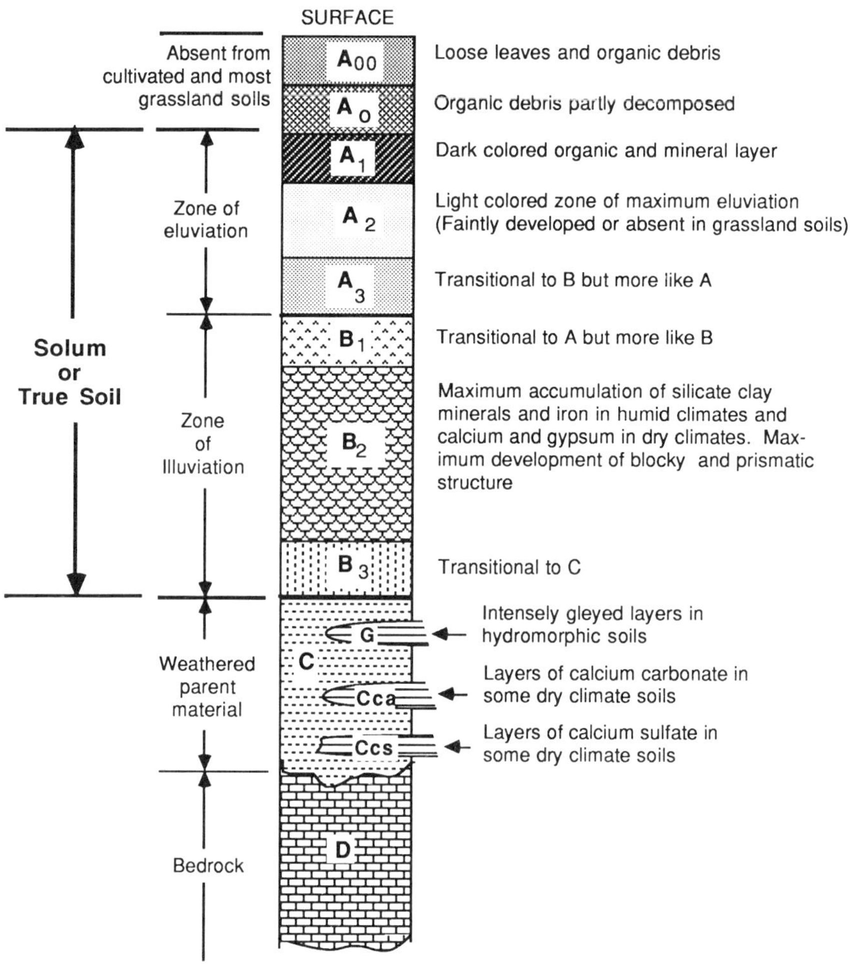

Fig. 7–1. A hypothetical soil profile having all the principal horizons. Very few soils would have all of the horizons shown. (From USDA, 1938, 1,169.)

classification. Horizons produced by soil-forming processes are called collectively the *solum*—the true soil lying above the unaltered parent material. A hypothetical soil profile is shown in figure 7-1. Many soils, however, will not have all the horizons depicted in the figure.

The nature of the soil profile is the most distinguishing feature of a soil. This makes it possible to separate soils into categories for classifying and mapping. Soil profiles are products of soil genesis, which in turn operates under a number of environmental influences and restraints. Among the major genetic factors determining types of soil are parent materials, climate, vegetation and other living organisms, topography, and time.

The O Horizon

The O horizon contains 20 percent or more organic matter consisting of fresh or partially decomposed leaf litter and other forest residue. This horizon occurs mostly in undisturbed forest areas; cultivation destroys it. It seldom occurs in grassland soils.

The A Horizon

The A horizon is a zone of leaching. As rain water percolates through the A horizon, soluble substances, such as calcium compounds, are carried to lower layers where they may be deposited or be completely leached from the profile. Other minerals, including compounds of iron, magnesium, potassium, and silicon, may be removed from the A horizon. Insoluble clay particles, formed by decomposition of larger mineral fragments, move out of the A horizon in suspension and lodge in lower horizons. Removal of materials from soil by percolating water, either in solution or suspension, is *eluviation*.

The A_1 horizon is the most active biologically; plant roots, small animals and insects, bacteria, and fungi are most concentrated in this zone. In most cases it has an accumulation of well-decomposed organic matter resulting from this concentration of biological activity. The organic matter coats the mineral particles and gives this horizon a darker color than the one below. In some forest soils, and in certain poorly drained grassland soils, the organic material has been removed from the A_2 by eluviation and the horizon is light colored. Otherwise, A_2 horizons are poorly developed in well-drained grassland soils.

The B Horizon

A common term for the B horizon is *subsoil*. B horizons are characteristic of soils with well-developed profiles, such as *planosols* and *gray brown podzolic soils* (see chapter 8). Because there is less biological activity, the organic content of the B horizon is lower than that of the A, and clay content often is high because of *illuviation*. Other minerals eluviated from the A horizon also may accumulate in the B horizon as stains, nodules, concretions, or *mineral pans*.[1]

The C Horizon

The C horizon occurs below what pedologists consider the true soil or solum, and often it is the parent material from which soil forms. Loose material in the C horizon may have weathered from the bedrock below, or it may have been transported into the area by water, wind, or glacial ice and deposited above the bedrock or other materials. The lower boundary of the C horizon may be the bedrock below, or it may be other buried unconsolidated materials. Generally, soil profile descriptions include the top foot or two below the solum.

Subhorizons

Sometimes it is desirable to designate subhorizons other than A_1, A_2, B_1, and so forth, within the framework of the major horizons. The profile model in figure 7-1 includes some symbols for the C horizon. Some subhorizon symbols may apply to any horizon, whereas others apply to a specific one. Symbols used in the present USDA soil taxonomy are listed in table 7-1. An example would be the *Bhir* horizon in a *spodosol* (*podzol* soil), indicating the presence of illuvial humus and illuvial iron in the B horizon.

Paleosols

Paleosols are soils buried below the modern land surface by materials laid down by glaciers, water, or wind. Today, some of these ancient soils are exposed by erosion or by cuts made by humans (fig. 7-2). Paleosols are normal soils formed on what was once the land surface. Since their characteristics reflect pedogenic processes operating at the time of their formation, paleosols give valuable clues to past environmental conditions of climate, vegetation, and drainage. Some, exposed at the surface through geologic erosion, serve as parent material for modern soils.

The Pedon

An individual soil is distinguished from other soils by its profile, and herein lies a problem in soil classification: How large an area does a single soil profile represent? Soil boundaries are seldom abrupt; rather, one soil grades into another over a distance of meters. Also, certain horizons in some soils are discontinuous, or they vary greatly in thickness within short horizontal distances. Consequently, two profiles taken a few meters apart may or may

[1] *Illuviation,* the opposite of *eluviation,* is the deposition of material from solution or suspension. A *pan* is a very compact horizon or layer in a soil resulting from a high clay content (*clay pan*) or from cementation of soil particles by iron oxide (*iron pan*), calcium carbonate (*caliche*), silica (*duripan*), or other minerals. Sometimes the general term *hard pan* is used to designate any such layer.

TABLE 7-1 Symbols for Designating Subhorizons in a Soil Profile

Symbol	Description
b	Buried soil horizon
ca	Accumulation of carbonate; commonly $CaCO_3$
cs	Accumulation of calcium sulfate ($CaSO_4$)
cn	Accumulation of concretions or nodules with sesquioxides
f	Soil frozen at time it is described—may or may not be permafrost
g	Strong gleying (reducing conditions)
h	Illuvial humus
ir	Illuvial iron
m	Strong cementation including ortstein, caliche, and duripan
p	Plowing or other disturbance (used only with prefix A)
sa	Accumulation of salts more soluble than $CaSO_4$
si	Cementation by siliceous material; *sim* if continuous (only with C)
t	Illuvial clay (used only with B)
x	Fragipan—a firm, brittle, high-density layer

Source: Soil Survey Staff, 1960, 28–29.

Fig. 7–2. Dark layers in a fresh railroad cut in Adair County in southwestern Iowa show the modern grassland (Mollisol) soil developed in the surface layer of a wind-deposited silt (loess), and a buried paleosol developed in Kansan glacial till. The paleosol was formed after the retreat of the glacier in what was then the surface layer before the loess was deposited over it. (Photo by W. E. Akin.)

not include all horizons, or they may show significant differences in vertical dimensions of some horizons.

A more workable concept would be to consider one soil in which the discontinuous A_2 horizon is an important characteristic for identification. A way around the difficulty encountered in relying on a two-dimensional profile is to include a three-dimensional body of soil with enough surface area to permit study of major variations in horizon shapes and relationships. Such a volume of soil is a *pedon* (Greek *pedon*, ground; rhymes with "head on"), defined as the smallest volume that can be called a soil. More precisely, the Soil Survey Staff of the USDA defines a pedon as follows:

> A pedon has three dimensions. Its lower limit is the vague and somewhat arbitrary limit between soil and "not-soil." The lateral dimensions are large enough to permit study of the nature of any horizons present, for a horizon may be variable in thick-

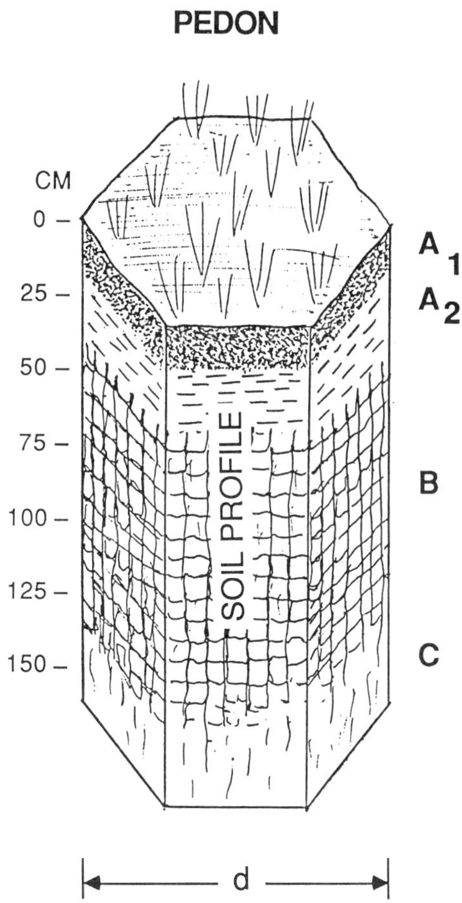

Fig. 7–3. A representative pedon in a Weller soil (Udalf). Weller soils developed under deciduous forest, but today they are mostly in crops. The diameter of the pedon is indicated by *d*, usually about 1 m. Horizons A_1, A_2, and B make up the solum.

ness or even discontinuous. Its area ranges from 1 to 10 square meters, depending on the variability in the horizons. Where horizons are intermittent or cyclic and recur at linear intervals of 2 to 7 meters (roughly 7 to 25 feet), the pedon includes one-half of the cycle. Thus each pedon includes the range of horizon variability that occurs within these small areas. Where the cycle is less than 2 meters or where all horizons are continuous and of uniform thickness, the pedon has an area of 1 square meter. Again, under these limits, each pedon includes the range of horizon variability associated with that small area. The shape of the pedon is roughly hexagonal.[2] One lateral dimension should not differ appreciably from any other. [Soil Survey Staff 1960]

Because of the arbitrary maximum size applied to a single pedon, it may require many contiguous pedons in an area to encompass a particular soil. A group of contiguous pedons having characteristics falling within the range of a single soil series has been termed a *polypedon*. The pedon model for a soil belonging to the Weller series is shown in figure 7-3.

Factors Influencing Soil Formation and Character

Parent Material

The breakup of the solid bedrock by chemical and mechanical weathering produces loose (unconsolidated) mineral material. This is the parent material within which a soil profile forms by *pedogenic processes*. If parent material develops from bedrock lying directly below, it is *residual* parent material (fig. 7-4); if it arrived at its present position through geologic agents, such as water, wind, or glacial ice, it is *transported* parent material. Transported parent material deposited by streams is called *alluvium*; it occupies flood plains, alluvial fans, and river deltas. Alluvium deposited in existing lakes is designated as *lacustrine*. Water velocity at the time of deposition sorts alluvial deposits according to particle size. *Colluvium* is material transported downslope by gravity through soil creep or landslides, and is unsorted as to particle size.

Figure 7-5 gives a highly generalized picture of the distribution of soil parent materials within the conterminous United States. Two types of transported materials, *glacial drift* and *loess*, are parent materials for some of the most productive soils in the American corn belt. *Glacial drift* is a collective term applied to any materials associated with glaciers; it includes *till* and *outwash*. Glacial till is deposited directly from melting glacial ice and is unsorted as to particle size and unstratified because water action is minimal. Glacial outwash, on the other hand, has been carried from the glacier by running water and is sorted and stratified, as is the case with any water-deposited materials. Loess is wind-deposited silt and clay-sized materials often associated with continental glaciation. Since the parent material from

[2] A hexagonal shape is used so that a pedon will have continuous boundaries with contiguous pedons with no intervening space, such as would occur if a circular shape were used.

Fig. 7–4. Bedrock-soil transition where weathering of bedrock has produced residual parent material and a thin soil. The soil is Lueders clay loam in Baylor County, Texas (Probable classification: *Order:* Mollisol; *Suborder:* Ustoll; *Great Group:* Argiustoll; *Subgroup:* Lithic Vertic Argiustoll). (U.S. Soil Conservation Service photo by C. A. Rogers, 1970.)

which an individual soil forms determines in part its texture and mineral composition, it is an important factor in distinguishing one type of soil from another.

Climate and Soil Formation

Soil formation, for the most part, is a process involving weathering of earth materials; it relates to climate directly through temperature and precipitation, indirectly through vegetation and topography. It is not surprising that the world climatic pattern reflects the broad pattern of world soils. Chemical weathering, particularly, is vigorous where there is a combination of high temperatures and high effective precipitation. Moisture is required for most chemical processes, and the rate at which most chemical reactions occur increases with increased temperature. Conversely, in cooler or drier climates, the soil forming processes lag.

Under chemical weathering, most primary minerals contained in the orig-

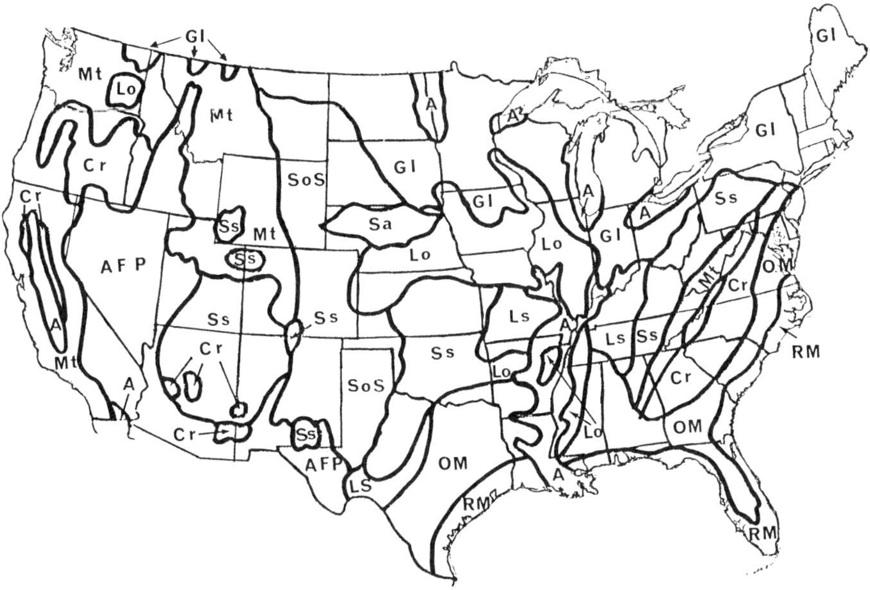

Fig. 7–5. Soil parent material of the conterminous United States. The map is highly generalized from various geologic and soil maps. Most divisions include areas of other parent materials, particularly alluvium, which are too small to show at this scale. Key:

RM *Recent marine deposits:* Sand, silts, some marls, and swamp and bog deposits that include some areas of peat.

OM *Older marine deposits of the inner Coastal Plain:* Includes marls and other weakly consolidated sediments.

A *Alluvium and lacustrine:* Includes the major alluvial valley of the Mississippi River and some tributaries, and the Central Valley of California. Lacustrine deposits are shown for the major glacial lakes. Many smaller areas of stream alluvium and lake deposits are not shown.

AFP *Alluvial fans and playa (dry lake) deposits of the Basin and Range Province:* Includes many mountain areas not differentiated on the map.

Lo *Loess:* Wind-deposited silts covering the outer margins of the glacial drift plains and adjacent areas, the Palouse of Washington, and the uplands adjacent to the lower Mississippi Valley.

Sa *Wind-deposited sand of the Sand Hills of Nebraska:* Many smaller areas of dune sand are not shown.

Gl *Glacial drift deposited by continental glaciers:* Includes till, outwash, and local lacustrine deposits that vary considerably in thickness and texture.

SoS *Soft sedimentary rock:* Poorly to moderately consolidated sediments of the western Great Plains.

Ss *Sandstones and shales:* Well-consolidated sedimentary rocks with some thinly bedded limestones.

Ls *Limestone:* Well-consolidated, generally thick marine limestones.

Cr *Crystalline rocks:* Includes metamorphic and igneous rocks, both ancient and recent volcanics.

Mt *Complex mountains:* Great variations of materials in short distances make mapping at this scale impossible. From Bear, Pritchard, and Akin 1986.)

inal parent material decompose and leach from the soil, or they recombine into secondary mineral compounds. Consequently, the resulting soil often bears little resemblance to the original parent material. The amount of water infiltrating the soil is greatly influenced both by microclimate, which determines local evapotranspiration, and by slope, which affects the amount of surface runoff. Although most water entering soil from precipitation eventually transpires and evaporates, it may remain in the soil long enough to participate in the process of soil formation.

Soils of the United States (see chapter 8) illustrate climatic effects on soil formation. In chapter 5 we saw the relationship between soil pH and rainfall (figure 5-13). In arid areas most soils are light colored because little organic matter accumulates under sparse vegetation, the solum is neutral to alkaline in reaction, and calcium carbonate ($CaCO_3$) sometimes accumulates at the bottom of the profile. In places the $CaCO_3$ accumulation is so intense that a cemented layer called *caliche* is formed.

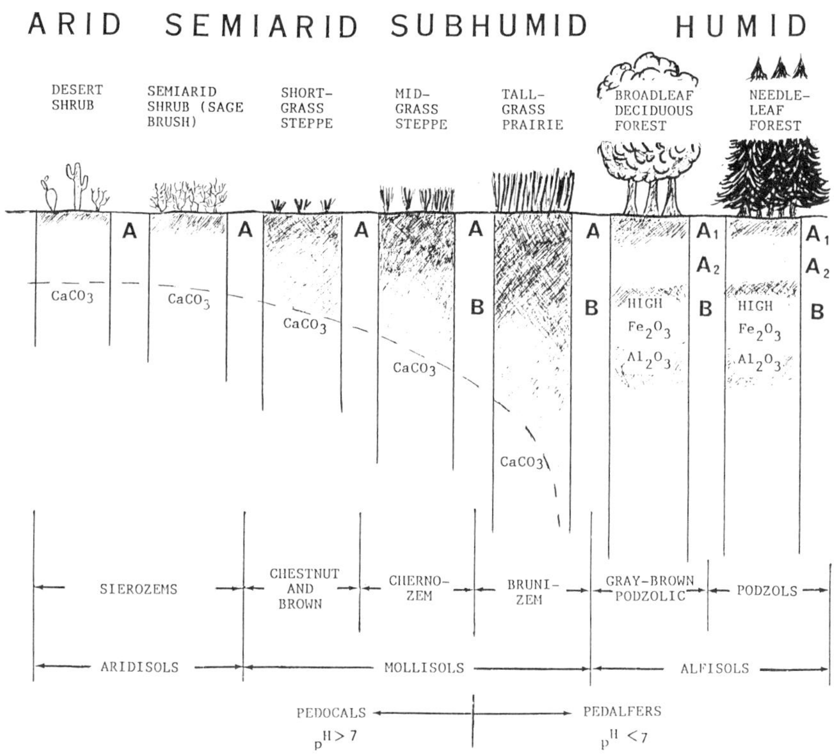

Fig. 7–6. Relationships among climate, vegetation, and soils along a line from the arid Southwest to the Great Lakes region of the United States. The drawings are not to scale. (From Bear, Pritchard, and Akin 1986.)

Eastward of the arid areas in the United States, as well as westward and northward, rainfall is sufficient to support denser vegetation cover and more vigorous weathering action. Thus, soils in these areas have deeper profiles and more organic matter and are more thoroughly leached of minerals than are soils of arid lands. In the humid eastern half of the United States, the $CaCO_3$ is leached completely from most soil profiles (fig. 7-6). C. F. Marbut, of the U.S. Department of Agriculture (see chapter 8), proposed a division of all soils into two categories: *pedocals,* for all soils with a Cca horizon (accumulation of $CaCO_3$), and *pedalfers,* for soils without such a horizon. Marbut's dividing line in the central United States coincides roughly with the 20 in (580 mm) isohyet in the north, 25 in (635 mm) in the central, and 35 in (889 mm) in the south; its location is shown in figure 5-14. Although the pedocal-pedalfer division is a useful generalization, it fails to classify soils that form from parent materials lacking $CaCO_3$ in their mineralogical makeup.

Temperature effects on soil formation can be seen by comparing soils of cool humid areas with those of warm and hot humid areas. Most obvious are differences in color: Soils in the cooler areas generally have gray or black A horizons and brown B horizons; as one travels equatorward, soils become progressively brighter and redder in color. Cooler soils owe their darker colors to the presence of organic humus, which decomposes slowly at low temperatures. On the other hand, higher temperatures accelerate decomposition of humus, and iron compounds become the chief coloring agents.

Vegetation and Other Living Organisms in Soil

Soils under almost identical conditions of parent material, slope, and climate may differ markedly if they formed under different vegetation covers In the central United States, a distinct contrast is visible between soils formed under tallgrass prairies and those formed under deciduous woodlands (figure 7-7). In fact, fundamental soil differences frequently occur between grasslands and forests. Within forested areas soils separate further into those formed under broadleaf deciduous forests and those formed under needle-leaf forests; in grasslands a similar contrast can be found between soils formed under sod-forming grasses, tall bunchgrass, and short bunchgrass. Generally, soils of middle latitude grasslands, are superior in natural fertility to those of forest.

The difference in natural fertility between soils of grasslands and soils of forests can be explained by several factors; of importance is the relationship between the quantity of organic matter produced and the level of nutrient demand. Most grass-produced organic matter dies annually and thus adds to soil humus; even extensive root systems of perennial grasses are replaced every few years. As a result, the upper horizons of grassland soils generally are much darker than those of forest soils. On the other hand, trees and their root systems live much longer than grasses and contribute less each year to

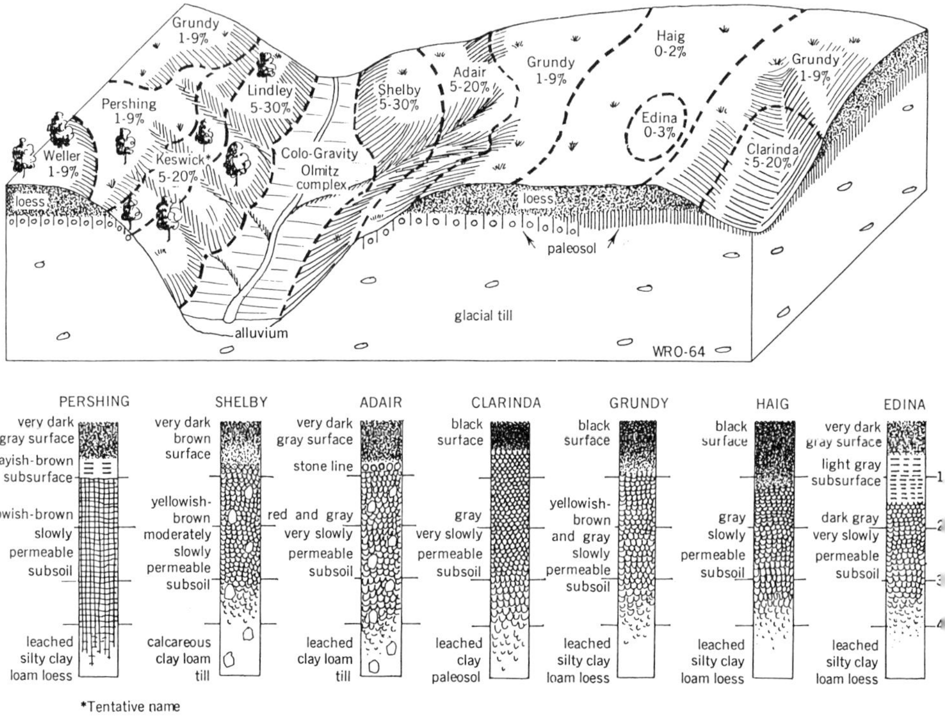

Fig. 7–7. Relationship of slope, vegetation, and parent material to soils of the Adair-Grundy-Haig and Grundy-Haig soil association areas in Iowa. Weller soils developed under a natural vegetation cover of broadleaf deciduous forest; Pershing soils formed under a mixture of grass and deciduous trees; all other soils in this association area formed under prairie grasses. (From W. R. Oschwald et al., *Soils of Iowa,* Ames: Department of Agronomy, Iowa State University, 1965.)

the accumulation of soil humus. Also, trees are less demanding of soil nutrients, particularly *bases,*[3] and so do not recycle these nutrients to the surface to the extent that grasses do.

Trees add organic matter to soil in the form of fallen leaves or needles. The decomposition of this litter produces organic acids which increase the leaching power of percolating water and hasten the removal of bases and other soluble minerals. Consequently, whereas most grassland soils are neutral to basic in reaction, most forest soils are acid. Most coniferous trees have lower base requirements than do deciduous trees, and soils developed under certain species may be intensely acid and low in fertility.

[3] *Bases* are substances containing an OH (hydroxide) radical, which forms an OH^- (hydroxyl) ion when dissolved in water; they react with acids to form salts including calcium carbonate, or lime ($CaCO_3$). Bases are hydroxides of positive elements.

In addition to vegetation, many other organisms live in and contribute to soil formation. Earthworms, burrowing rodents, and insects contribute by mixing and opening the soil for easier air and water penetration. In grasslands their activity prevents formation of well-differentiated A and B horizons because of the downward mixing of organic matter. Microorganisms, including bacteria, fungi, and *actinomycetes*,[4] contribute to this decomposition by converting complex organic compounds into water, CO_2, and simple inorganic compounds that add to plant nutrients. We described some of these processes in the section on soil and plant nutrition in chapter 5.

Topography and Soils

Whereas climate exerts the major control on broad global soil patterns, topography and parent material control the details of the local soil landscape. Of greatest importance is the degree of slope and the relative position of soils on that slope. Generally, there is a sequence of soils ranging from flat summits, down the side slopes of valleys to lower slopes, and then to valley bottoms. This local arrangement of soils is called a *toposequence*. If all soils in a toposequence are derived from similar material and are formed under the same vegetation cover, the slope sequence is a *catena* (Latin "chain").

Figure 7-8 illustrates the significance of slope position on the development of soil profiles. Slope affects soil formation directly, not only by its influence on runoff and infiltration by water, but also by erosion and deposition of materials. For the most part, under similar climatic conditions, soils with very little slope are much wetter than soils on steep slopes. Exceptions occur where soil materials are very coarse, thus causing excessive drainage even on flat surfaces, or in areas with excessive rainfall, where water-logged soils may be found even on steep slopes. Where water tables are high enough at least some time each year to saturate the solum, soil development is profoundly affected. With a given combination of climate, parent materials, and vegetation, a specific relationship exists between soil slope, slope position, and soil genesis. Nevertheless, the relationship between topography and soils under one combination may not apply to a different combination of materials, vegetation, and climate.

Figure 7-7 illustrates a toposequence in the Adair-Grundy-Haig soil association: Edina and Haig soils are on the flat summits, Grundy on more gentle side slopes, Adair and Shelby soils on steeper lower slopes, and alluvial soils in valley bottoms. Only Edina, Haig, and Grundy soils form a catena, being formed from the same parent material. In the toposequence on the opposite valley slope in figure 7-7, the Weller, Grundy, and Pershing soils occupy similar topographic positions, but they were formed under dif-

[4]*Actinomycetes* are a group of organisms that are sometimes classified as bacteria and sometimes as fungi. A few cause plant diseases; others cause diseases in humans and animals. Some produce antibiotic substances—for example, *Streptomyces griseus* produces streptomycin, and *Penicillium notatum* is the source of penicillin.

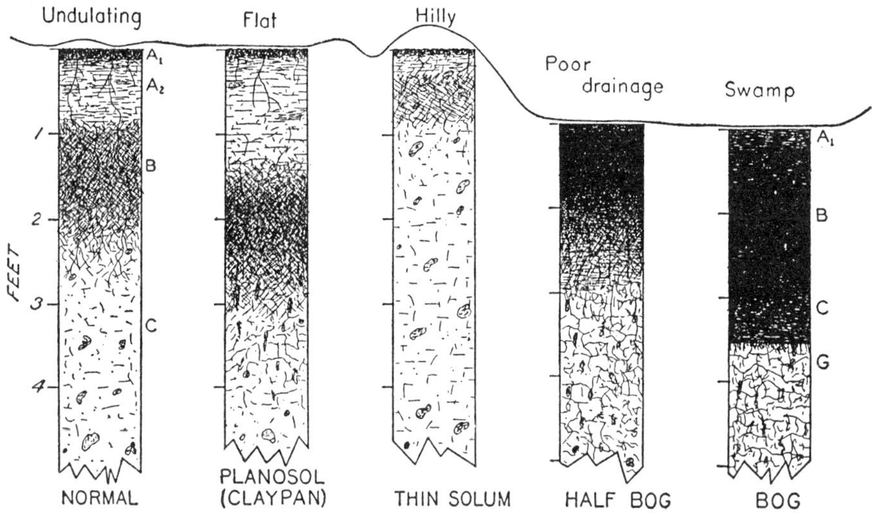

Fig. 7–8. Soil profiles developed from similar parent materials but at different topographic positions. Shallow soil is developed on hilly land because excessive erosion prevents formation of a deep profile, whereas soils on flat uplands have a thick profile with leached A horizons and a dense clay pan in the B horizon. Profiles represent, from left to right, Miami, Crosby, Rodman, Brookston, and Carlisle soil series. (From USDA, 1938, 967.)

ferent vegetation covers; therefore, they belong to a different soil series and they do not form a catena. Figure 7-9 shows the location of this soil association in Iowa, where there forest-prairie transition was located at the time of European settlement.

Time and Soil Formation

Soil-forming processes require time to alter parent material into soil with a well-developed profile. Soils exist in all stages of development, from very young to extremely old. Landscape terms, such as *young, mature,* and *old,* sometimes indicate relative progress of soil development through time. Generally speaking, young soils are fertile since the nutrient loss by leaching is minimal. A soil may require hundreds or even thousands of years to progress from youth to maturity, finally acquiring a well developed solum and strongly differentiated A and B horizons. Some remain in perpetual youth because of rapid erosion of surface horizons or constant deposition of new materials. A good example of soil rejuvenation is the constant renewal of fertility to some tropical soils by volcanic ash, making them much more productive than many mature tropical soils under the same climatic conditions.

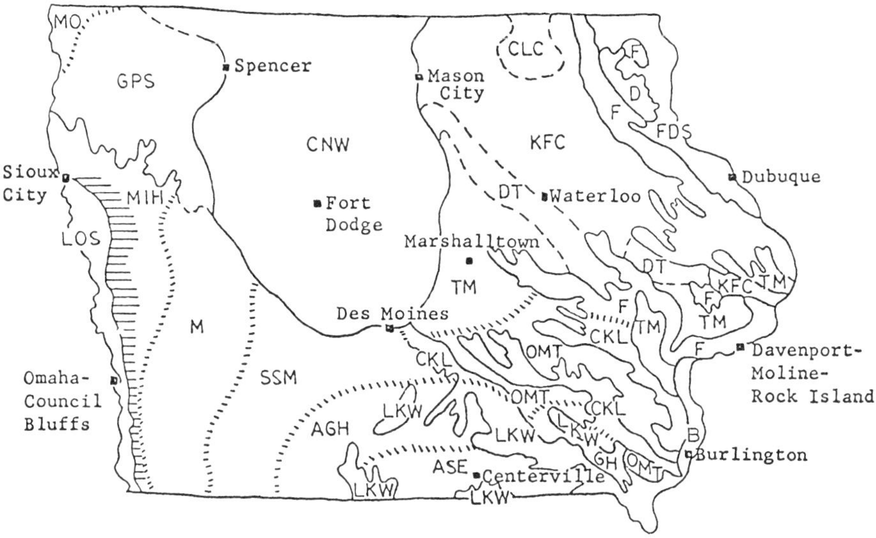

Western Iowa bluffs area — Tentative boundary — Gradational boundary — Abrupt boundary

AGA	Adair-Grundy-Haig	GH	Grundy-Haig
ASE	Adair-Seymoure	KFC	Kenyuon-Floyd-Clyde
B	Soils of Mississippi Bottoms	LKW	Lindley-Keswick-Weller
CKL	Clinton-Keswick-Lindley	LOS	Luton-Onawa-Salix
CLC	Cresco-Lourdes-Clyde	M	Marshall
CNW	Clarion-Nicollet-Webster	MIH	Monona-Ida-Hamburg
D	Downs	MO	Moody
DT	Dinsdale-Tama	OMT	Otley-Mahaska-Tantor
F	Fayette	SSM	Shelby-Sharpsburg-Macksburg
FDS	Fayette-Dubuque-Stonyland	TM	Tama-Muscatine
GPS	Galva-Primghar-Sac		

Fig. 7–9. Principal soil association areas of Iowa. (From Akin and Dideriksen, 1967. Used with permission of the *Journal of Soil and Water Conservation.*)

As soils approach maturity, pedogenic processes decline, and the profile development assumes a nearly steady state, maintained by conversion of parent material into soil at a rate equal to loss by erosion and leaching. Differentiation between A and B horizons continues during maturity, but the rate slows. Soil fertility may not be as high as during early youth, but the release of nutrients by weathering of primary minerals may maintain relatively high fertility.

A soil attains old age when most primary minerals have weathered and no new source of nutrients is available to replace those lost by leaching. Eluviation continues, and the B horizon may accumulate so much clay that movement of air and water is restricted and plant roots may have difficulty with penetration. Old soils are among the most infertile and are destined to remain so until geologic processes, such as glaciation, deposition of new materials, and accelerated erosion, interrupt soil development and set the stage for a new cycle. In part, this explains the high fertility of relatively young glacial, loessial, alluvial, and volcanic ash soils in many parts of the world, and it also explains the infertility of old upland soils in many tropical areas where for millions of years glaciation or other geologic disturbances did not interrupt pedogenesis.

Physical Characteristics of Soil

The most important physical characteristics of soil are texture, structure, and color. Parent material determines texture, but soil-forming processes greatly influence distribution of particles of different sizes throughout the solum. A close relationship exists between structure and color and pedogenesis.

Soil Texture

Soil texture is described according to the relative abundance of sand, silt, and clay, as a percentage of total weight. Sizes grade from very coarse sand, particles of which have a diameter as great as 2 mm (0.08 in), through finer sands, silts, and clay. Table 7-2 gives the size limits of soil texture classes used by the USDA. These soil textural classifications are based on a range of mixtures of various sized particles determined in the laboratory or estimated in the field. Figure 7-10 is a graphic device for determining textural classifications of a soil when percentages of various sized particles become known. Soil textural terms and class names used by the USDA are given in table 7-3, and examples of the makeup of typical texture classes are shown in figure 7-11.

It is possible to estimate soil texture in the field without laboratory analysis. Such obvious classes as coarse sand and very fine clay are easy to observe, but long experience with soils is necessary if one is to differentiate texture with

Table 7-2 Size Limits (in Millimeters) of Soil Separates from Two Schemes of Analysis

U.S. Department of Agriculture Scheme		International Scheme	
Name of Separate	Diameter (range)	Fraction	Diameter (range)
Very coarse sand[a]	2.0 –1.0		
Coarse sand	1.0 –0.5	I	2.0 –0.2
Medium sand	.5 –0.25		
Fine sand	.25–0.10	II	0.20–0.02
Very fine sand	.10–0.05		
Silt	.05–0.002	III	0.02–0.002
Clay	Below 0.002	IV	Below 0.002

[a]Prior to 1947 this separate was called fine gravel. Now fine gravel is used for coarse fragments from 2 mm to ½ inch in diameter.
Source: USDA 1938, 207.

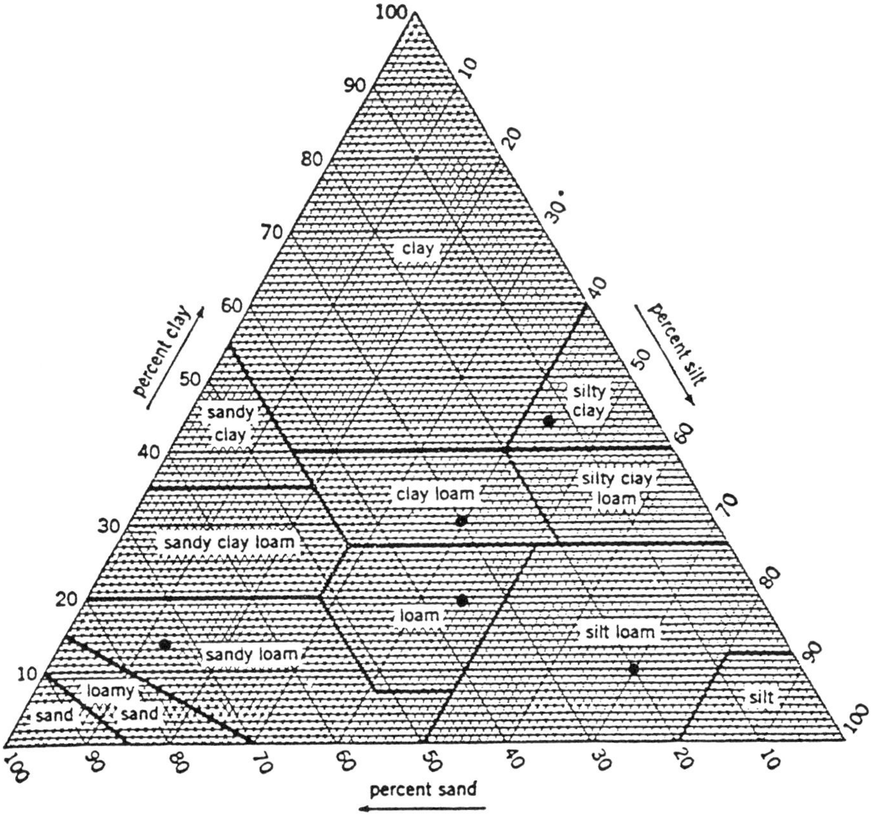

Fig. 7–10. Graph for determining soil textural classes from percentages of clay, silt, and sand contained in a soil. Dots on the diagram indicate the composition of soils of various textures illustrated in figure 7-11. (From USDA, *Soil Survey Manual*, 1961.)

TABLE 7-3 Soil Textural Terms and Class Names

General Terms	Relative Terms	Basic Soil Textural Class Names
Sandy soils	coarse-textured soils	sands loamy sands
	moderately coarse-textured soils	sandy loam fine sandy loam very fine sandy loam
Loamy soils	medium-textured soils	loam silt loam silt
	moderately fine-textured soils	clay loam sandy clay loam silty clay loam
Clayey soils	fine-textured soils	sandy clay silty clay clay

Source: Soil Survey Staff, *Soil Survey Manual*, USDA, 1951, 213.

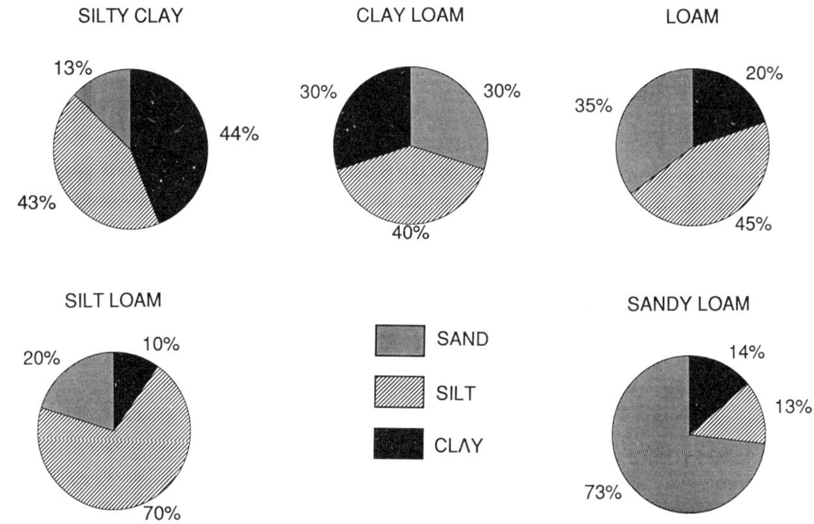

Fig. 7–11. Textural compositions of five soils according to the USDA criteria for textural classes. These soils are positioned in the soil texture graph in Figure 7-9.

any degree of accuracy. Guidelines for determining soil textural class by field observation and feel of the soil are given in an appendix of *Soil Taxonomy* (Soil Survey Staff 1975). Although today laboratory analysis takes precedence over subjective field classifications, these guidelines are useful for describing the nature of the textural classes and for field observations.

Soil Structure

Soil structure is determined by the natural aggregation of primary soil particles (sand, silt, and clay) into clusters called *peds*. Fine-textured particles such as clay and humus stick together (*flocculate*) and act as agents for binding coarser textured materials into peds. Other soil fragments, such as clods formed when soils are disturbed, are not peds. Soil structure may be radically different in soils having the same texture. In some very sandy soils, and in deflocculated clays, structure may be entirely lacking. In fine textured soils with abundant clay, structure is important, especially since it may increase permeability and may facilitate penetration by plant roots.

Peds take several forms, and they may vary in size, producing fine, medium, and coarse structure (fig. 7-12). The five structural categories are: (a) structureless, (b) granular, (c) platy, (d) blocky, and (e) prismatic (table 7-4).

Structureless Soils. This type includes *single grain* soils, particularly those with high sand and low clay content, and *massive* soils, in which the entire

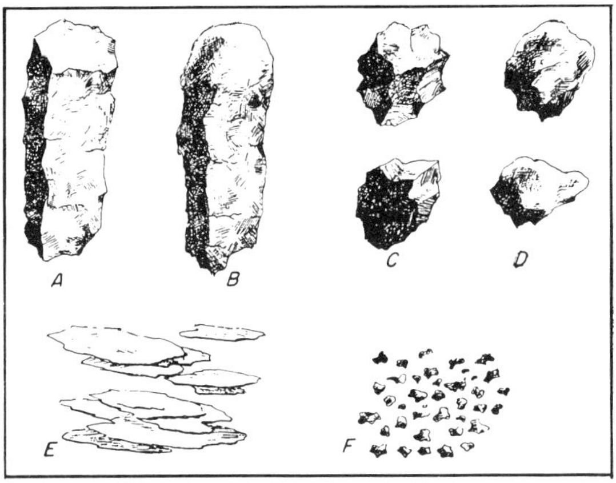

Fig. 7–12. Drawings indicating shapes of peds associated with several types of soil structure: A, prismatic; B, columnar; C, angular blocky; D, subangular blocky; E, platy; and F, granular. (From USDA, *Soil Survey Manual*, 1961.)

soil mass adheres without forming recognizable peds. Massive structure is typical of some soils with high clay content in which clay particles are not flocculated.

Granular Structure. In this category, peds are roughly spheroidal and do not fit together tightly. *Crumb* structure is a particularly porous form of granular structure. Water, air, and plant roots can readily penetrate this soil, making it particularly well suited for agriculture. Some tropical soils, however, have such well developed spheroidal peds that their structure is too loose and lack of moisture holding capacity limits productivity.

Platy Structure. Particles of platy soils are arranged in planes, usually horizontal, by lateral movement of water through soil. Platy structure occurs most frequently in A_2 horizons, or at the soil surface.

Blocky Structure. This type of soil structure is produced by peds having roughly equal horizontal and vertical dimensions. They are distinguished from granular peds because they fit snugly against adjacent peds, leaving little space between them. Blocky peds are *angular blocky* if corners and edges are sharp, and *subangular blocky* if they are rounded. Angular blocks are more characteristic of B horizons, while subangular blocks may be found in either A or B horizons.

Prismatic Structure. In this category, soil particles are arranged around a vertical axis, forming peds with vertical dimensions greater than the horizontal. As in blocky soils, the peds fit snugly against faces of adjacent peds. If the tops of the prisms are rounded by eluviation, the structure becomes *columnar*. Prismatic structure occurs most frequently in B horizons of well-developed soils, although occasionally it may occur in A horizons. Columnar structure characterizes B horizons of very strongly developed soils, usually indicating old age or high sodium content.

Development of Structure. Structure develops during the soil-forming pro-

TABLE 7-4 Types and Classes of Soil Structure

Class	Type (Shape and Arrangement of Peds)						
	platelike with one dimension (the vertical) limited and greatly less than the other two; arranged around a horizontal plane; faces mostly horizontal	prismlike with two dimensions (the horizontal) limited and considerably less than the vertical; arranged around a vertical line; vertical faces well defined; vertices angular		blocklike; polyhedronlike, or spheroidal, with three dimensions of the same order of magnitude, arranged around a point			
				blocklike; blocks or polyhedrons having plane or curved surfaces that are casts of the molds formed by the faces of the surrounding peds		spheroids or polyhedrons having plane or curved surfaces which have slight or no accommodation to the faces of surrounding peds	
		without rounded caps	with rounded caps	faces flattened; most vertices sharply angular	mixed rounded and flattened faces with many rounded vertices	relatively non-porous peds	porous peds
	platy	prismatic	columnar	(angular) blocky[a]	subangular blocky[b]	granular	crumb
Very fine or very thin	very thin platy; <1 mm	very fine prismatic; <10 mm	very fine columnar; <10 mm	very fine angular blocky; <5 mm	very fine subangular blocky; <5 mm	very fine granular; <1 mm	very fine crumb; <1 mm
Fine or thin	thin platy; 1–2 mm	fine prismatic; 10–20 mm	fine columnar; 10–20 mm	fine angular blocky; 5–10 mm	fine subangular blocky; 5–10 mm	fine granular; 1–2 mm	fine crumb; 1–2 mm
Medium	medium platy; 2–5 mm	medium prismatic; 20–50 mm	medium columnar; 20–50 mm	medium angular blocky; 10–20 mm	medium subangular blocky; 10–20 mm	medium granular; 2–5 mm	medium crumb; 2–5 mm
Coarse or thick	thick platy; 5–10 mm	coarse prismatic; 50–100 mm	coarse columnar; 50–100 mm	coarse angular blocky; 20–50 mm	coarse subangular blocky; 20–50 mm	coarse granular; 5–10 mm	
Very coarse or very thick	very thick platy; >10 mm	very coarse prismatic; >100 mm	very coarse columnar; >100 mm	very coarse angular blocky; >50 mm	very coarse subangular blocky; >50 mm	very coarse granular; >10 mm	

[a]Sometimes called *nut*. The word *angular* in the name can ordinarily be omitted.
[b]Sometimes called *nuciform*, *nut*, or *subangular nut*. Since the size connotation of these terms is a source of great confusion to many, they are not recommended.
Source: Soil Survey Staff 1975, 475.

cess when wetting and drying cause swelling and shrinking of soil materials. This shifting of soil particles produces planes of weakness in the soil mass, which gradually develop into ped faces. Freezing and thawing, penetration by plant roots, and animal activity eventually may increase separation between peds. Also important are the cementing agents produced by plants, animals, and microorganisms that help stabilize peds (Thompson and Troeh 1978, 53). Plowing or compaction by heavy equipment can destroy structure in some soils if they are wet. This undesirable condition is called *puddling* because the larger pores collapse and particles run together as the peds disintegrate.

Soil Color

Color is one of the most obvious and easily distinguishable of soil characteristics. Although it has little direct influence on soil quality, it may be a good indicator of other characteristics. Color may range from black to white through many shades of gray, brown, yellow, and red. Since verbal descriptions of soil color are highly subjective, pedologists use a color chart based on the Munsell color system.

This chart consists of 175 colored "chips" systematically arranged on cards according to *hue, value,* and *chroma*—three variables that combine to give all colors. Hue is the dominant spectral color based on the wavelength of reflected light. The Munsell color system includes all colors of the visible spectrum, but only the red (R), yellow-red (YR), and yellow (Y) portions of the spectrum are significant for describing most soil colors. Value is the relative darkness or lightness of the color on a scale of 0 to 10, pure black being 0 and pure white, 10. Chroma is the relative purity of the spectral color, and it increases with decreasing grayness, a pure color having a value of 20 and gray being 0 or neutral (N). Soils seldom have chromas exceeding 8. Various combinations of value and chroma produce a range of soil colors within each hue (table 7-5). Hues used in soil color classification are 10R, 2.5YR, 5YR, 7.5YR, 10YR, 2.5Y, and 5Y. As numbers increase within each letter combination, hue becomes increasingly yellow and less red.

TABLE 7-5 Effect of Hue on Descriptive Soil Color with a Constant Value and Chroma of 5/6

Munsell Color	Descriptive Color
10R 5/6	red
2.5YR 5/6	red
5YR 5/6	yellowish red
7.5YR 5/6	strong brown
10YR 5/6	yellowish brown
2.5Y 5/6	light olive brown
5Y 5/6	olive

Soil colors are described by Munsell letter-number combinations from color charts, and also by corresponding color names. Thus, a soil matching the chip on card 10YR numbered 3/2 is a "very dark gray brown (10YR 3/2)"; the hue is 10YR, value 3, chroma 2. Soil colors always are given for moist conditions, the usual field situation; colors for dry conditions are recorded whenever possible.

Most soil minerals are white or gray, but soil color often is affected strongly by the humus and iron compounds coating the particles. An exception is the A_2 horizon in some soils in which eluviation removes humus and iron minerals. Generally, humus gives a dark color to soil; as little as 2–3 percent may be enough to produce black soil. All humus, however, is not black; some high humus soils are brown or reddish in color.

Unweathered iron compounds, generally dark colored, oxidize into iron oxides with a range of colors from red to bluish gray; which color predominates is dependent on the type of iron oxide present. *Ferric oxide* or *hematite* (Fe_2O_3) is red; *hydrated ferric oxide* or *limonite* ($Fe_2O_3 \cdot nH_2O$) is yellowish brown; and *ferrous oxide* (FeO) is bluish gray. The ferric iron compounds—also called *trivalent*—and aluminum oxides (Al_2O_3) are *sesquioxides*. The type of iron oxide produced by soil weathering depends on availability of oxygen and moisture conditions. Oxidation produces ferric oxides in well-drained, well-aerated soils; hydrous ferric oxides occur where oxygen is still plentiful but the soil usually is moist, the variable amount of H_2O being indicated by the "n" in the formula; ferrous oxides form where drainage is poor and oxygen supply limited. Upon exposure to air, the ferrous oxides will oxidize further into ferric forms; sometimes in just a few minutes soil colors can change from bluish gray to brown or red.

Soil Mineralogy, Chemistry, and Fertility

In chapter 5 we examined some sources of plant nutrients found in soils, along with the processes by which plants obtain nutrients through exchange of cations and anions with the soil solution and with colloidal particles of clay and humus. If we exclude saline and some highly weathered soils, over 99 percent of the cations are *adsorbed*[5] on the colloidal surfaces. Anion exchange primarily takes place between plant roots and the soil solution; it involves negatively charged plant nutrient ions (table 7-6). Although a few cations may be present in the soil solution, most are adsorbed by colloids. The positively charged cations are held on surfaces of negatively charged soil particles called *micelles* (fig. 7-13). Since colloidal particles are extremely small (0.2 μm to 1.0 μm, or < 0.001 mm), they present an aggregate surface area many times that of coarser soil particles; this is an important factor in their ability to adsorb vast numbers of cations. The external surface area of

[5] *Adsorption* (not to be confused with *absorption*) is the concentration of ions or molecules on a surface; in soils, this contributes to cation and anion exchange capacity.

TABLE 7-6 Important Soil Elements

Element[a]	Symbol	Principal Ions
Aluminum	Al	Al^{+++}
Boron*	B	$B_4O_7^{--}$
Calcium*	Ca	Ca^{++}
Carbon*[b]	C	CO_3^{--}
Chlorine*[c]	Cl	Cl^-
Cobalt*	Co	Co^{++}
Copper*	Cu	Cu^{++}
Hydrogen*[b]	H	H^+, OH^-
Iron*	Fe	Fe^{++}, Fe^{+++}
Magnesium*	Mg	Mg^{++}
Manganese*	Mn	Mn^{++}, MnO_4^-
Molybdenum	Mo	MoO_4^{--}
Nitrogen*	N	NH_4^+, NO_2^-, NO_3^-
Oxygen[b]	O	(combined)
Phosphorus*	P	$H_2PO_4^-$, HPO_4^{--}
Potassium*	K	K^+
Silicon	Si	Si^{++++}
Sodium*[c]	Na	Na^+
Sulfur*	S	SO_4^{--}
Zinc*	Zn	Zn^{++}

[a]Starred elements are important plant nutrients; underlined elements are macronutrients.
[b]Carbon, hydrogen, and oxygen are obtained mostly from air and water.
[c]Formerly thought not essential to plants; now known to be essential in minute quantities. Sodium may partially substitute for potassium.

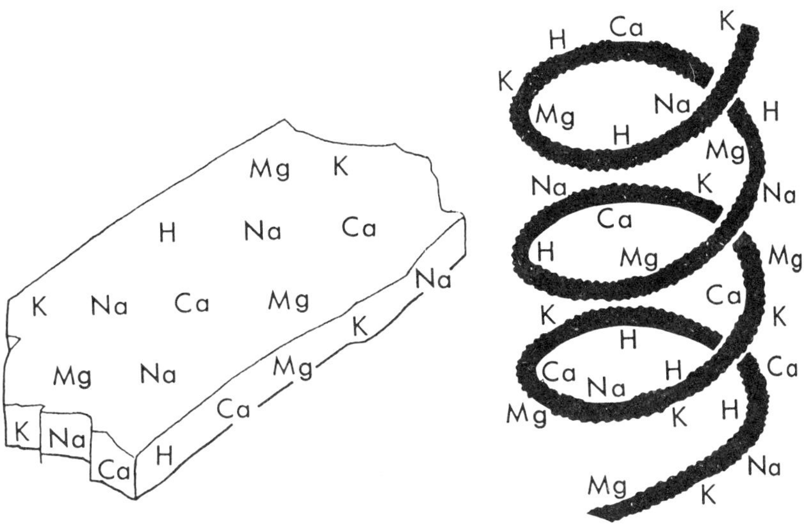

Fig. 7–13. On the left is a negatively charged colloidal clay particle (*micelle*) with attached (*adsorbed*) cations. On the right is organic matter, which serves a similar function. These materials act as giant anions adsorbing nutrient cations from the soil solution and holding them in exchangeable form available to growing plants through cation exchange, but not subject to loss through leaching (see fig. 5-10). (From Berger 1972.)

1 g of colloidal clay is approximately one thousand times that of 1 g of coarse sand, so that a layer of soil seven inches deep and an acre in extent has forty times the surface area of the state of Illinois (Brady 1974). Thus, most chemical activity in soil occurs in colloidal materials, and it may involve organic colloids or clay minerals.

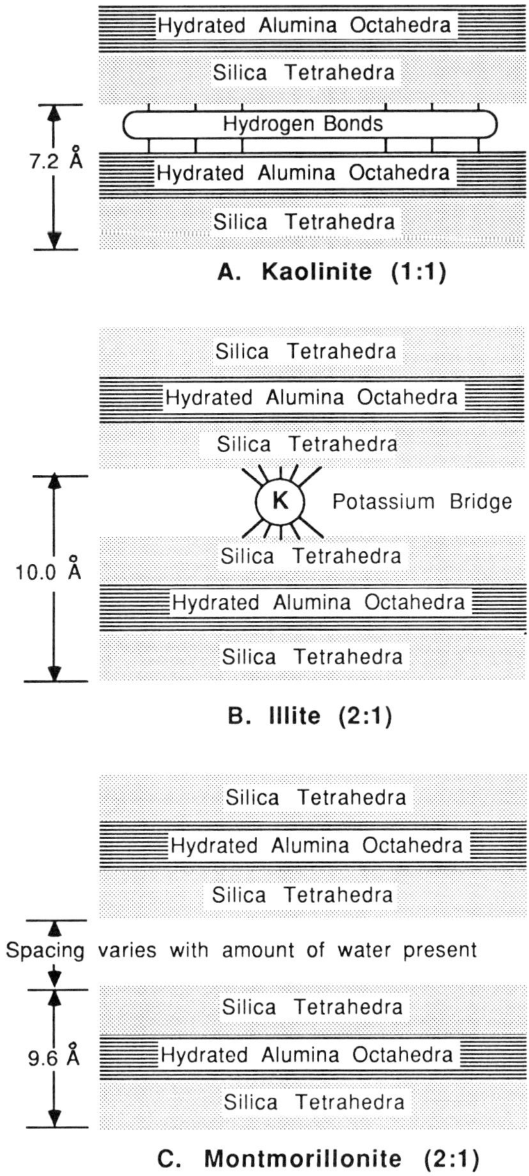

Fig. 7–14. Arrangement of alumina and silica sheets in 1:1 and 2:1 layer silicates: kaolinite, illite, and montmorillonite. (From USDA 1957.)

Clay Minerals in Soil

Not all clay particles are strictly colloidal in size since they may range up to 0.002 mm, but the colloidal fraction dominates. At one time it was thought that all clay minerals were *amorphous* (without crystalline structure), but development of x-ray analysis and the electron microscope demonstrate that most clay particles are definitely crystalline. Clay minerals fall into two groups; *silicate clays,* and iron and aluminum *hydrous oxide clays*. From the standpoint of plant nutrition, the most important are silicate clays. Among these, the ability to attract and hold cations varies greatly, depending on the mineralogical structure of the clay crystals.

Building blocks of clay crystals are *silica tetrahedra* and *alumina octahedra* arranged in sheets. A silica tetrahedron consists of a single silicon atom surrounded by four oxygen atoms forming a four-sided figure. When arranged in the same plane, they form a silica sheet. An alumina octahedron contains one aluminum atom surrounded by six oxygen or hydroxol ions to form an eight-sided figure. When joined in a plane they form an *alumina sheet*. Silica and alumina sheets bond through common oxygen atoms to form clay minerals and micas. Some clay minerals, like *kaolinite,* consist of alternating silica and alumina sheets in a 1:1 ratio, whereas others, such as *montmorillonite* and *illite,* consist of silica and alumina sheets bonded in a 2:1 ratio (figs. 7-14 and 7-15). The particular arrangement of the sheets determines the nature of

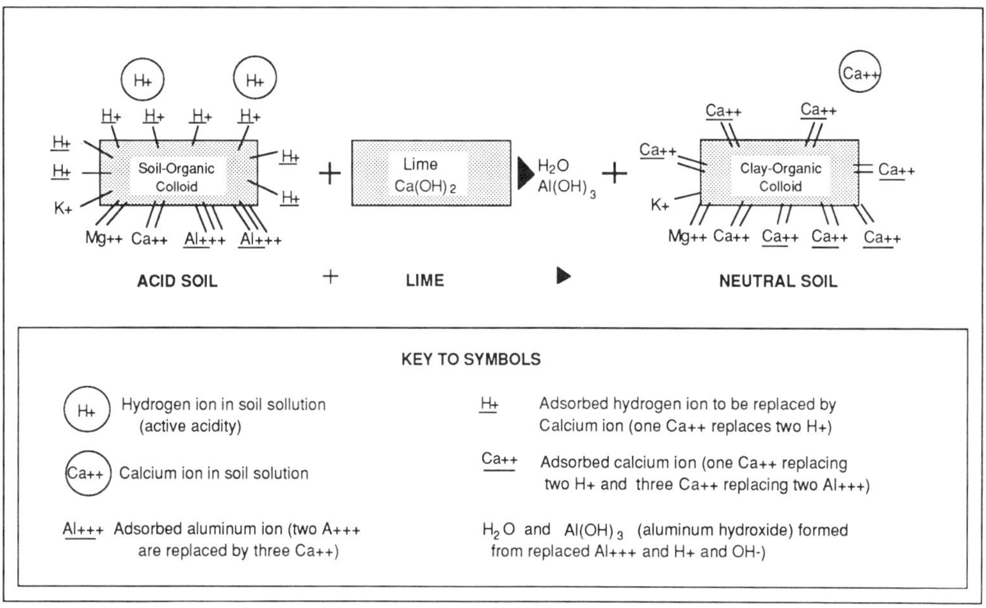

Fig. 7–15. Cation exchange reaction when an acid soil is limed. The hydrogen ions held by the clay-organic colloidal material are replaced with calcium ions from the lime. (From USDA 1957.)

the clay crystal and its ability to attract and hold nutrient cations in the soil.

Most clay micelles have a negative charge and they act as giant anions, attracting positively charged metallic ions, such as Ca^{++} and K^{+}, important in plant nutrition. This negative charge comes from *isomorphous substitution* in the crystal lattice, a process in which higher valence ions may be replaced in the crystal structure by lower valence ions of about the same size. For example, when Al^{+++} replaces Si^{++++} in the otherwise neutral silicate layer, the charge is unbalanced. This happens because the three positive charges of the aluminum ion cannot neutralize the four negative charges of the surrounding oxygen ions.[6] When Mg^{++} replaces Al^{+++} in the alumina layer, a like negative charge exists. This negative charge is satisfied by a swarm of cations that attach themselves to the clay crystal (fig. 7-13). These attached (adsorbed) cations can be replaced by others, either from the soil solution or plant roots.

Cation-Exchange Capacity in Soil

The total potential for exchanging cations in a soil, including both clay and organic colloids, determines its *cation-exchange capacity*. The term *base-exchange capacity* also may be used for this potential. The cation-exchange capacity of a soil is expressed quantitatively by *milliequivalents* per hundred grams of soil (meq/100g); a milliequivalent is the amount of material that will combine with or can replace a milligram of hydrogen. This measurement is valuable for comparing cation-exchange capacity of different soils.

Cation-exchange capacity varies considerably from one clay mineral to another, depending on the arrangement of the silica and alumina sheets and their manner of bonding. The major clay minerals are *kaolinite, chlorite, illite, montmorillonite, allophane,* and *vermiculite*. Kaolinites and montmorillonites are most abundant.

Kaolinite structure is such that bonding is tight enough to prevent water and ions from entering between layers of the crystal (fig. 7-14); therefore kaolinite is a *nonexpanding clay*. Also, the alumina and silica sheets tend to be neutral in charge and furnish few negatively charged attachment points for cation adsorption. Most adsorption occurs along broken edges of crystals where about equal numbers of positive and negative charges are unsatisfied because of disruption of the sheets. Kaolinite has the lowest cation-exchange capacity of any crystalline clay (table 7-7).

Montmorillonite is an *expanding clay mineral* (actually a group of clay minerals varying slightly in their ionic makeup). The theoretical formula is $Al_2Si_4O_{10}(OH)_2 \cdot {}_nH_2O$, but the actual formula is variable; some of the Al^{+++} may be replaced by Fe^{++} and Mg^{++}, and some Si^{++++} may be replaced by Al^{+++}. Note that in all of these substitutions a deficiency in positive charge results (2 for 3 and 3 for 4); therefore, the layers are negatively charged,

[6] Although the Si^{++++} is surrounded by 4 O^{--}, with a total negative charge of 8, four are shared with adjacent tetrahedra, thus binding them into the silica sheet.

TABLE 7-7 Cation exchange capacity of clay minerals and humus

Material	Crystal Lattice	Cation Exchange Capacity (meq/100g)
Kaolinite	nonexpanding	3–15
Chlorite	partly expanded	20–40
Illite	nonexpanding	20–40
Montmorillonite	expanding	60–100
Allophane	indeterminate	50–200
Vermiculite	partly expanded	100–200
Humus	amorphous	100–300

Source: Adapted from Thompson and Troeh 1978, 151.

giving montmorillonite a relatively high capacity to adsorb and hold cations. Note in figure 7-14 that spacing between layers of a montmorillonite crystal is variable, causing soils containing large quantities of montmorillonite clay to swell and shrink considerably as wetting and drying occur. This characteristic is the result of the very weak attraction of oxygen to oxygen between the sheets, called *van der Waal's bonding,* which is not strong enough to prevent water from entering.

Other clay minerals occur in varying amounts in soils, generally being mixed with kaolinite or montmorillonite. Illite is particularly common in association with the montmorillonite. Allophane is an ultrafine clay fraction that does not have an identifiable structure. Although it may occur in all soils, it is particularly prevalent in young volcanic soils, and it has a relatively high cation-exchange capacity. Table 7-7 lists some characteristics of crystalline clay minerals.

Humus is the other important source of cation-exchange capacity. Weight for weight, the exchange capacity of humus greatly exceeds that of clay minerals, but the percentage of organic matter is much less in mineral soils, seldom exceeding 5 percent. In some tropical soils composed mostly of oxide clays, $Al_2O_3 \cdot {}_nH_2O$ and $Fe_2O_3 \cdot {}_nH_2O$, most cation-exchange capacity is in organic matter. In some acid organic soils a high percentage of the exchange sites may be occupied by hydrogen cations rather than by plant nutrients.

Percent Base Saturation in Soils

Cation-exchange capacity is only one indicator of a soil's fertility or potential fertility. Actual fertility is determined by the number of exchange sites occupied by plant nutrients, such as Ca^{++}, Mg^{++}, and K^{+}, compared to those occupied by H^{+} and Al^{+++}, or other nonessential ions. A good measure of the actual storage of plant nutrient ions in a soil is its *percent base saturation,* defined as the percentage of the cation-exchange capacity actually occupied by metallic nutrient cations, excluding aluminum and acid-forming hydrogen ions (table 7-8). As one would expect, the pH of the soil solution

TABLE 7-8 Cation Exchange Capacity (CEC), Percent Base Saturation, and pH in the Plow Layer (Ap) for Selected Soils Under Diverse Environmental Conditions

Soil Order[a]	Location	Natural Vegetation	Precipitation		CEC in meq/100g	Percent Base Saturation	Total Nutrient Cations (meq/100g)	pH
			in	mm				
(1) Mollisol[b]	Minnesota	Prairie	28	711	49.9	92	45.90	7.2
(2) Mollisol	Iowa	Prairie	33	838	26.1	65	16.96	5.6
(3) Alfisol	Indiana	Deciduous Forest	45	1,143	13.9	68	9.45	6.9
(4) Alfisol	Washington	Mixed Forest	44	1,118	25.3	20	5.06	5.5
(5) Oxisol	Puerto Rico	Tropical Forest[c]	77	1,956	17.7	12	2.00	3.8
(6) Oxisol	Puerto Rico	Tropical Forest[c]	77	1,956	19.2	7	1.34	4.5
(7) Ultisol	Puerto Rico	Tropical Forest[c]	60	1,524	14.2	8	1.14	4.2
(8) Ultisol	Alabama	Mixed Forest[c]	55	1,397	3.2	34	1.09	5.7

Source: Soil Survey Staff 1960.
[a] See chapter 8 for definition of soil orders.
[b] Soil sample from a poorly-drained position; soil has a very high Ca content.
[c] Estimate of vegetation cover by author, based on climate and geographic location; no natural vegetation given.

is related to percent base saturation, with increasing pH indicating increasing percent base saturation.

Anion Exchange in Soils

Some nutrient ions form anions (negatively charged) in the soil solution, and their utilization by plants involves anion exchange. *Anion exchange capacity* is measured in the same way as cation-exchange capacity, except that the value is expressed in milliequivalents of anions per hundred grams of soil. The most prevalent nutrient ions in soils are NO_3^-, Cl^-, SO_4^{--} and $H_2PO_4^-$.

Developmental Age of Soils and Soil Minerals

As weathering and pedogenesis progress, soil minerals tend to disappear in a predictable order. In young soils, many primary minerals inherited from parent material are still present, but soluble salts, such as gypsum and calcite, may have been lost. In early maturity clay-sized mineral particles such as hornblende, biotite, and plagioclase feldspar, containing calcium, magnesium, and iron, have disappeared from the A horizon. Meanwhile, silicate clays, such as montmorillonite and illite, are forming and accumulating in B horizons. By late maturity most primary minerals are weathered from the clay and silt fraction, and the more complex silicate clays are changing into kaolinite, and iron and aluminum oxides are increasing. In old soils A horizons have lost almost all clay and silt, leaving mostly quartz sand; the B horizons contain concentrations of kaolinite and oxide clays. In general terms, as soil ages, subsequent progressive alteration and loss of soil minerals occurs and natural fertility and potential productivity decline.

Soil-Forming Processes

The most important soil-forming processes are (1) *podzolization*, (2) *calcification*, (3) *ferralization* or *latosolization*, (4) *salinization*, (5) *alkalization*, and (6) *gleization* and the formation of organic or *peat* soils. Some processes are reversible; *desalinization* and *dealkalization* can occur in some saline and alkaline soils. Some of these soil-forming processes are related closely to climate and vegetation, whereas parent material, topography, and drainage are more important in others.

Podzolization

Podzolization operates in areas with plentiful moisture and a vegetation cover of forest or heath. Mild podzolization also may occur in humid grasslands. Two phases are involved. First, surface accumulation of a mat of organic matter, called *mor*, appears. Organic acids form in this layer and facilitate leaching of clays and iron compounds from upper horizons. Second, the leached materials are deposited in lower horizons (fig. 7-16). For intense podzolization to occur, abundant moisture and loose, sandy parent material must be present under a forest cover—ideally pine—or under heaths (see

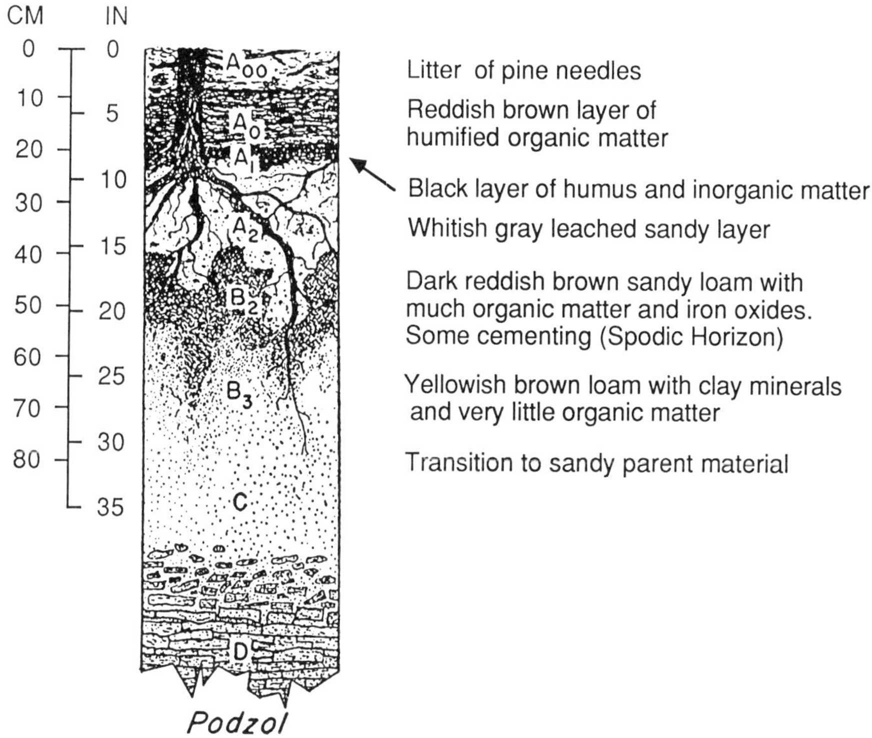

Fig. 7–16. A soil produced by podzolization in a cool, humid climate under a forest or heath. This soil is a Spodosol in the new soil taxonomy. (From USDA, *Climate and Man, Yearbook of Agriculture, 1941,* 273.)

chapter 5). Intense podzolization causes soil to become depleted of bases, grow more acid, and develop a sandy leached A_2 horizon from which most clay and organic matter has disappeared (fig. 7-16). It is this gray sandy A_2 horizon, often mixed with flecks of black humus from the A_1 horizon, which gives the appearance of ashes to a freshly plowed field; for this reason, the Russian word *pod,* meaning "under," and *zola,* "ash," combine to give *podzol* (fig. 7-17). The sesquioxides and eluviated colloids accumulate in the B horizon, giving it a brown color and sometimes cementing it into a hardpan called *ortstein.*

We should make a distinction between podzol soils and the process of podzolization: the true podzol is limited to certain combinations of vegetation and parent material, whereas the process is active in many soils that do not develop the true podzol profile. Such soils are termed *podzolic* rather than podzol. In less intensely podzolized soils, the gray sand layer is missing and the mor layer is replaced by a *mull* layer in which organic matter is more

Fig. 7–17. An area of heath podzol (Spodosol) on the sandy plains of western Denmark, south of Viborg in the valley of the Karup Aa. White sand from the bleached A_2 horizon has drifted in the furrows and in the lee of the planted windbreak. (Photo by W. E. Akin, March 1962.)

decomposed and mixed with mineral matter in the A_1 (fig. 7-7). In a podzolic soil the precipitated humus layer in the B horizon is not present, the accumulation of sesquioxides is less prominent, and silicate clays are more plentiful than in a true podzol. In contrast to the inherent low fertility of intensely podzolized soils (true podzols), moderately podzolized soils (podzolic) are likely to be much more fertile. If podzolization is defined in terms of acid leaching of bases, colloids, and sesquioxides from upper horizons, and their subsequent deposition (illuviation) in the lower profile, it is much more widespread than most descriptions indicate.

Often too much emphasis is placed on cool climates as requisite for podzolization and formation of true podzols. It is true that cool conditions favor the accumulation of mor, the acid surface horizon of litter and humus in a forest soil that aids podzolization. Nevertheless, where parent material and vegetation cover are favorable, and high rainfall assures a water surplus and moist soil conditions throughout the year, high temperatures do not prevent

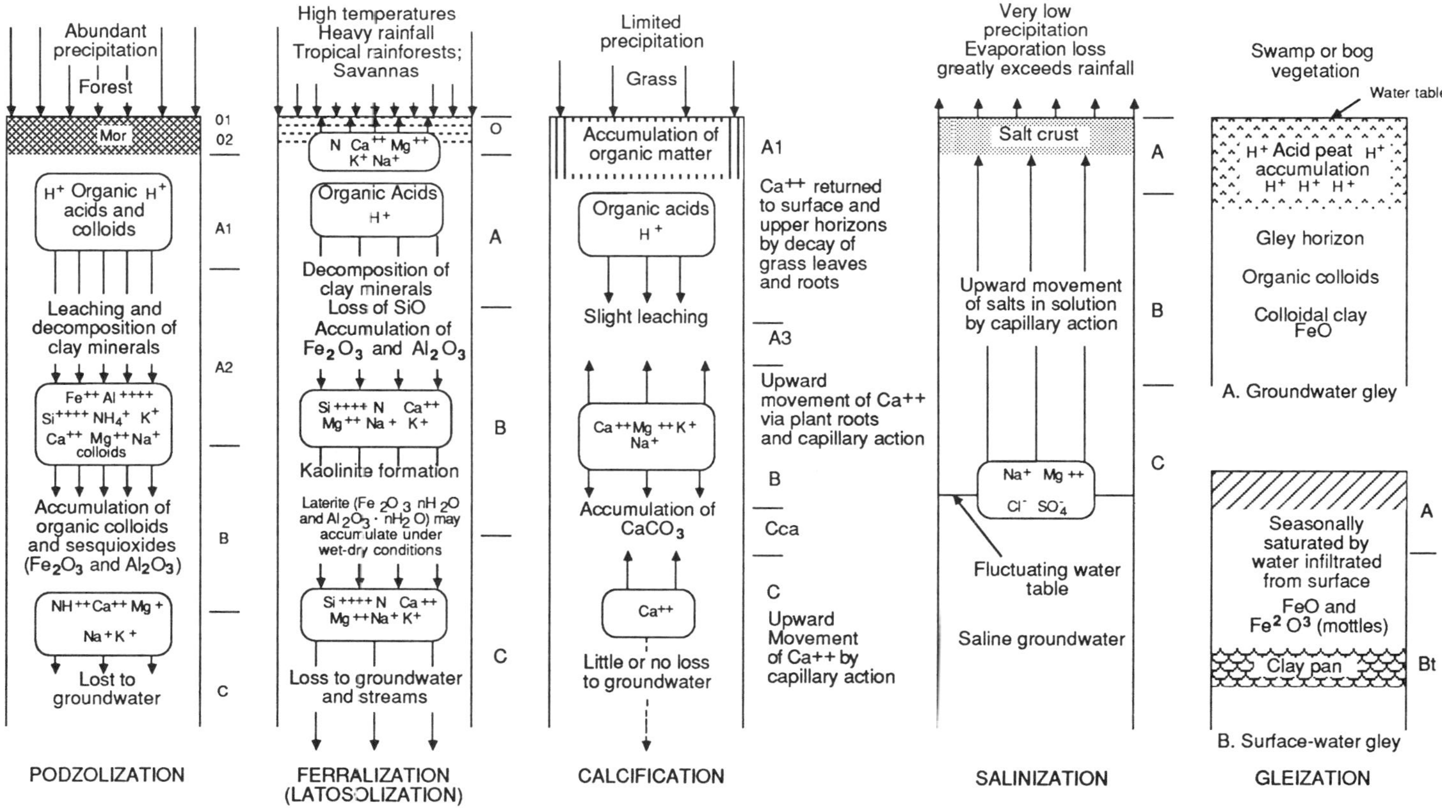

Fig. 7–18. The soil-forming processes.

podzolization, or even the development of true podzols. An example is the so-called "giant podzol" of Surinam which has an A_2 horizon consisting of almost 3 m (10 ft) of gray to white sand. This area has a mean annual temperature of 26°C (79°F) and mean annual precipitation of 2,300 mm (92 in). The present vegetation is savanna, although the original vegetation probably was forest (Buol, Hole, and McCracken 1980, 269).

Podzolization is most intense on sandy soils under certain types of vegetation cover, such as hemlock (*Tsuga canadensis*) or heather (*Calluna vulgaris*), that has low nutrient requirements. Consequently, unused nutrients leach rather than recycle through plants and do not return to the surface. In the boreal forest where true podzols are most extensive, they are often best represented under pine on sandy outwash materials (fig. 7-18). Spruce is more demanding of soil nutrients and dominates areas less sandy and more fertile. Under spruce the rate of podzolization is slower, and many soils are more properly podzolic rather than true podzols. Even more nutrient-demanding are broadleaf deciduous forests where generally soils are podzolic but not podzols. Some of the most intense podzolization has occurred on cleared and cultivated land abandoned to heaths, such as those in western Denmark described in chapter 5. Many of these heath podzols have strongly developed iron pans (*ortstein*) in the B horizon.

Calcification

Under drier conditions, grasses and shrubs replace forests, and in middle latitudes calcification replaces podzolization (fig. 7-18). As the term implies, calcification produces a soil sufficiently high in calcium to give a neutral to basic reaction. Both climate and vegetation contribute to this process. Low rainfall means less leaching of upper horizons, and bases removed from the A horizon concentrate in the B or C horizons instead of disappearing from the profile. Consequently, an accumulation of $CaCO_3$ appears at a depth marking the average penetration of percolating rain water. The more soluble gypsum ($CaSO_4 \cdot 2H_2O$) sometimes accumulates in drier soils at a depth greater than the calcium carbonate. Only under very dry conditions will the highly soluble sodium chloride (NaCl) accumulate in the soil horizons. Under some conditions the $CaCO_3$ layer may be dense enough to form a hardpan (*caliche*).

Vegetation plays a role in the process of calcification. Grasses, in particular, are efficient in absorbing calcium cations and then returning them to the upper horizons. When the plant dies, calcium returns to the soil, thus counteracting the leaching process, and the humus-rich A_1 horizon generally is neutral or, at most, mildly acid. In humid grasslands, although soils are acid in reaction and the pedogenic process is akin to podzolization, the return of calcium to the upper part of the solum by grasses assures a higher pH and higher fertility.

Ferralization (Latosolization)

Pedogenic processes in tropical areas still are not understood completely and remain controversial. Confusion occurs, in terminology—not only in those terms used to describe processes, but also in terms for the soils they produce. *Laterization, latosolization,* and *ferralization* are used to describe the process, and *laterite, latosol, ferralite,* and *oxisol* are terms applied to soils. Ferralization refers to the processes that produce and concentrate sesquioxides of iron and aluminum in soil. Within the tropics, a high concentration of iron oxide and a low organic content result in a dominance of soils with red or yellow colors. The term *ferralization* seems preferable to *laterization* for describing this process, since the latter implies the eventual formation of a *laterite* (Latin, "brick"), a material that, upon exposure, hardens irreversibly. The term *laterite* is synonymous with the newer term *plinthite.* Laterite is by no means present or in the process of developing in all tropical soils where ferralization is occurring.

Under heavy rainfall and high temperatures of the tropics, weathering processes are intense, producing deep soil profiles from which most plant nutrients are leached or, as in the case of the phosphorus described above, adsorbed by iron and aluminum oxides or hydroxides and made unavailable to plants. Initially, the pedogenic process would be akin to podzolization, with decomposition of clay minerals and leaching of by-products from the A horizon. But because of high temperatures, heavy rainfall, and lack of a dormant period with frozen soil, the processes are much more intense. In extensive tropical areas, particularly in Africa and South America, soils on the incredibly old, stable geomorphic landscapes have not experienced the rejuvenating effects of glaciation or recent volcanic activity. Silica has been removed from clay minerals, and the remaining aluminum and iron have oxidized into sesquioxides that make up a significant percentage of soil materials. Clay minerals weather into kaolinite, or they may lose all of their silica to form oxide clays (fig. 7-18). In some tropical soils the concentration of oxides may be so high that they can be used as iron and aluminum (*bauxite*) ores.

Where severe dry seasons occur, soils are dry for part of the year and strongly leached for the remainder, with water tables fluctuating accordingly. Plinthite or laterite accumulates in the horizon of the fluctuating water table as a soft material high in sesquioxides. If erosion subsequently exposes the laterite horizon to air, it will harden to a permanent laterite crust. Most laterite crusts occupying surface positions in some tropical areas may be ancient subsurface horizons now exposed by erosion.

Tropical soils have a reputation for low fertility—a reputation that, with some important exceptions, seems deserved. Oxide clays and kaolinite have low cation-exchange capacity, and intense leaching removes most available nutrients from the soil solution. In chapter 6 we discussed the rapid cycling of plant nutrients within the biomass of the rainforest, from decaying organic matter to living plants, before loss through leaching. Therefore, most exchangeable nutrient ions in many tropical soils are bound up in organic matter. When vegetation is removed to make way for cultivation, the cycle is broken, organic matter quickly oxidizes, and nutrients are lost through leaching. Experience shows that with clearing and cultivation, fertility in tropical soils usually declines rapidly. This has led to the development of shifting cultivation in which a clearing is made, farmed a few years, and then abandoned. Which is more important in forcing the primitive agriculturist to move to a new field, decline in fertility or rapid invasion of weeds? Fortunately, small clearings heal in a few years and the forest regenerates. Extensive clearing, however, is another matter. In many areas, after extensive clearing, cultivation, and eventual abandonment, regeneration does not occur.

Tropical soils often have good structure permitting tillage soon after heavy rainfall. This characteristic causes some planners to speak with eloquent optimism about the potential role of tropical soils in world agriculture. But the fact that these soils are tillable almost immediately after tropical downpours is the result, in part, of the absence of expanding clay minerals such as montmorillonite. Moreover, the soils lack the high cation-exchange capacity that these clay minerals would impart. Fertilizers can be added, but the high content of sesquioxides tends to render phosphates in fertilizers insoluble and unavailable to plants. The process is called *phosphate fixation*, illustrated by the following reaction between soluble aluminum and phosphate (Buckman and Brady 1974):

$$\underset{}{Al^{+++}} + \underset{\text{(soluble)}}{H_2PO_4^-} + 2H_2O \leftrightarrows 2H^+ + \underset{\text{(insoluble)}}{Al(OH)_2H_2PO_4}$$

A similar reaction occurs between phosphate and iron cations in solution, as well as with insoluble hydrous oxides. These reactions take place over a wide range of pH, but they may be reversed by the presence of certain humic acids from organic matter. However, rapid loss of organic matter makes this reverse process less significant in the tropics than in middle latitudes.

Relatively fertile tropical soils may be found where new materials, such as alluvium or volcanic ash, accumulate. Young materials may have enough base concentration to weather into montmorillonite, giving a much higher cation-exchange potential than is characteristic of many older tropical soils. Commercial tropical agriculture often concentrates on such soils as these. Unfortunately, large areas within the tropics have very old land surfaces, and over millions of years the pedogenic processes have taken their toll on fertility.

Salinization

Saline soils are those with a high content of soluble salts, including those of sodium (chloride, sulphate, bicarbonate) and calcium (sulphate and chloride), and a pH between neutral and 8.5. Salts move upward in the soil profile through capillary action and may form a white crust on the surface; thus, they have been called *white alkali* soils (fig. 7-18). More recently, the Russian term *solonchak* has been applied to saline soils. A good rain may temporarily remove the salt crust by solution. If the water table is deep, capillary movement may not reach the surface, and salts may accumulate in a *salic* horizon within the profile.

Since these white salts form neutral solutions, pH may be surprisingly low in some soils, and there may be a low concentration of exchangeable sodium ions. Colloidal clays and organic matter usually are well flocculated in saline soils, and if the water table is lowered the porous structure assures good internal drainage. If drainage is improved, the excess salts may leach through application of irrigation water or from increased precipitation. The process of salt removal in solution is *desalinization.*

Alkalization

If sodium carbonate is present, exchangeable sodium ions associate with colloids and the colloids disperse; thereby, deflocculation destroys the soil structure, producing a structureless, impervious soil. Because of dispersed organic matter, the soil may be black in color, accounting for the term *black alkali,* a name sometimes applied to these soils. The Russian term *solonetz* is preferred. A soil horizon in which sodium ions dominate the clay-humus complex is a *natric* horizon.

Because of their impervious character, solonetz soils leach slowly and reclamation is difficult. With improved drainage, however, and with increased irrigation or precipitation, a solonetz soil can become leached enough to permit successful cultivation. The process by which the sodium ions are removed from the clay-humus complex is *solodization*; the soils produced range from a *solodized solonetz* to a *soloth.* In a solodized solonetz the hard clayey layer, (natric horizon) is overlain by a light-colored leached layer, which may be neutral or slightly acid in reaction. The end product of solodization is the soloth, a soil with a light-colored acid surface horizon over a heavy dark brown B horizon.

Peat Accumulation and Gleization

Where the water table is on or above the surface, as it is in swamps and bogs, the remains of plants build up an accumulation of organic debris known as peat. Because *anaerobic* conditions prevail (free oxygen is not present), decomposition of vegetable matter is retarded and peat layers of great thickness may accumulate. When this progresses to a more advanced state of decomposition with a higher mineral content, it is called *muck.* Collectively, in older classifications these highly organic soils are known as *bog soils*; they are *histosols* in the new USDA soil taxonomic system (see chapter 8).

Where the profile consists of a peaty layer over a mineral layer, the soil is a half-bog soil. The character of the mineral horizon in a half-bog soil is produced by the process of *gleization* in which the ferric iron oxide (Fe_2O_3) is reduced to the ferrous compound (FeO), giving the material a gray to bluish-gray color. Gleying may occur in upland soils in which internal drainage is retarded by clay pans, causing the upper horizons to be saturated for long periods (fig. 7-18). The gley horizon in these soils usually is mottled by rust-colored stains where air penetrates along roots and cracks during times when the soil is not saturated, and it has oxidized the FeO to Fe_2O_3. Upland soils with gleyed horizons have been called *wisenboden, humic-gley,* and *planosols*. In the USDA taxonomy they are classified under a number of moist suborders of *Alfisols, Mollisols,* and *Inceptisols* (see chapter 8).

The general problem of classification . . . may be stated as follows: To provide that things shall be thought of in such groups, and those groups in such an order, as will best conduce to the remembrance and to the ascertainment of their laws.

JOHN STUART MILL, *A System of Logic*

Chapter 8

CLASSIFICATION AND GLOBAL DISTRIBUTION OF SOILS

Most modern soil classification systems derived from concepts developed in Russia in the late nineteenth century. In 1914, European soil scientists became aware of these concepts when a summary appeared, written in German, by K. D. Glinka. This text was translated into English in 1927 by C. F. Marbut, an official of the U.S. Department of Agriculture, thus breaching the linguistic barrier between Russian and American soil scientists (Glinka 1927).

Marbut, educated both as a geologist and as a geographer, joined the U.S. Soil Survey in 1910, first as consultant, and later as its director. He published his system of soil classification in 1927, and developed it more fully in 1936. We introduced his basic concept of pedocals and pedalfers in chapter 7 (Marbut 1936). Extensive modification of the system followed in 1938 (Baldwin, Kellog, and Thorpe 1938). Modification continued in subsequent years (Thorp and Smith 1949) and, until the present Comprehensive System of Soil Taxonomy replaced it in 1965, it remained the basis for all soil classification made by the U.S. Department of Agriculture (table 8-1).

It is not surprising that Russian soil scientists were the first to formulate broad concepts of climate-vegetation-soil interrelationships. The vast expanse of old imperial Russia with its great variety of physical environments furnished an ideal laboratory for highly trained soil scientists to observe environmental impacts on soil genesis. Only the United States, and possibly China, possess a comparable diversity of climate, vegetation, and soil landscapes; undoubtedly, this mutual diversity has been a contributing factor in American soil scientists having rivaled the Russians in pedological contributions. Canadians and Australians have made notable advances in soil classification, since their countries also encompass large territories with varied landscapes. British soil scientists always have maintained a global interest in soils, particularly those of the tropics, in part because of once vast colonial interests.

In this chapter we consider two soil classification systems: the 1938 classification by the U.S. Soil Survey, as modified up to 1965, and the 1965 *Comprehensive System of Soil Taxonomy* (formerly the *7th Approximation*), now used by the Soil Survey Staff. Although in the United States the older system has been replaced by the newer one, the older one, incorporating concepts and terminology inherited from the Russian school, is still recognized and used as the basis for some current classifications. Geographers find the 1938 system compatible with many of their global concepts of climate, vegetation, and soils, so they often are reluctant to abandon it in favor of the newer, more complex taxonomy. Some current textbooks and other publications still contain the 1938 system, and knowledge of its terminology and concepts is essential to comprehend the literature on soils and physical geography published before 1965.

In the same vein, since the new system is gaining worldwide recognition, a familiarity with its broader aspects is desirable; for example, some names of mapping units used for the FAO/UNESCO World Soil Map derive from the new U.S. system, others from the 1938 system, and still others are unique. The Canadian, British, and Australian systems continue to use some terminology based on Russian concepts of pedology, and they use many categories of the 1938 system (table 8-2). It should be stressed, however, that definitions of terms common to all systems may vary between systems.

The 1938–49 Classification

The 1938 soil classification system of the U.S. Department of Agriculture is summarized in *Soils and Men, The 1938 Yearbook of Agriculture,* and an updating of modifications to the system appeared in 1949 (Riecken and Smith 1949, 107–15; Thorp and G. D. Smith 1949, 117–26). Further modifications were made before the introduction of the new system in 1965.

The 1938 scheme, an outgrowth of Marbut's earlier system, draws heavily on Russian concepts and uses much Russian terminology. Soils are classified according to *order*, *suborder*, *great soil group*, *family*, *series*, *type*, and *phase*. While the system was in use, the family category was never fully developed. Soils may be grouped into *associations*, when they are found in the same area, or into *catenas*, if all associated soils have developed from the same parent materials (see chapter 7 for examples of associations and catenas).

Table 8-1 summarizes the higher categories of the 1938 classification through the level of great soil groups, as modified through 1959. Lower categories, too numerous to list here, consist of some ninety-five hundred soil series currently recognized in the United States. A single soil type, such

TABLE 8-1 The 1938 Soil Classification System of the United States Department of Agriculture, Soil Conservation Service, with Revisions and Additions through 1959

Order	Suborder	Great Soil Group
Zonal soils	1. Soils of the cold regions	Tundra soils
	2. Light-colored podzolized soils of the timbered regions	Podzol soils
		Gray wooded soils
		Brown Podzolic soils
		Gray-Brown Podzolic soils
		Red-Yellow Podzolic soils
		Terra Rossa
		Sols Bruns Acides
	3. Laterite soils of forested warm-temperate and tropical regions (Latosols)	Reddish-Brown Lateritic soils
		Yellowish-Brown Lateritic soils
		Laterite soils
		Latosols
		Humic Ferruginous Latosols
		Humic Latosols
		Low Humic Latosols
		Latosolic Brown Forest soils
	4. Soils of the forest-grassland transition	Degraded Chernozem
		Noncalcic Brown soils
	5. Dark-colored soils of semiarid, subhumid, and humid grasslands	Chestnut soils
		Reddish Chestnut soils
		Chernozems
		Brunizems (Prairie soils)
		Reddish Prairie soils
	6. Light-colored soils of arid regions	Desert soils
		Red Desert soils
		Sierozems
		Brown soils
		Reddish-Brown soils
Intrazonal Soils	1. Halomorphic (saline and alkali) soils of imperfectly drained arid regions and littoral deposits	Solonchaks (Saline soils)
		Calcium Carbonate Solonchaks[a]
		Solonetz soils
		Soloth soils
	2. Calcimorphic soils	Calcisols
		Rendzina soils
		Brown Forest soils
		Calcium Carbonate Solonchaks[a]
	3. Volcanic ash soils	Ando soils[b]
	4. Dark-colored expanding clay soils of wet-dry climatic regimes	Grumusols
	5. Hydromorphic soils of marshes, swamps, seep areas, and flats	Humic Gley soils (wisenboden)
		Alpine Meadow soils
		Bog soils

TABLE 8-1 (continued)

Order	Suborder	Great Soil Group
		Half-Bog soils
		Low-Humic Gley soils
		Planosols
		Ground-Water Podzol soils
		Ground-Water Laterite soils
		Hydrol Humic Latosols
Azonal Soils	(no suborders)	Lithosols
		Regosols (includes Dry Sands)
		Alluvial soils

Sources: Thorp and Smith 1949, 118; Soil Survey Staff 1975, 433–35.

[a] Calcium carbonate solonchaks are included under both halomorphic and calcimorphic suborders.

[b] Ando soils have been considered a zonal soil group but the dark color seems to be the result of a combination of allophane, a common weathering product of volcanic ash, with humus to form dark-colored material. Since parent material rather than climate determines soil characteristics, and since Ando soils occur in several climatic regimes, they probably should be considered intrazonal.

TABLE 8-2 Soil Mapping Units for the FAO/UNESCO Soil Map of the World

Major Units	Soil Characteristics
Fluvisols	Recent alluvial soils with little or no diagnostic horizons
Rhegosols	Soils formed from unconsolidated materials except recent alluvium, with little or no diagnostic horizons
Arenosols	Sandy soils
Gleysols	Soils with gley horizons
Rendzinas	Shallow soils over limestone
Rankers	Shallow soils over noncalcareous materials
Andosols	Volcanic ash soils with dark epipedons
Vertisols	Cracking clay soils
Ermosols	Desert soils; called Yermosols in later revisions
Xerosols	Dry soils of semiarid regions
Halosols	Solonetz and Solonchak; natric and salic horizons
Planosols	Soils with abrupt textural change at A-B contact
Castanozems	Chestnut colored surface; soils of the steppes
Cherozems	Black colored, high humus soils of steppes and prairies
Phaeozems	Dark soils that are more leached than Chernozem or Castanozem
Cambisols	Soils that have been greatly changed by weathering processes
Luvisols	Soils with argillic B horizons and medium to high base saturation
Acrisols	Highly weathered, low base status, and argillic horizons
Paramasol	Soils of the high Andean grasslands or *paramos*
Podzol	Podzol soils
Ferralsols	Soils with high content of sesquioxides in the clay fraction
Histosols	Organic soils
Lithosols	Shallow soils over hard bedrock

Source: Dudal 1968.

as a Marshall silt loam of western Iowa, would be classified from higher to lower category as follows:

Order: Zonal soil
Suborder: Dark colored soil of semiarid, subhumid, and humid grasslands
Great Soil Group: Brunizem (prairie soil)
Series: Marshall
Type: Marshall silt loam

The same soil can be further classified according to phase if it deviates from the normal type because of erosion, slope, stoniness, or soluble salt content, for example, Marshall silt loam, steep phase.

Soil Orders (1938): Zonal, Intrazonal, and Azonal

Zonal Soils. These soils are associated primarily with climate and vegetation regions (fig. 8-1). They occur where local features, such as drainage and slope, permit maximum influence of climate and vegetation on soil-

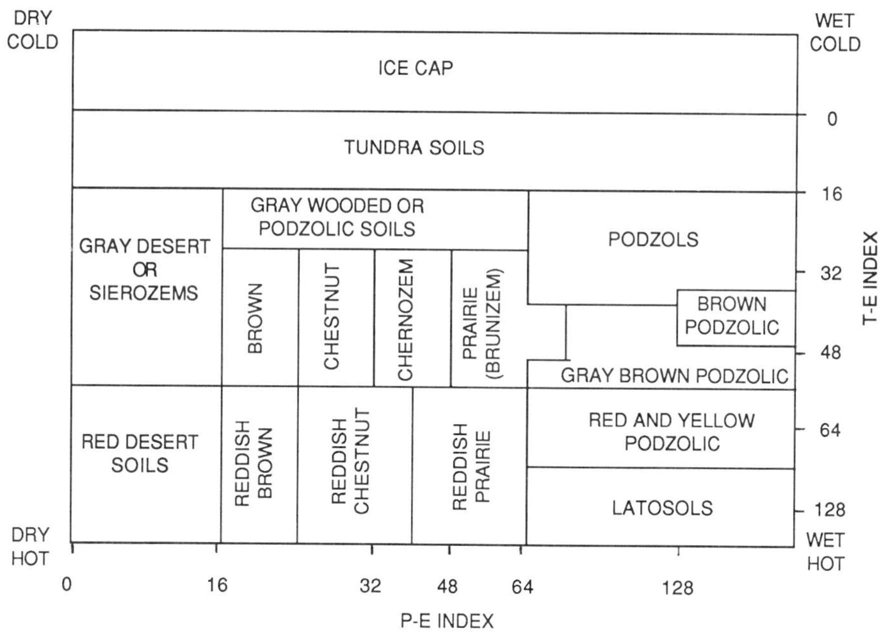

Fig. 8–1. A schematic continent showing the relationship between great soil groups of the revised 1938 soil classification system and Thornthwaite's precipitation and temperature efficiency indices. (Modified from D. I. Blumenstock and C. W. Thornthwaite, "Climate and the World Pattern," in *Climate and Men, Yearbook of Agriculture, 1941,* Washington, D.C.: USDA, 1941:98–127.)

forming processes. Most zonal soils occupy upland positions with just enough slope to ensure good drainage, but not enough to produce rapid erosion that would prevent the development of a mature profile. Zonal soils occupy extensive areas; they may include soils formed from a variety of parent materials, which have been subjected to the pedogenic processes. The use of terms that emphasize either climate or vegetation reflects this concept of zonality; for example, "soils of the cold zone" and "soils of the forested warm-temperate and tropical regions" (table 8-1).

Intrazonal Soils. These soils form where the soil-forming process is dominated by some local condition, such as poor drainage (*hydromorphic soils*), the presence of saline or alkali conditions (*halomorphic soils*), or high lime content of the parent material (*calcimorphic soils*). Such soils are not confined to a single zone but extend across zonal boundaries.

Azonal Soils. These soils have not developed profiles in which horizons are well differentiated. Horizon differentiation may be lacking in very young soils formed in alluvium, or on steep slopes where rapid erosion prevents horizon development. Thin soils over bedrock are *Lithosols* (fig. 7-4), while young soils formed in deep unconsolidated materials, such as dry sand or loess, are Regosols (from *regolith*, the loose material overlying solid bedrock). Alluvial soils, found on flood plains of rivers, are important to agriculture, especially in extensive flood plains, such as those of the lower Mississippi River Valley and the Central Valley of California (fig. 8-2).

Zonal and Intrazonal soil orders subdivide into suborders, but suborders are not recognized in Azonal soils. Each suborder contains one or more great soil groups, which can be associated with one of the soil-forming processes outlined in chapter 7. The names of some of these great soil groups derive from the chief pedogenic process that is thought to produce the soil; for example, Podzol and Podzolic (podzolization), and Latosol (laterization). Other great soil groups have names relating to the climate or vegetation under which they developed—Prairie soils, Red Desert soils, and Tundra soils. Soil color is indicated in other names, as in Chernozem (black soil), Brunizem (brown soil), Chestnut, Reddish-Brown. This lack of consistency in naming soil groups is a weakness of the 1938 system.

Zonal Soils and Their Distribution

Tundra Soils of Cold Regions

Tundra soils are associated with the treeless, but ice-free, arctic and subarctic regions, and some high alpine areas. Native vegetation consists of low-growing lichens, mosses, flowering plants, and shrubs. Subsoil is often permanently frozen, internal drainage is poor, and alternate freezing and thawing of the horizons above permafrost cause mechanical mixing of soil materials. Under these conditions the major soil-forming process is gleization (fig. 7-18). The presence of permafrost distinguishes tundra soils from

Fig. 8–2. A gully wall in a flood plain in Ellsworth County, Kan. reveals a sequence of soil horizons buried by repeated deposition of sediments during floods. The dark bands are minimal A_1 horizons that have had their development interrupted by deposition of new sediment. Minimal horizon development is a characteristic of the order of Entisols in the new soil taxonomy of the Soil Conservation Service. This soil is classified as a Typic Ustifluvent. In the 1938 system of soil classification it is an Alluvial soil. (U.S. *Soil Conservation Service photo by Mc Lean,* n.d.)

their Intrazonal counterparts; thus, where permafrost is absent, the soils in areas of tundra vegetation are Intrazonal.

Generally, tundra soils have a peaty layer covering a bluish gray gley horizon, and they are very similar to Humic Gley and Half-Bog soils classified in the Intrazonal order. Tundra soils occupy extensive areas in Canada, Alaska, and the Soviet Union where limited portions are pastures for reindeer and muskox, a few hardy crops, and hunting and trapping of wild game, but they contribute little to world agricultural production.

Podzolized Soils of Cool Regions

Soils that are wholly or partly the product of podzolization have wide distribution. Podzolization is most intense in areas of cool, humid climates, under a vegetation cover of coniferous trees, particularly pine, or under heath. Exceptions occur however; well-formed Podzols have been reported even in hot, humid tropical lowlands (see chapter 7). The most extensive areas of true Podzols occur in boreal forests where climate is subarctic (*Dc* or *Dd*). Even here, one finds soils that formed beneath more nutrient-demanding species, such as some varieties of spruce or broadleaf trees, or on clay soils, more akin to Podzolic soils than to true Podzols. Extensive areas of Intrazonal and Azonal soils are common within the zone occupied by Podzols; especially common are Peat and Bog soils of the hydromorphic suborder.

Heath Podzols of northwestern Europe are human-induced (anthropic); that is, when people cleared the forest from sandy outwash plains for cultivation and pastures and then abandoned them, instead of the original forest returning, heath vegetation moved in (see chapter 5).

Podzolized Soils of Temperate Forested Regions

Less drastically podzolized soils occur in areas of humid continental (*Da* and *Db*) and marine (*Cb*) climates, where the natural vegetation is broadleaf deciduous forests. The great soil group division here is the Gray-Brown Podzolic soils, typical in large areas of the United States east of the Mississippi River and north of the Ohio River. They also are found in some western states, including western Montana and northern Idaho, and the coastal and valley sections of the northern Pacific coast, south to San Francisco. Gray-Brown Podzolic soils occur in most of Europe except along parts of its eastern, northern, and southern margins; they are the major zonal soils of the central region of China and northern Japan.

Although only moderately fertile under natural conditions, the Gray-Brown Podzolic soils respond well to good management and can be very productive. Climates under which these soils are most extensively developed are those favorable for agriculture; notably, in the eastern corn belt of the United States, in the highly productive agricultural areas of Western Europe, and in areas of intensive agriculture in China and Japan. In some parts of Asia and Europe, centuries of intensive cultivation have altered the structure of the soils until now they bear little resemblance to original zonal soils. Generally, in Western Europe good soil management practices, such as liming and manuring, have improved the cultivated soils, and maintenance of permanent grass pastures has produced soils often more akin to those of grasslands than of forests. The artificially constructed soils of Asian rice paddies are a separate case, which cannot be classified in this system.

Soils of the Humid Subtropics

Under warmer conditions in the humid subtropics (*Ca*), and in some areas of the marginal tropics (*Am*, *Aw*, *Af*), where the native vegetation consists of broadleaf deciduous trees, broadleaf evergreens, and mixed forests, soils have been subjected to pedogenic processes transitional between those of the cooler middle latitudes and those of the tropics. Consequently, soils have characteristics transitional between those produced by podzolization and laterization. In the 1938 classification, these are the Red and Yellow Podzolic soils, the dominant color being determined by low humus content and the presence of iron oxides. In Red Podzolic soils, iron oxides are less hydrolized than in Yellow Podzolic soils.

Red and Yellow Podzolic soils are typical of the old cotton belt of the southeastern United States. Natural fertility is considerably lower than that of most Gray-Brown Podzolic soils, but under good management and fertilization these soils may be quite productive. In addition to cotton, they support a wide variety of crops, including corn, soybeans, tobacco, peanuts, and sweet potatoes. Today, large areas of worn-out cotton-growing soils are in pastures and woodlots; others are in natural or planted commercial forests of pine for lumber, pulp, and naval stores (see chapter 6).

Soils of the Humid Tropics

As world population increases, more attention is directed toward the vast expanses of uncultivated land lying between the tropics of Cancer and Capricorn. The humid tropics (*Af*, *Am*, *Aw*) contain large areas with low population densities, in particular the Amazon basin of South America, Equatorial Africa, and some areas on the mainland and islands of Southeast Asia. Some locations within the humid tropics support high population densities, notably the Indian Peninsula, the island of Java, and the great river deltas of Burma, Thailand, and Indochina, and some areas of West Africa. At least a part of this uneven distribution of population within the humid tropics can be attributed to differences in soil productivity; but historical, cultural, political, and economic differences may be of equal importance. The exact contribution of soils to the agricultural development of the tropics is difficult to ascertain because tropical soils are poorly understood and even more poorly mapped.

According to the 1938 system, soils of the humid and subhumid tropics are grouped under the suborder of Latosols—that is, soils formed by the process of laterization (ferralization) described in chapter 7. Mature soils have low cation-exchange capacity and low base saturation, and they often are red or yellow in hue. Once these soils have been cleared of natural vegetation, they rapidly lose both plant nutrients through leaching and humus by oxidation. Artificial fertilizers, which are applied so successfully to other soils, may be less effective here. In addition to the lack of cation-exchange capacity for holding nutrients out of solution and thus retarding loss by leaching, there is a tendency for phosphates to combine with the sesquioxides to form insoluble iron and aluminum phosphates, making phosphorus unavailable to plants. In contrast to their low fertility, most tropical Latosols have good structure because they are low in plastic clays and almost devoid of expanding clays; thus, many tropical soils can be cultivated immediately after heavy and prolonged rainfall without difficulty or damage to their structure.

Most world soil maps based on the 1938 system show a large part of the humid tropics occupied by Latosols or Lateritic soils (fig. 8-3). This is an oversimplification resulting from lack of detailed knowledge and a tendency to equate Latosols and Laterites with humid tropical climates and tropical forests and savannas. As more is learned about tropical soils, it is becoming apparent that they show as much variability as soils outside the tropics. For example, soil surveys in the Amazon region of Brazil indicate the presence of yellow Latosols, dark red Latosols, red and yellow quartz sands, soils with lateritic concretions, red and yellow Podzols, *Terra Preto do Indio*, *Terras Roxas*, Grumusols, Ground Water Laterites, Gley, Low-Humic Gley, alluvial soils, Solonchaks, Solonetz, and organic soils (Falesi, 1974).[1] All of these Zonal, Azonal, and Intrazonal soils occur within the area of South America mapped as Latosolic soils in figure 8-3.

Within large areas of the humid tropics, traditional farmers practice shifting cultivation in which they clear small areas in rainforests, burn the debris and plant among the stumps. Because soil fertility declines rapidly and weeds become a problem, they abandon clearings after a fallow period of one to four years and prepare new ones. Abandoned clearings require a fallow period of ten to twelve years; during this time fertility slowly rebuilds under second-growth forests. It is obvious that such a system will not support a dense population, but some researchers believe that this may be the only method

[1] *Terra Preto do Indio*, "black soil of the Indian," is a distinctive area of black soil, possibly an accumulation of organic debris associated with ancient Indian settlements. There is some evidence for this since many artifacts are found associated with these soils. *Terras Roxas*, "purple soils," derive from basaltic lavas and generally are reddish violet in color and fertile.

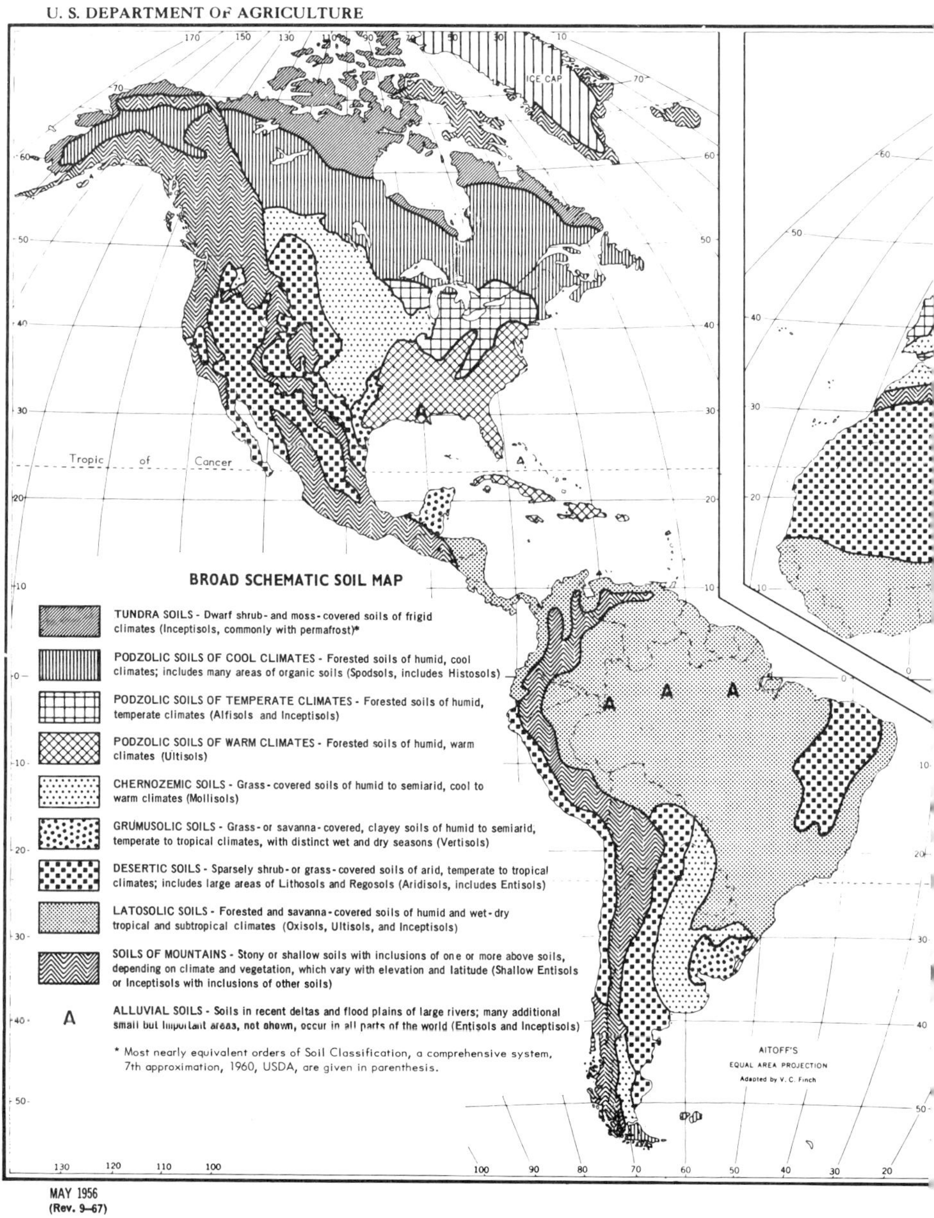

Fig. 8–3. Generalized and schematic soil map of the world according to the revised 1938 system of soil classification. The most nearly equivalent orders of the new soil taxonomy are given in parentheses. (Courtesy U.S. Soil Conservation Service.)

SOIL CONSERVATION SERVICE

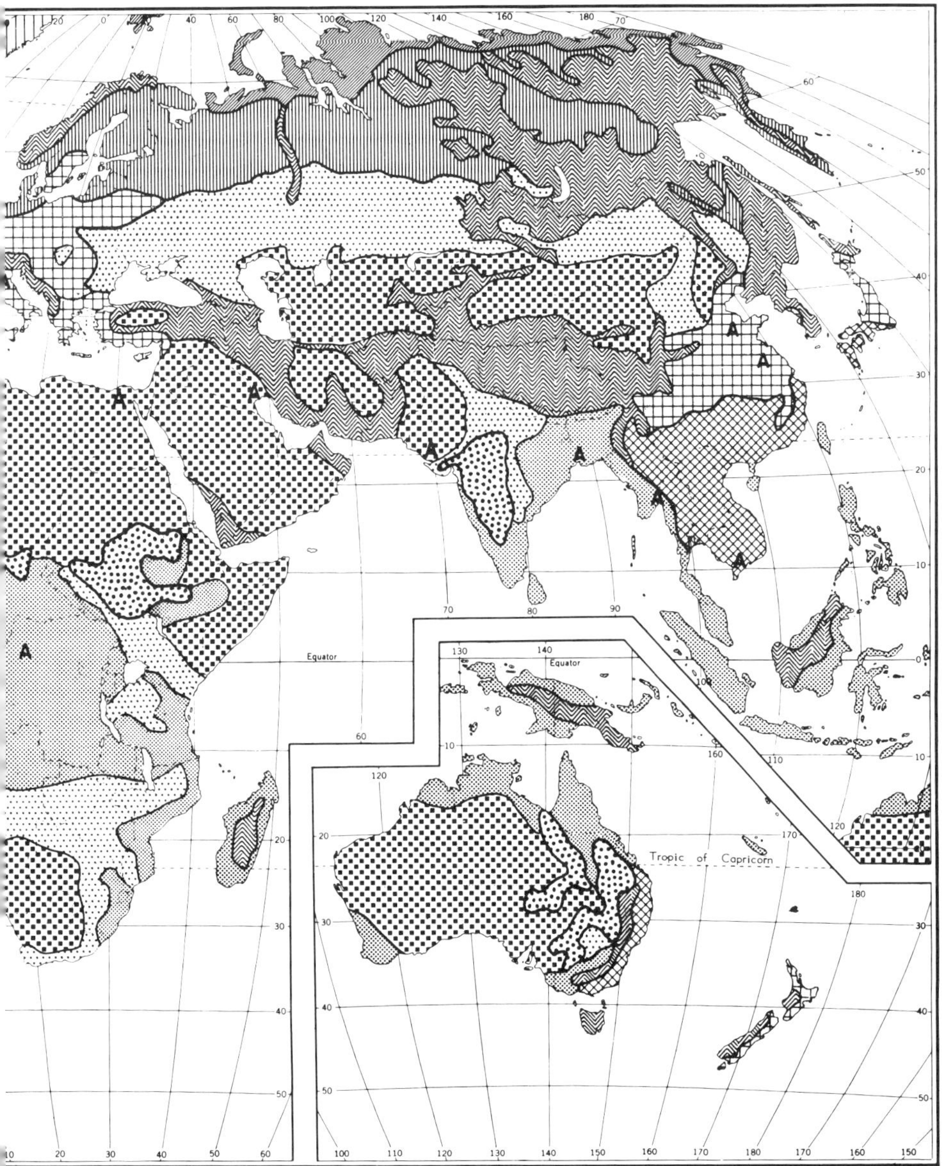

USDA-SCS-HYATTSVILLE, MD. 1967

for cultivating the more fragile tropical soils without catastrophic soil erosion and degradation of the forest biome. More extensive clearing and more intensive use may permanently degrade these soils and recovery to original forest may not occur.

Soils of the Middle Latitude Grasslands and Deserts

Middle latitude grassland soils account for a large share of the world's commercial production of wheat, corn, and soybeans. Along humid margins of grasslands are the most productive corn-producing soils in the world, and in warmer areas of humid and subhumid grasslands, cotton is important. In subhumid, semiarid, and arid areas, small grains, sorghum, cotton, and forage crops are produced either by dry-farming or irrigation.

In cooler middle latitudes (*Da*, *Db*, *BSk*, and *BWk*), moister areas have soils developed under tall prairie grasses, the Brunizems (table 8-3). Soils formed under slightly drier subhumid climates with mid- and shortgrass steppes, belong to the suborder Chernozems, rich, black soils of the wheat lands of the North American Great Plains and the steppes of the Soviet Union. Under increasingly drier conditions, vegetation ranges from short bunchgrass to desert scrub, and corresponding soils range from Chestnut, through Brown, to Sierozem. Profile development and horizon thickness decrease with diminished precipitation, but the main horizon sequence (A_1, A_3, B_2 and Cca) remains the same, the only exception being lack of a Cca (calcium carbonate) horizon in Brunizems, the result of more vigorous leaching under a more humid climate. Where temperatures are higher (*Ca*, *BSh*, *BWh*), the sequence of grassland and desert soils is Reddish-Prairie, Reddish-Chestnut, Reddish-Brown, and Red Desert. More intense weathering under higher temperature regimes accounts for these reddish colors, a result of more iron oxides and more rapid decomposition of organic matter than is the case in cooler climates.

Intrazonal Soils

Major characteristics of Intrazonal soils are determined largely by some local environmental factor other than climate. The five orders of Intrazonal soils are: (1) hydromorphic, (2) halomorphic, (3) calcimorphic, (4) volcanic ash soils, and (5) expanding-clay soils. The characteristics of Azonal soils are determined by the nature of parent material, or by steep slopes; thus, influences of pedogenic processes are minimal because of the youthfulness of parent materials, resistance of parent materials to weathering, or possibly interruption of soil-forming processes by erosion.

Hydromorphic Soils

Because of low-lying positions or poor internal drainage, excess soil moisture is characteristic of hydromorphic soils. In low areas, where water tables are

TABLE 8-3 Characteristics and Classification of Representative Soils of the Clarion-Nicolet-Webster Soil Association Area of Iowa

Soil Type	Slope Percent		Land Position	Parent Material	Original Vegation	Corn Yield Potential (Bu/Acre)[b]	1938 SCS Classification[b]	Classification by 7th Approximation[b]	
	Typical	Range						Order	Suborder
Hayden	2–5	2–50	upland ridges & side slopes	glacial till	forest	84	Gray Brown Podzolic	Alfisols	Udalfs
Ames	0–2	0–3	upland flats & depressions	glacial till	forest	72	Gray Brown Podzolic	Alfisols	Udalfs
Clarion	2–5	2–3	upland highs and ridges	glacial till	prairie grasses	90	Brunizem	Mollisols	Udolls
Nicollet	1–3	0–5	upland intermediate highs	glacial till	prairie grasses	95	Brunizem	Mollisols	Udolls
Webster	0–2	0–3	upland flats	glacial tille, reworked till	prairie grasses	91	Humic Gley	Mollisols	Aquolls
Glenco	0	0	upland depressions	local alluvium	sedges and prairie grasses	90	Humic Gley	Mollisols	Aquolls
Storden	5–50	5–50	upland knobs & side slopes	glacial till	prairie grasses	60	Regosol	Entisols	Orthents

Source: (cols. 1–7): W. R. Oschwald et al., *Principal Soils of Iowa,* Cooperative Extension Service, Iowa State University, 1965.

[a]Corn yield potentials in 1965 with proper land management, including conservation practices and drainage where necessary. Corn is not recommended on steeper slopes.

[b]Classification made by the author.

near the surface, soils may be Ground-Water Podzols, Half-Bog, or Bog soils, in order of increasingly poor drainage. Ground-Water Podzols have a thick cover of undecomposed organic matter and a cemented layer in the B horizon called *ortstein*, and they develop where the groundwater table is high but fluctuating. In Half-Bog soils a gley horizon is present, whereas in the true Bog, peats and mucks predominate.

Wisenboden—sometimes called Humic Gley—are hydromorphic soils associated with humid grasslands and Zonal Prairie soils. Periodically wet conditions are the result of flat topography, rather than of poor internal drainage caused by clay pans, and these soils are very productive when properly drained. Similar soils are commonest on young glacial drift plains of middle latitudes.

On older glaciated areas of the central United States, where parent materials are clay-rich loess, Planosols have developed on the flat topographic divides. An example of a Planosol is the Grundy soil depicted in figure 7-8 and listed in Table 8-4. These soils, developed under prairie grasses, range all the way from southern Iowa and Illinois to northeastern Oklahoma. Farther east, they are found under broadleaf deciduous forests in southern Indiana and Ohio (fig. 8-4). Generally, Planosols are difficult to drain because of clay pans, and they are less productive than their Zonal associates, Prairie or Gray-Brown Podzolic soils.

Hydromorphic soils occur in association with all Zonal soils; for example, under poor drainage conditions in the tropics, they include Ground-Water Laterites and Hydrol Humic Latosols.

Halomorphic Soils

Halomorphic soils occur in arid regions where drainage is poor and where an abnormal concentration of salt is present in the upper soil horizons; the term *halomorphic* comes from the mineral *halite* (NaCl). Zonal associates of halomorphic soils are Chernozems, Chestnut, and Brown soils, along with various desert soils. High concentrations of soluble salts of sodium, calcium, magnesium, and potassium produce saline soils or Solonchaks. Some halomorphic soils can be made productive by improving drainage and by leaching or flushing excess salts from the soil. Also, some crops, including cotton, date palm, sugar beet, and some forages, are salt tolerant to some degree, but even these will not survive with high salt concentrations.

Calcimorphic, Volcanic Ash, and Expanding-Clay Soils

Three suborders of Intrazonal soils are separated on the basis of parent materials. In these suborders the influence of parent materials is stronger than that of zonal climate in determining the nature of the soil. Calcimorphic soils owe their uniqueness to high lime content in their parent materials. Brown

Table 8-4 Characteristics and Classification of Representative Soils of the Adair-Grundy-Haig Soil Association Area of Iowa

Soil Type	Slope Percent		Land Position	Parent Material	Original Vegation	Corn Yield Potential (Bu/Acre)[a]	1938 SCS Classification[b]	Classification by 7th Approximation[b]	
	Typical	Range						Order	Suborder
Weller	2–5	1–9	upland	loess	forest	65	Gray-Brown Podzolic	Alfisols	Udalfs
Lindley	14–25	5–30	upland and side slopes	glacial till	forest	55	Gray-Brown Podzolic	Alfisols	Udalfs
Shelby	14–25	5–30	upland and side slopes	glacial till	prairie grasses	60	Brunizem	Mollisols	Udolls
Adair	9–14	5–20	upland and side slopes	reddish paleosol, till	prairie grasses	50	Brunizem	Mollisols	Udolls
Grundy	2–5	1–9	upland ridges & side slopes	loess	prairie grasses	78	Planosol/ Brunizem	Mollisols	Udolls
Haig	0–1	0–2	upland flats	loess	prairie grasses	77	Humic Gley	Mollisols	Udolls/ Aquolls
Edina	0–2	0–3	upland depressions	loess	prairie grasses, sedges	71	Planosol	Mollisols	Albolls

Source: (cols. 1–7): W. R. Oschwald et al., *Principal Soils of Iowa,* Cooperative Extension Service, Iowa State University, 1965.
[a]Corn yield potentials in 1965 with proper land management, including conservation practices and drainage where necessary. Corn is not recommended on steeper slopes.
[b]Classifications made by the author

Fig. 8–4. Soil profile of a Humeston soil, Appanoose County, Iowa. This soil has a mollic epipedon (A_1) over a light-colored albic (A_2) horizon. The light color results from the removal of organic matter by eluviation. The B horizon has a high clay content and a blocky structure (argillic horizon), which has low permeability to percolating water. Parent material is loess. Humeston soils were called Planosols in the 1938 classification and Argialbolls in the new soil taxonomy. (Soil Conservation Service photo by J. R. Culver, 1969.)

Forest soils, formed under trees in humid climates, make up one of the great soil groups of the calcimorphic suborder; their grassland counterparts are the Rendzinas. Soils formed on young volcanic ash deposits have dark color and considerable reserves of unweathered minerals, and they contain a higher proportion of 2:1–lattice clay minerals and allophane than their associated Zonal soils. Called Ando soils or Andosols, they sometimes are grouped with Zonal soils. Nevertheless, since their dark color seems to be a product of soil mineralogy rather than organic humus, and since Ando soils occur in a variety of climates, including marine, humid subtropical, and tropical, and under vegetation ranging from broadleaf to coniferous forests, their inclusion in the Intrazonal order seems to follow the logic of the 1938 system. In the humid tropics, when compared with Zonal Latosols, these young volcanic soils may be exceptionally fertile.

A third group of Intrazonal soils owes its characteristics to a high content of expanding clay (40–70 percent), and to a wet-dry climatic regime. Montmorillonitic clays swell during the wet season and shrink in dry weather, producing deep cracks into which surface particles fall (fig. 8-5). In time this process causes the profile to become inverted; thus, the name Vertisol is applied to these soils in the present USDA soil taxonomy, replacing the older name, Grumusol. Natural fertility is high in Grumusols because of their good cation-exchange capacity and high base saturation. If tilled when wet, however, these clay soils may become unworkable under the weight of farm machinery or work animals. In India and Sudan, Grumusols are important for cotton, maize, millet, and sorghum; in Texas and parts of Latin America they produce cotton or are in pastures.

Azonal Soils

Within the Azonal soil order (there are no Azonal suborders) are three great soil groups: *Alluvial soils*, *Lithosols*, and *Regosols*. Because of extreme youth or lack of weatherable minerals, Azonal soils show very little profile development. Alluvial soils are on relatively recent flood plains and usually have very shallow profiles, which often consist only of a darkened surface layer of accumulated organic matter. Often, a series of buried profiles will indicate a frequent addition of materials to the surface by flood waters (fig. 8-2). With good texture and drainage, alluvial soils may be more fertile than Zonal soils in any given area. Fertile flood-plain soils of the humid tropics are a good example.

Lithosols are thin soils in contact with bedrock. They are found principally where erosion on steep slopes prevents deep accumulations of regolith (fig. 8-6), or in arid or cold regions where weathering processes are very slow. The Regosols are azonal soils on thick, unconsolidated materials, such as loess and wind-blown sands, and on steep slopes in glacial till where erosion prevents formation of deep profiles. The Storden soil profile in figure

Fig. 8–5. Profile of Zita soil, Crosby County, Tex. This soil has deep cracks that develop because of swelling and shrinking of expanding clay minerals, such as montmorillonite. This soil was classified as a Grumusol in the revised 1938 system and a Vertisol in the new taxonomy. (U. S. Soil Conservation Service photo by W. M. Koos, n.d.)

Fig. 8–6. A shallow soil over limestone in Howard County, Iowa. Classed as a Lithosol in the 1938 classification system, it is now placed in the order of Entisols, suborder of Orthents, great group of Udorthents, and subgroup of Lithic Udorthents. (U. S. Soil Conservation Service photo by K. F. Kinion, n.d.)

8-7 is an example of Regosols formed on steeply sloping glacial till. The Sand Hills of Nebraska represent an unusually large area of Regosols (fig. 7-5).

Shortcomings of the 1938 System

The 1938 USDA system of soil classification is based largely on genetic processes, and at least in theory Zonal soils closely correlate with climate. Since many genetic processes are not clearly understood, it is often difficult to assign a definite process to an individual soil. The tendency has been to assume that a certain process is operating in a given climatic and vegetation environment. It is true that each of the Zonal soils seems to be most extensively represented within a certain range of climatic and ecological conditions, but it is also true that a particular Zonal soil may occur in widely different climatic environments.

Fig. 8–7. A shallow Storden soil developed on glacial till in north central Iowa. Relatively rapid erosion on steep slopes has prevented the formation of a mature soil profile; only an A horizon about 15 cm (6 in) thick is present over a C horizon. Storden soils are Regosols in the 1938 classification and Entisols (Typic Udorthent) in the new soil taxonomy. (U. S. Soil Conservation Service photo by Jones, n.d.)

In Puerto Rico one finds soils morphologically similar to Gray-Brown Podzolic soils of middle latitudes, and soils closely resembling the Reddish-Brown Lateritic soils of the tropics occur in parts of Virginia and Oregon. We could cite many other examples. Thorp and Smith (1949) suggested that such anomalies might be inherited from environmental conditions that no longer exist, and they summarized the dilemma resulting from these apparent anomalies as follows (p. 125):

> However the matter may be explained, the fact remains that there [is] a very definite overlap of soils commonly associated with cooler climates. Soil scientists are so accustomed to finding soils of certain great soil groups within a limited climatic environment that some are inclined to overlook this overlap. When we classify soils according to their characteristics, the present writers feel that *climatic environment, although relevant, should not be considered strictly as a soil characteristic. Greatest emphasis in soil classification should be given to the profile characteristics of soils themselves* [italics added].

Even then, plenty of geographic and genetic bias will remain in our classification.

The new system developed by the Soil Conservation Service to replace the 1938 system follows suggestions made by Thorp and Smith. It stresses profile characteristics that can be measured in the field or laboratory—independent, for the most part, of genetic processes, climate, or vegetation. This is not to say that soil genesis is not important or that soil-forming processes will no longer be a subject of research; in fact, many diagnostic features in the new system are selected because they relate directly to soil genesis. Nevertheless, *if a soil is described independently of process, its classification will remain the same even though ideas about its origin may change.* Except for the fact that soil temperatures and soil wetness and dryness are used as diagnostic characteristics, the new system is relatively free of climatic bias. There is far from complete correlation between distribution patterns of new soil orders and suborders and patterns of the 1938 system; nevertheless, considerable paralellism remains.

In his foreword to the first published version of the new taxonomy in 1960, Charles E. Kellogg, then assistant administrator for soil survey, USDA, commented on the need for a new system of classification:

> We needed one that would include as many of the new data as possible, provide for the incorporation of data that would become available, and furnish a basis for predicting how the various soils of the world would respond to modern management.
>
> This is not to say that the older systems are not useful. *We will continue to have need for the concepts of zonal, intrazonal, and azonal soils, even though we leave them out of the system of classification.* But we felt a growing need for a new system that would get away from the bias of genetic factors outside the soil itself and from the extreme emphasis on virgin soils. [Soil Survey Staff 1960, iv; italics added]

THE COMPREHENSIVE SOIL TAXONOMY OF THE U.S. DEPARTMENT OF AGRICULTURE

The system of soil classification used today in the United States developed through a series of stages called "approximations." Each stage was evaluated by the Soil Survey Staff of the USDA and by leading soil scientists from many countries. In 1960, the *7th Approximation* was published, and after 1965 it became the basis for soil classification in the United States. Supplements to the 1960 classification followed in 1964, 1966, 1967, and 1968, leading to the current revised system, published in a single volume in 1975 (Soil Survey Staff 1975). This is the source for most of the technical material included in this chapter.

The classification system based on the new soil taxonomy has advantages over the 1938 system; some of these advantages, as summarized by Guy D. Smith (1963), are:

1. Classification of soils, rather than soil-forming processes, is the focus.
2. Definitions stress soil and its properties, rather than geology and climatology.

3. It is possible to classify soils of unknown genesis since only observable soil properties are considered.
4. The classification can be applied with reasonable uniformity by many soil scientists; classification is not influenced by differences in opinion as to soil formation.

Properties used for defining taxa in the new classification include many related to soil genesis, although they are observable and measurable without reference to genetic processes. They incorporate all physical, biologic, and chemical properties of soil already outlined in chapter 7: (a) soil moisture and temperature regimes; (b) soil color, texture, and structure; (c) humus content and the presence or absence of clay minerals and iron and aluminum oxides; and (d) the levels of cation-exchange capacity, base saturation, salinity, and pH. In higher categories of the system, classification depends not only on the presence or absence of specific soil properties, but also on diagnostic horizons. The following brief definitions of major soil properties will provide a general idea of the criteria that are used in classification.

Diagnostic Horizons

Diagnostic horizons are of two classes—*epipedons* (Greek *epi*, over, and *pedon*, soil) normally formed at the soil surface, and other horizons formed below the soil surface. An epipedon includes the upper part of a soil profile darkened by organic matter, or eluviated, or possibly both. Epipedon is not necessarily synonymous with A horizon, but it may include part or all of the illuvial B horizon, if darkening by organic matter extends from the surface into the B. The six epipedons are summarized in table 8-5.

Major diagnostic subsurface horizons and some of their characteristics ap-

TABLE 8-5 Diagnostic Surface Horizons (Epipedons) Used in the Comprehensive Soil Taxonomy of the U.S. Department of Agriculture[a]

Epipedon	Major Characteristics
Mollic (L. *mollis,* soft)	Thick, dark colored, humus-rich, high base saturation (>50%), strong structure, soft when dry
Anthropic (Gr. *anthropos,* human)	Resembles Mollic; results from long period of human occupation and cultivation
Umbric (L. *umbra,* shade)	Same as Mollic but with a lower base saturation (<50%)
Histic (Gr. *histos,* tissue)	High organic content; peat or muck if it has not been plowed; saturated with water during part of the year
Plaggen (Ger. *plaggen,* sod)	Manmade surface layer produced by long-continued manuring; has a high organic content, dark color, and commonly contains artifacts and spade marks
Ochric (Gr. *ochros,* pale)	Light colored with low organic content; may be hard and massive when dry

[a] Soil Survey Staff 1975.

TABLE 8-6 Diagnostic Subsurface Horizons Used in the Comprehensive Soil Taxonomy of the U.S. Department of Agriculture[a]

Horizon	Major Characteristics
Argillic (L. *argilla*, white clay)	Accumulation of silicate clays
Agric (L. *ager*, field)	Illuvial horizon of clay and humus formed under cultivation just below the plow layer (Ap)
Natric (L. *natrium*, sodium)	Argillic horizon high in sodium with columnar or prismatic structure
Sombric (Fr. *sombre*, shadow)	Contains illuvial humus with low cation-exchange capacity and low base saturation; restricted to cool moist soils of highland tropics or subtropics
Spodic (Gr. *spodos*, wood ash)	Organic matter and aluminum and iron oxides; has a high cation-exchange capacity
Placic (Gr. *plax*, flat stone)	Thin, black to dark reddish pan cemented by iron, sometimes with manganese or organic matter
Cambic (L. *cambiare*, change)	Altered by physical movement or aggregation of particles, or by chemical reactions
Oxic (oxides)	Residual concentration of 1:1 lattice clays and hydrated oxides of Fe and Al (sesquioxides), with very low cation-exchange capacity
Duripan (L. *durus*, hard)	Hard pan cemented by silica, largely restricted to volcanic areas in Mediterranean or arid climates
Fragipan (L. *fragilis*, brittle)	Brittle layer with a hard consistency when dry, and weakly brittle when wet; probably formed by binding by clay and weak cementation by a variety of agents
Albic (L. *albus*, white)	A residual concentration of quartz and other primary minerals with gray to white color; often forms an A_2 horizon
Calcic (calcium)	Accumulation of appreciable amounts of calcium carbonate, sometimes with magnesium carbonate
Petrocalcic (Gr. *petra*, rock)	Hard, massive, rock-like continuous horizon cemented by calcium carbonate
Gypsic (gypsum)	Accumulation of appreciable amounts of gypsum ($CaSO_4 \cdot 2H_2O$)
Petrogypsic (Gr. *petra*, rock)	A gypsic horizon strongly cemented by gypsum so that roots cannot enter
Salic (L. *sal*, salt)	Horizon containing secondary enrichment of salts more soluble in cold water than gypsum
Sulfuric (sulfur)	Sulfur-rich horizon formed by oxidation of sulfide-rich mineral or organic matter in artificially drained soils; highly toxic to plants and free of living roots

[a]Soil Survey Staff 1975.

pear in table 8-6. Some are considered to be B horizons; others are part of A horizons. Generally, subsurface horizons represent accumulations of materials such as clay, gypsum, calcium carbonate, sesquioxides, and salts. Some are cemented layers, or pans, which restrict movement of water and penetration of subsoil by plant roots.

Organic Soil Materials

In organic soils (Histosols), some layers do not meet the definition of soil horizons because they were formed in different environments as they accumulated. For soil to be considered organic, more than half the upper 80 cm (32 in) must be organic matter. Organic soil materials are *fibric*, *hemic*, and *sapric*, according to the degree of decomposition of the original plant materials.

Fibric Soil Materials. Taking their name from the Latin word *fibra*, fiber, these are the least decomposed of organic materials—they were formerly "peat." With the exclusion of living roots, fibers are fragments of plant tissue that retain recognizable cellular structure of the original plant. Minimum fiber content ranges between two-fifths and three-fourths of soil volume in soil materials classified as fibric.

Hemic Soil Materials. These organic materials (from Greek *hemi*, half, implying intermediate decomposition) are intermediate between fibric and more decomposed sapric materials. Formerly, these were "mucky peat" or "peaty muck."

Sapric Soil Materials These materials (from Greek *sapros*, rotten), formerly called "muck," are the most highly decomposed organic materials. Recognizable fiber content of sapric materials is less than one-sixth of soil volume, excluding coarse fragments and mineral layers.

In addition to the organic fraction test, organic soil materials are extracted in a solution of sodium pyrophosphate. With the solution placed on white chromatographic paper, each class of organic soil materials must yield a color within a certain range of value and chroma. The color of the solution is tested against a Munsell color chart.

Other Diagnostic Soil Characteristics

In addition to soil properties listed in tables 8-5 and 8-6, others are important for defining soil taxa. A brief description of some follows.

Abrupt textural change may occur, from an ochric epipedon or an albic horizon to an argillic horizon.

Coefficient of linear extensibility (COLE) is the ratio of the difference between the moist length and the dry length of a clod. A high COLE value is characteristic of soils rich in montmorillonitic clays (expanding clays).

Durinods (Latin *durus*, hard, and *nodus*, knot) are weakly to strongly cemented nodules. The chief cementing agent is silica.

Gilgai is the microrelief typical of clayey soils having a high coefficient of expansion (Vertisols); it consists of a succession of microbasins and microknolls in nearly level areas, or of microvalleys and microridges running up and down slopes. Relief in gilgai varies from a few centimeters to one meter; occasionally, it may approach two meters.

Lithic contact is a boundary between soil and underlying rock material within 50 cm of the surface. The underlying material must have a hardness exceeding 3 on the Mohs scale.[2] Diagnostic soil horizons, such as duripans or petrocalcic horizons, are not considered lithic contacts.

Paralithic contact is similar to a lithic contact, except the material below the contact is softer than 3 on the Mohs scale.

Particle-size classes, described in chapter 7, are important for defining some taxa according to amounts of sand, silt, clay, and rock fragments in selected horizons, or within arbitrary depth limits.

Permafrost is a layer in which temperature always is below 0° C (32°F). Ice may or may not be present. If not, it is called "dry permafrost" and its consistency is loose.

Petroferric contact (Latin *petraus*, rock, and *ferrum*, iron) is the boundary between soil and a continuous layer of iron-cemented material having little or no organic matter. Indurated (hardened) horizons containing significant amounts of organic matter, such as spodic horizons (ortstein), are not considered petroferric horizons. Petroferric layers are well developed in some tropical and subtropical areas; sometimes they are called hardened laterite.

Plinthite (Greek *plinthos*, brick) is an iron-rich, humus-poor mixture of clay with quartz and other materials; usually it occurs as dark mottles. Plinthite changes irreversibly into ironstone hard pan or into irregular aggregates with repeated wetting and drying, especially when directly exposed to sun. In moist soil, plinthite is soft and can be cut by a spade. After irreversible hardening, it is no longer called plinthite, but *ironstone*. Plinthite and unhardened laterite are the same.

Soil Moisture Regimes

Soil moisture regimes are important for determining some soil-forming processes. A soil moisture regime can be a product of climate, but not always. Soils in arid climates may be wet or seasonally wet, depending on topography and drainage, high water tables, and water sources other than local rainfall. Soil moisture regimes are defined in terms of groundwater level and presence or absence of soil moisture in the soil section above the water table.

Aquic Moisture Regimes. This type of moisture regime (from Latin *aqua*, water) requires soil saturation by groundwater or by the capillary fringe for several days each year, or long enough for dissolved oxygen to be removed. The lack of oxygen produces a reducing environment and favors gley formation. If groundwater is always at surface level, or very close, soils have a *peraquic* regime (not used as a formative element in naming taxa).

[2]The Mohs scale determines the hardness of a mineral by comparing it to hardness of one of a group of ten selected minerals on a scale of 1 (softest) to 10 (hardest). The scale is as follows: 1, talc; 2, gypsum; 3, calcite; 4, fluorite; 5, apatite; 6, orthoclase; 7, quartz; 8, topaz; 9, corundum; 10, diamond.

Aridic and Torric Moisture Regimes. These regimes (Latin *aridus*, dry, and *torridus*, hot and dry) are the same, except they refer to different categories of the taxonomy. Soils with an aridic (torric) moisture regime are dry more than half the time during which soil temperature at a depth of 50 cm (20 in) is above 5°C (41°F), and they are never moist for as long as ninety consecutive days during which soil temperature at 50 cm is above 8°C (46.4°F). Such soils are normal in arid climates, but they may occur in semiarid climates if their physical properties keep them dry; for instance, the presence of a crusty surface that excludes water, or bedrock that is very close to the surface. Leaching is almost absent in these moisture regimes, and soluble salts may accumulate.

Udic Moisture Regimes. These regimes (Latin *udus*, humid) are characteristic of well-drained soils where rainfall is very evenly distributed, or where enough rain falls in summer to assure stored moisture approximately equal to or exceeding the amount lost through evapotranspiration. In most years the soil is moist for a total of ninety days or more, not necessarily consecutive ones. In cool climates with marked seasonal fluctuation of temperature, more than forty-five consecutive days are moist in the four months following the summer solstice (summer rainfall regime). If precipitation exceeds evapotranspiration in all months of most years, the regime is *perudic* (not used as a formative element in soil taxa names).

Ustic Moisture Regimes. Taking their name from Latin *ustus*, burnt, implying dryness, these regimes are intermediate between aridic (torric) and udic moisture regimes. Moisture is limited, but present when temperature conditions are suitable for plant growth. The designation is not applied to very cold (*pergelic*) or cool (*cryic*) regimes; these we will define later.

Xeric Moisture Regimes. These regimes (Greek *xeros*, dry) are typical of Mediterranean (*Cs*) climates with moist, cool winters and hot, dry summers. In six years out of ten, soil is dry for forty-five or more consecutive days in summer and early fall, and is moist for forty-five or more consecutive days during winter and early spring.

Soil Temperature Regimes

Soil temperature, along with soil moisture, is an important factor in plant growth, in other biological activity, and in soil formation. Each pedon has a characteristic temperature regime related to mean annual soil temperature, seasonal fluctuations from that mean, and a seasonal soil temperature gradient within the main root zone. For defining classes at various categoric levels in the taxonomy, several temperature regimes are used.

Pergelic Temperature Regimes. The name (Latin *per*, throughout, and *gelare*, to freeze) connotes permanent frost. Soils with a pergelic temperature regime have a mean annual temperature below 0°C (32°F). If moist, these soils have a permafrost layer; if excess water is not present, they have dry permafrost.

Cryic Temperature Regimes. Soils with a cryic temperature regime (Greek *kryos*, coldness) have mean annual temperatures above 0°C (32°F), but lower than 8°C (47°F). If cryic soils have an aquic moisture regime they are commonly churned by frost action. In organic soils, frost action produces characteristic mounds.

Frigid Temperature Regimes. Frigid soils are warmer in summer than the cryic, but their mean annual temperature is below 8°C (47°F), and the difference between mean winter and mean summer temperatures at a depth of 50 cm (20 in) exceeds 5°C (9°F).

Mesic Temperature Regimes. These occur in soils with mean annual temperatures between 8°C and 15°C (47°F and 59°F), and the difference between mean winter and mean summer temperatures exceeds 5°C (9°F) at a depth of 50 cm (20 in).

Thermic Temperature Regimes. These are found in soils in which the mean annual temperature is between 15°C and 22°C (59°F and 72°F), and the mean annual range of soil temperature at 50 cm (20 in) exceeds 5°C (9°F).

Hyperthermic Temperature Regimes. This designation pertains to mean annual soil temperatures exceeding 22°C (72°F), with the difference between mean winter and mean summer soil temperatures exceeding 5°C (9°F) at 50 cm (20 in). If the mean summer and mean winter temperatures differ less than 5°C (9°F) at 50 cm (20 in), the prefix *iso* is used with the above terms: i.e., *isofrigid, isomesic, isothermic.*

Categories of the System

There are six categories of classification; from the broadest to the most specific they are: (1) order, (2) suborder, (3) great group, (4) subgroup, (5) family, and (6) series. Categories of the system through the level of great groups are listed in table 8-7. Lower categories of subgroups, families, and series are much too numerous to consider individually; we will summarize only the principles used to differentiate these lower categories.

Order. There are ten orders, based primarily on those aspects of soil morphology thought to be associated closely with soil genesis. Thus, the orders of the new classification most closely correlate with the suborders of the revised 1938 system (table 8-8); again, it should be emphasized that morphology rather than genesis is the basis for classification, and correlation with the older system is only approximate.

Suborder. There are some forty-seven suborders recognized, but the number for any single order ranges from two in the Aridisols to seven in the Mollisols (table 8-7). Suborders are subdivisions of the orders into groups of soils that have common properties, but whose characteristics differ from those of other suborders of the same order. For most, these diagnostic characteristics involve wetness, temperature, and vegetation. Exceptions are the Entisols and Inceptisols, where soil materials and topographic position also differentiate suborders.

Table 8-7 Names of Orders, Suborders, and Great Groups

Order	Suborder	Great group
Alfisols	Aqualfs	Albaquals
		Duraqualfs
		Fragiaqualfs
		Glossaqualfs
		Natraqualfs
		Ochraqualfs
		Plinthaqualfs
		Tropaqualfs
		Umbraqualfs
	Boralfs	Cryoboralfs
		Eutroboralfs
		Fragiboralfs
		Glossoboralfs
		Natriboralfs
		Paleboralfs
	Udalfs	Agrudalfs
		Ferrudalfs
		Fragiudalfs
		Fraglossudalfs
		Glossudalfs
		Hapludalfs
		Natrudalfs
		Paleudalfs
		Rhodudalfs
		Tropudalfs
	Ustalfs	Durustalfs
		Haplustalfs
		Natrustalfs
		Paleustalfs
		Plinthustalfs
		Rhodustalfs
	Xeralfs	Durixeralfs
		Haploxeralfs
		Natrixeralfs
		Palexeralfs
		Plinthoxeralfs
		Rhodoxeralfs
Aridosols	Argids	Durargids
		Haplargids
		Nadurargids
		Natrargids
		Paleargids
	Orthids	Caliorthids
		Camborthids
		Durorthids
		Gypsiorthids
		Paleorthids
		Salorthids

TABLE 8-7 (*continued*)

Order	Suborder	Great group
Entisols	Aquents	Cryaquents
		Fluvaquents
		Haplaquents
		Hydraquents
		Psammaquents
		Sulfaquents
		Tropaquents
	Arents	Arents
	Fluvents	Cryofluvents
		Torrifluvents
		Tropofluvents
		Udifluvents
		Ustifluvents
		Xerofluvents
	Orthents	Cryorthents
		Torriorthents
		Troporthents
		Udorthents
		Ustorthents
		Xerorthents
	Psamments	Cryopsamments
		Quartzpsamments
		Torripsamments
		Tropopsamments
		Udipsamments
		Ustipsamments
		Xeropsamments
Histosols	Fibrists	Borofibrists
		Cryofibrists
		Luvifibrists
		Medifibrists
		Sphagnofibrists
		Tropofibrists
	Folists	Borofolists
		Cryofolists
		Tropofolists
	Hemists	Borohemists
		Cryohemists
		Luvihemists
		Medihemists
		Sulfihemists
		Sulfohemists
		Tropohemists
	Saprists	Borosaprists
		Cryosaprists
		Mediasaprists
		Troposaprists

TABLE 8-7 (*continued*)

Order	Suborder	Great group
Inceptisols	Andepts	Cryandepts
		Durandepts
		Dystrandepts
		Eutrandepts
		Hydrandepts
		Placandepts
		Vitrandepts
	Aquepts	Andaquepts
		Cryaquepts
		Fragiaquepts
		Halaquepts
		Haplaquepts
		Humaquepts
		Placaquepts
		Plinthaquepts
		Sulfaquepts
		Tropaquepts
	Ochrepts	Cryochrepts
		Durochrepts
		Dystrochrepts
		Eutrochrepts
		Fragiochrepts
		Ustochrepts
		Xerochrepts
	Plaggepts	Plaggepts
	Tropepts	Dystropepts
		Eutropepts
		Humitropepts
		Sombritropepts
		Ustropepts
	Umbrepts	Cryumbrepts
		Fragiumbrepts
		Haplumbrepts
		Xerumbrepts
Mollisols	Albolls	Argialbolls
		Natralbolls
	Aquolls	Argiaquolls
		Calciaquolls
		Cryaquolls
		Duraquolls
		Haplaquolls
		Natraquolls
	Borolls	Argiborolls
		Calciborolls
		Cryoborolls
		Haploborolls
		Natriborolls
		Paleborolls
		Vermiborolls

TABLE 8-7 (*continued*)

Order	Suborder	Great group
	Rendolls	Rendolls
	Udolls	Argiudolls
		Hapludolls
		Paleudolls
		Vermudolls
	Ustolls	Argiustolls
		Calciustolls
		Durustolls
		Haplustolls
		Natrustolls
		Paleustolls
		Vermustolls
	Xerolls	Argixerolls
		Calcixerolls
		Durixerolls
		Haploxerolls
		Natrixerolls
		Palexerolls
Oxisols	Aquox	Gibbsiaquox
		Ochraquox
		Plinthaquox
		Umbraquox
	Humox	Acrohumox
		Gibbsihumox
		Haplohumox
		Sombrihumox
	Orthox	Acrorthox
		Eutrorthox
		Gibbsiorthox
		Haplorthox
		Sombriorthox
		Umbriorthox
	Torrox	Torrox
	Ustox	Acrustox
		Eutrustox
		Sombriustox
		Haplustox
Spodosols	Aquods	Cryaquods
		Duraquods
		Fragiaquods
		Haplaquods
		Placaquods
		Sideraquods
		Tropaquods
	Ferrods	Ferrods
	Humods	Cryohumods
		Fragihumods
		Haplohumods
		Placohumods
		Tropohumods

TABLE 8-7 (*continued*)

Order	Suborder	Great group
	Orthods	Cryorthods
		Fragiorthods
		Haplorthods
		Placorthods
		Troporthods
Ultisols	Aquults	Albaquults
		Fragiaquults
		Ochraquults
		Paleaquults
		Plinthaquults
		Tropaquults
		Umbraquults
	Humults	Haplohumults
		Palehumults
		Plinthohumults
		Sombrihumults
		Tropohumults
	Udults	Fragiudults
		Hapludults
		Paleudults
		Plinthudults
		Rhodudults
		Tropudults
	Ustults	Haplustults
		Paleustults
		Plinthustults
		Rhodustults
	Xerults	Haploxerults
		Palexurults
Vertisols	Torrerts	Torrerts
	Uderts	Chromuderts
		Pelluderts
	Usterts	Chromusterts
		Pellusterts
	Xererts	Chromoxererts
		Pelloxererts

Source: Soil Survey Staff 1975, 86–87.

Great Group. There are about 185 great groups recognized in the United States. These are subdivisions of the suborders based on the assemblage of diagnostic horizons, and the most significant properties of the entire soil. A few suborders have no great groups because information is lacking about the soils, or there is little variation within soils of the suborder.

Subgroup. There are almost a thousand subgroups recognized in the United States—obviously, too many to list. Subgroups are divisions within a great

TABLE 8-8 Present Soil Orders and Approximate Equivalents in the Revised 1938 Classification

Present Order	Approximate Equivalents
Alfisols	Gray-Brown Podzolic, Gray Wooded soils, Noncalcic Brown soils, Degraded Chernozem, and associated Planosols and Half-Bog soils
Aridisols	Desert, Reddish Desert, Sierozem, Solonchak, some Brown and Reddish Brown soils, and associated Solonetz
Entisols	Azonal soils, some Low Humic Gley soils
Histosols	Bog soils
Iceptisols	Ando, Sol Brun Acide, some Brown Forest, Low Humic Gley, and Humic Gley soils
Mollisols	Chestnut, Chernozem, Brunizem (Prairie), Rendzinas, some Brown, Brown Forest, and associated Solonetz and Humic Gley soils
Oxisols	Lateritic soils, Latosols
Spodosols	Podzols, Brown Podzolic soils, and Ground-Water Podzols
Ultisols	Red-Yellow Podzolic soils, Reddish-Brown Lateritic soils of the United States, and associated Planosols and Half-Bog soils
Vertisols	Grumusols, and some Rendzinas, Regosols, and Alluvial soils

Source: Soil Survey Staff 1960, 13.

group around a central concept identified as *typic* (typical). Other subgroups have characteristics that are transitional (intergrade) between those of the typic subgroup and those of another great group.

Family. Soils within a subgroup having similar physical and chemical properties that are important to plant growth and affect their response to management are assigned to a family. Such properties include particle size distribution within the zone of major biologic activity, mineralogy of those horizons, soil temperature regimes, soil thickness penetrated by roots, pH, and so forth. There are about forty-five hundred families recognized in the United States.

Series. This is the lowest category of the taxonomy; about 10,500 soil series are recognized in the United States. Some series names have been in use for many years, carried over from the older classification. The name of a series derives from a place or a natural feature near where it was first recognized; thus, to someone not familiar with its diagnostic characteristics, a series name tells nothing about soil characteristics.

NOMENCLATURE OF THE SOIL TAXONOMY

Most criticism of the new soil taxonomy, particularly from some geologists and geographers, is leveled at the terminology used to designate soil taxa. Long exposure to old terminology, based largely on Russian words with a sprinkling of Greek and Latin roots, may make terms like *Chernozem* seem natural, whereas *Mollisol* may seem alien. It is doubtful, however, that anyone unfamiliar with either system would find one term more difficult than the other. For a general treatment of world soils, the new system, through levels of order and suborder, seems as useful as the older one, and the new system has this advantage: each part of the soil name conveys a concept of soil character or soil environment. Admittedly, below the level of suborder, classes become more complex and numerous. We do not suggest that one memorize lesser categories; it is sufficient to comprehend the logic of the system, which by use of a key permits an analysis of characteristics of each taxon. A parallel to soil taxonomy is found in better-known and accepted plant and animal taxonomies; a botanist, for instance, would not be expected to know the names of all plants classified within the system.

Naming the Categories of the Taxonomy

In all categories above series, the name of each taxon indicates its class. The name of orders all end in *sol* (Latin *solum*, soil), e.g., Molli*sol* (Latin *mollis*, soft). Suborder names all have two syllables. The first refers to a diagnostic quality of the soil, such as *aqu-* (Latin *aqua*, water) in *Aqu*ents, *Aqu*olls, or *Aqu*alfs. The second syllable is the formative element from the name of the order to which the soil belongs; in the above examples, we have *-ent* from *Ent*isol, *-oll* from M*oll*isol, and *-alf* from *Alf*isol. See tables 8-9 and 8-10 for a list of formative elements in the names of soil orders and suborders. The name of a great group consists of one or two formative elements suggesting diagnostic properties (table 8-11). Aquents having a cryic temperature regime are *Cry*aquents, and those with a sandy texture are called *Psamm*aquents (Greek *psammos*, sand, plus aquent).

Subgroup names consist of the name of a great group modified by one or more adjectives (table 8-12). For a subgroup that typifies the central concept

TABLE 8-9 Formative Elements in Names of Soil Orders

Name of Order	Formative Element in Name of Order	Derivation of Formative Element
Alfisol	Alf	Meaningless syllable
Aridisol	Id	L. *aridus,* dry
Entisol	Ent	Meaningless syllable
Histosol	Ist	Gr. *histos,* tissue
Inceptisol	Ept	L. *inceptum,* beginning
Mollisol	Oll	L. *mollis,* soft
Oxisol	Ox	F. *oxide,* oxide
Spodosol	Od	Gr. *spodos,* wood ash
Ultisol	Ult	L. *ultimus,* last
Vertisol	Ert	L. *verto,* turn

Source: Soil Survey Staff 1975, 87.

TABLE 8-10 Formative Elements in Names of Suborders

Formative Element	Derivation	Mnemonicon	Connotation
Alb	L. *albus,* white	Albino	Presence of albic horizon
And	Modified from ando	Ando	Andolike
Aqu	L. *aqua,* water	Aquarium	Aquic moisture regime
Ar	L. *arare,* to plow	Arable	Mixed horizons
Arg	Modified from argillic horizon; L. *argilla,* white clay	Argillite	Presence of argillic horizon
Bor	Gr. *boreas,* northern	Boreal	Cool
Ferr	L. *ferrum,* iron	Ferruginous	Presence of iron
Fibr	L. *fibra,* fiber	Fibrous	Least decomposed stage
Fluv	L. *fluvius,* river	Fluvial	Flood plain
Fol	L. *folia,* leaf	Foliage	Mass of leaves
Hem	Gr. *hemi,* half	Hemisphere	Intermediate stage of decomposition
Hum	L. *humus,* earth	Humus	Presence of organic matter
Ochr	Gr. base of *ochros,* pale	Ocher	Presence of ochric epipedon
Orth	Gr. *orthos,* true	Orthophonic	The common ones
Plagg	Modified from Ger. *plaggen,* sod		Presence of plaggen epipedon
Psamm	Gr. *psammos,* sand	Psammite	Sand texture
Rend	Modified from Rendzina		High carbonate content
Sapr	Gr. *sapros,* rotten	Saprophyte	Most decomposed stage
Torr	L. *torridus,* hot and dry	Torrid	Torric moisture regime
Ud	L. *udus,* humid	Udometer	Udic moisture regime
Umbr	L. *umbra,* shade	Umbrella	Presence of umbric epipedon
Ust	L. *ustus,* burnt	Combustion	Ustic moisture regime
Xer	Gr. *xeros,* dry	Xerophyte	Xeric moisture regime

Source: Soil Survey Staff 1975, 88.

TABLE 8-11 Formative Elements in Names of Great Groups

Formative element	Derivation	Mnemonicon	Connotation
Acr	Modified from Gr. *akros,* at the end	Acrolith	Extreme weathering
Agr	L. *ager,* field	Agriculture	An agric horizon
Alb	L. *albus,* white	Albino	An albic horizon
And	Modified from ando	Ando	Andolike
Arg	Modified from argillic horizon; L. *argilla,* white clay	Argillite	An argillic horizon
Bor	Gr. *boreas,* northern	Boreal	Cool
Calc	L. *calcis,* lime	Calcium	A calcic horizon
Camb	L.L. *cambiare,* to exchange	Change	A cambic horizon
Chrom	Gr. *chroma,* color	Chroma	High chroma
Cry	Gr. *kryos,* icy cold	Crystal	Cold
Dur	L. *durus,* hard	Durable	A duripan
Dystr, dys	Modified from Gr. *dys,* ill; dystrophic, infertile	Dystrophic	Low base saturation
Eutr, eu	Modified from Gr. *eu,* good; eutrophic, fertile	Eutrophic	High base saturation

TABLE 8-11 (*continued*)

Formative element	Derivation	Mnemonicon	Connotation
Ferr	L. *ferrum,* iron	Ferric	Presence of iron
Fluv	L. *fluvus,* river	Fluvial	Flood plain
Frag	Modified from L. *fragilis,* brittle	Fragile	Presence of fragipan
Fragloss	Compound of fra(g) and gloss		See the formative elements frag and gloss
Gibbs	Modified from gibbsite	Gibbsite	Presence of gibbsite in sheets or nodules
Gyps	L. *gypsum,* gypsum	Gypsum	Presence of a gypsic horizon
Gloss	Gr. *glossa,* tongue	Glossary	Tongued
Hal	Gr. *hals,* salt	Halophyte	Salty
Hapl	Gr. *haplous,* simple	Haploid	Minimum horizon
Hum	L. *humus,* earth	Humus	Presence of humus
Hydr	Gr. *hydor,* water	Hydrophobia	Presence of water
Luv	Gr. *louo,* to wash	Ablution	Illuvial
Med	L. *media,* middle	Medium	Of temperate climates
Nadur	Compound of na(tr) and dur		See the formative elements natr and dur
Natr	Modified from *natrium,* sodium		Presence of natric horizon
Ochr	Gr. base of *ochros,* pale	Ocher	Presence of ochric epipedon
Pale	Gr. *paleos,* old	Paleosol	Excessive development
Pell	Gr. *pellos,* dusky		Low chroma
Plac	Gr. base of *plax,* flat stone		Presence of a thin pan
Plagg	Modified from Ger. *plaggen,* sod	sod	Presence of plaggen epipedon
Plinth	Gr. *plinthos,* brick		Presence of plinthite
Psamm	Gr. *psammos,* sand	Psammite	Sand texture
Quartz	Ger. *quarz,* quartz	Quartz	High quartz content
Rhod	Gr. base of *rhodon,* rose	Rhododendron	Dark red color
Sal	L. base of *sal,* salt	Saline	Presence of salic horizon
Sider	Gr. *sideros,* iron	Siderite	Presence of free iron oxides
Sombr	F. *sombre,* dark	Somber	A dark horizon
Sphagn	Gr. *sphagnos,* bog	Sphagnum	Presence of Sphagnum
Sulf	L. *sulfur,* sulfur	Sulfur	Presence of sulfides or their oxidation products
Torr	L. *torridus,* hot and dry	Torrid	Torric moisture regime
Trop	Modified from Gr. *tropikos,* of the solstice	Tropical	Humid and continually warm
Ud	L. *udus,* humid	Udometer	Udic moisture regime
Umbr	L. base of *umbra,* shade	Umbrella	Presence of umbric epipedon
Ust	L. base of *ustus,* burnt	Combustion	Ustic moisture regime
Verm	L. base of *vermes,* worm	Vermiform	Wormy, or mixed by animals
Vitr	L. *vitrum,* glass	Vitreous	Presence of glass
Xer	Gr. *xeros,* dry	Xerophyte	A xeric moisture regime

Source: Soil Survey Staff, 1975, 89.

TABLE 8-12 Adjectives in Names of Extragrades and Their Meaning

Adjective	Derivation	Mnemonicon	Connotation
Abruptic	L. *abruptum,* torn off	Abrupt	Abrupt textural change
Aeric[a]	Gr. *aerios,* air	Aerial	Aeration
Anthropic	Modified from Gr. *anthropos,* man	Anthropology	An anthropic epipedon
Arenic	L. *arena,* sand	Arenose	Sandy epipedon between 50 cm and 1 m thick
Cumulic	L. *cumulus,* heap	Accumulation	Thickened epipedon
Epiaquic[a]	Gr. *epi,* over, above and aquic		Surface wetness
Glossic	Gr. *glossa,* tongue	Glossary	Tongued horizon boundaries
Grossarenic	L. *grossus,* thick and L. *arena,* sand		Thick sandy epipedon >1 m thick
Hydric	Gr. *hydor,* water	Hydroponics	Presence of water
Leptic	Gr. *leptos,* thin		A thin soil
Limnic	Modified from Gr. *limn,* lake	Limnology	Presence of limnic layer
Lithic	Gr. *lithos,* stone	Lithosphere	Presence of a shallow lithic contact
Pachic	Gr. *pachys,* thick	Pachyderm	A thick epipedon
Paralithic	Gr. *para,* beside, and lithic		Presence of a shallow paralithic contact
Pergelic	L. *per,* throughout in time and space, and L. *gelare,* to freeze		Permanently frozen or having permafrost
Petrocalcic	Gr. *petra,* rock and calcic from calcium		Presence of a petrocalcic horizon
Petroferric	Gr. *petra,* rock and L. *ferrum,* iron		Presence of a petroferric contact (ironstone)
Plinthic	Modified from Gr. *plinthos,* brick	Plinthite	Presence of plinthite
Ruptic[a]	L. *ruptum,* broken	Rupture	Intermittent or broken horizons
Superic	L. *superare,* to overtop	Superimpose	Presence of plinthite at the surface
Terric	L. *terra,* earth		A mineral substratum
Thapto[a]	Gr. *thapto,* buried		A buried soil

[a]Not strictly an extragrade. Name used to indicate a special departure from the typic subgroup.
Source: Soil Survey Staff 1975, 90.

of the great group, the adjective *typic* is used; thus, the Torrifluvent that typifies that great group, is called a *Typic* Torrifluvent. A Torrifluvent having many, but not all, the properties of a Vertisol superimposed on the complete set of diagnostic properties of a Torrifluvent, is called a *Vertic* Torrifluvent. Such a soil is said to be *intergrade* between Vertisols and Entisols. *Extragrade* subgroups have characteristics that are not representative of the great group,

TABLE 8-13 Examples of Names of Taxa in Each Category

Order	Suborder	Great group	Subgroup	Family[a]	Series
Alfisols					
Aridisols					
Entisols	Aquents				
	Arents				
	Fluvents	Cryofluvents	Typic Cryofluvents	Coarse-loamy, mixed, acid	Susitna
		Torrifluvents	Typic Torrifluvents	Fine-loamy, mixed (calcareous), mesic	Jocity and Youngston
			Vertic Torrifluvents	Clayey over loamy, mixed, (calcareous) hyperthermic	Glamis
	Orthents	Cryorthents	Typic Cryorthents	Loamy-skeletal, carbonatic	Swift Creek
			Pergelic Cryorthents	Loamy-skeletal, mixed (calcareous)	Durelle
Histosols					

[a]The complete name of a family consists of the name of the subgroup modified by adjectives to describe certain defined properties.
Source: Soil Survey Staff 1975, 84.

but do not indicate transitions to any other known soil; e.g., a Cryorthent having a permafrost layer is called a Pergelic Cryorthent. Table 8-13 gives some examples of names of taxa in each category. Examples of the derivation of names of taxa from order through subgroup are as follows, with formative elements in italics:

*Alf*isol (order)
Bor*alf* (suborder)
Cryo*boralf* (great group)
Typic *Cryoboralf* (subgroup)

Sp*od*osol (order)
Aqu*od* (suborder)
Cryaqu*od* (great group)
Pergelic Cryaqu*od* (subgroup)

The syllable *alf* identifies each of the lower categories as members of the Alfisol order; in the same way, the syllable *od* identifies members of the Spodosol order. Only the subgroup name is required for placing a soil in its proper great group, suborder, and order.

Family names consist of the name of a soil subgroup along with three or more adjectives to indicate particle size, mineralogy, temperature regime, and other characteristics. In table 8-13, the second example under the great group subdivision is a family of *fine-loamy* (indicating particle size distribution), *mixed calcareous* (indicating mineralogy), *mesic* (describing soil temperature regime), *Typic Torrifluvents* (designating the subgroup); thus, this soil belongs to a family of fine-loamy, mixed (calcareous), mesic, Typic Torrifluvents.

Soil Orders and Their Suborders

The USDA soil taxonomy recognizes ten orders. Each has from two to seven suborders, and each suborder is divided into a number of great groups. The names of the orders, suborders, and great groups are given in table 8-7. Only orders and suborders are described here, but the general idea of the nature of any great group or subgroup can be determined by following the key of formative elements furnished in the tables in this chapter, according to procedures described in the section on nomenclature. For example, the great group *Tropohumults* belong to the order of Ultisols, from the formative element, *ult*; in the same manner, the suborder *Hum*ults indicates Ultisols rich in humus, and the formative prefix *tropo-* indicates humid tropical by temperature and moisture characteristics. Figure 8-8 shows distribution of orders, suborders, and great groups within the United States; figure 8-9 shows world distribution. A brief description of the nature and occurrence of each of the ten orders and their suborders follows.

Fig. 8–8. Generalized soil map of the United States according to the new soil taxonomy. See figure 8-9 for key to the letter designations for soil orders and suborders of the system, with a brief description of each. (Courtesy U.S. Department of Agriculture—Soil Conservation Service.)

Alfisols

Alfisols have a moderate accumulation of silicate clay minerals in the B horizon, producing an argillic horizon, and a medium to high base saturation; a mollic epipedon or a spodic horizon are absent. Surface horizons usually are ochric or umbric epipedons, ranging in color from gray to brown (table 8-5). Under warmer or drier climatic conditions, surface horizons may be yellowish to reddish in hue. Water is available to mesophytic plants more than half the year, or at least for three consecutive months during the warm season. Because of their high base status and good moisture supplies, most Alfisols are intensely cultivated. Moisture may be a limiting factor in some Xeralfs and Ustalfs, and clay-rich horizons may cause difficulty in cultivating some Alfisols.

Alfisols include most soils classified in the older system as Noncalcic Brown, Gray-Brown Podzolic, and Gray-Wooded. Some Planosols, Half-Bog soils, and solodized Solonetz also are included with Alfisols. Alfisols formed under boreal forests, deciduous broadleaf forests, some tallgrass prairies, and, where there is a marked dry season, scattered xerophytic trees and grasses. Thus, they arc more varied in climatic and vegetation associations than their counterparts in the 1938 system.

Suborders of Alfisols are:

Aqualfs: Aquic moisture regime (wet) unless artificially drained.
Boralfs: Cool to cold; well drained; under broadleaf deciduous forests.
Udalfs: Udic moisture regime (moist); warm; broadleaf deciduous forests. (fig. 7-4).
Ustalfs: Ustic moisture regime (winter dry); warm; deciduous or xerophytic woodlands, savannas.
Xeralfs: Xeric moisture regime (summer dry); grasses, forbs, and shrubs.

In the United States (fig. 8-8), some of the best agricultural soils are Udalfs in the eastern corn belt. Aqualfs are common in the young glacial plains of the upper Midwest; Boralfs occur at higher elevations in the Rocky Mountains and in the western Great Lakes area; Ustalfs are present in Texas, Oklahoma, and eastern New Mexico; and Xeralfs are found mostly in California. Boralfs cover wide areas of Canada and northern Eurasia, from the Baltic into Russia and Siberia; Xeralfs are encountered in many areas of Mediterranean climate (*Cs*), including southern Europe, Middle Chile, southern South Africa, and southwestern Australia.

Aridisols

Aridisols are found mostly in dry climates where water is unavailable to mesophytic plants for prolonged periods. Moisture is unavailable three months or more out of a year, and soil is warm. Because of the sparse cover of ephemeral grasses and forbs and scattered xerophytic plants, surface horizons have low organic content and generally a light color; that is, they have ochric epipedons. Other pedogenic horizons commonly found in Aridisols are accumulations of silicate clays (argillic), calcium carbonate (calcic), gypsum (gypsic), or more soluble salts (salic). Horizons cemented by calcium carbonate (petrocalcic), or silica (duripan), often are found.

Temperature regimes of Aridisols range from cryic to isohyperthermic, and moisture regimes are dominantly aridic, except for some saline Aridisols having ustic or xeric moisture regimes. Temperature and moisture regimes are indicated at the subgroup level. Because of sparse vegetation cover and dry conditions of surface soil, wind erosion is prevalent, often removing fine materials from the surface, and creating a *desert pavement* from the gravel and stones left behind.

There are two suborders of Aridisols, classified on the basis of presence or absence of an argillic horizon. The *Argids* have an illuvial horizon in which silicate clays accumulate; this is indicative of their position on older, more stable land surfaces. The *Orthids* develop on geologically younger materials. On the youngest surfaces in dry regions, such as recent alluvium, and on steep slopes, the soils usually develop no diagnostic horizons and thus are Entisols. Aridisols include most soils classified as Desert, Reddish

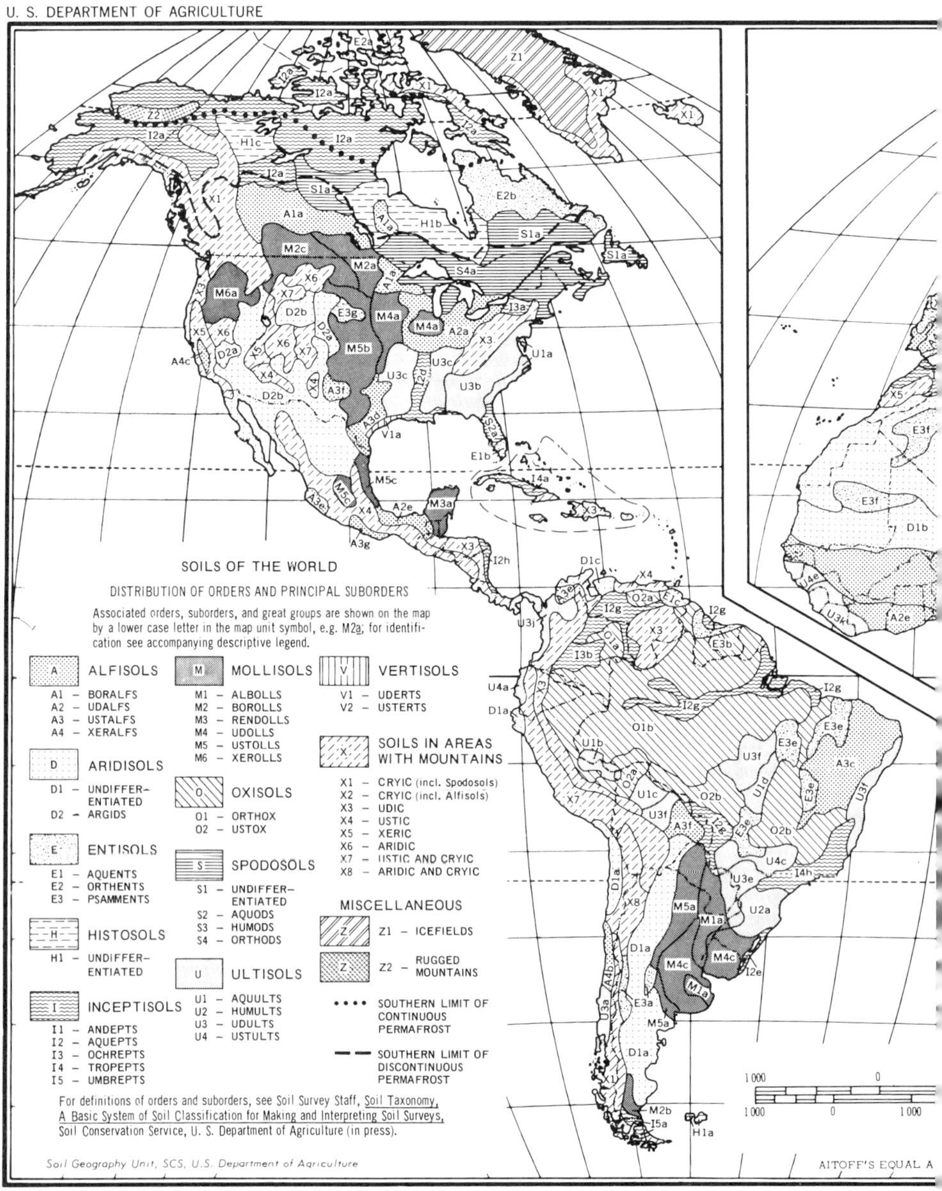

Fig. 8–9. Soils of the world. Distribution of orders and principal suborders according to the new soil taxonomy. Lower-case letters, referring to associated orders and suborders, are not defined in the map key. (Courtesy U. S. Department of Agriculture, Soil Conservation Service.)

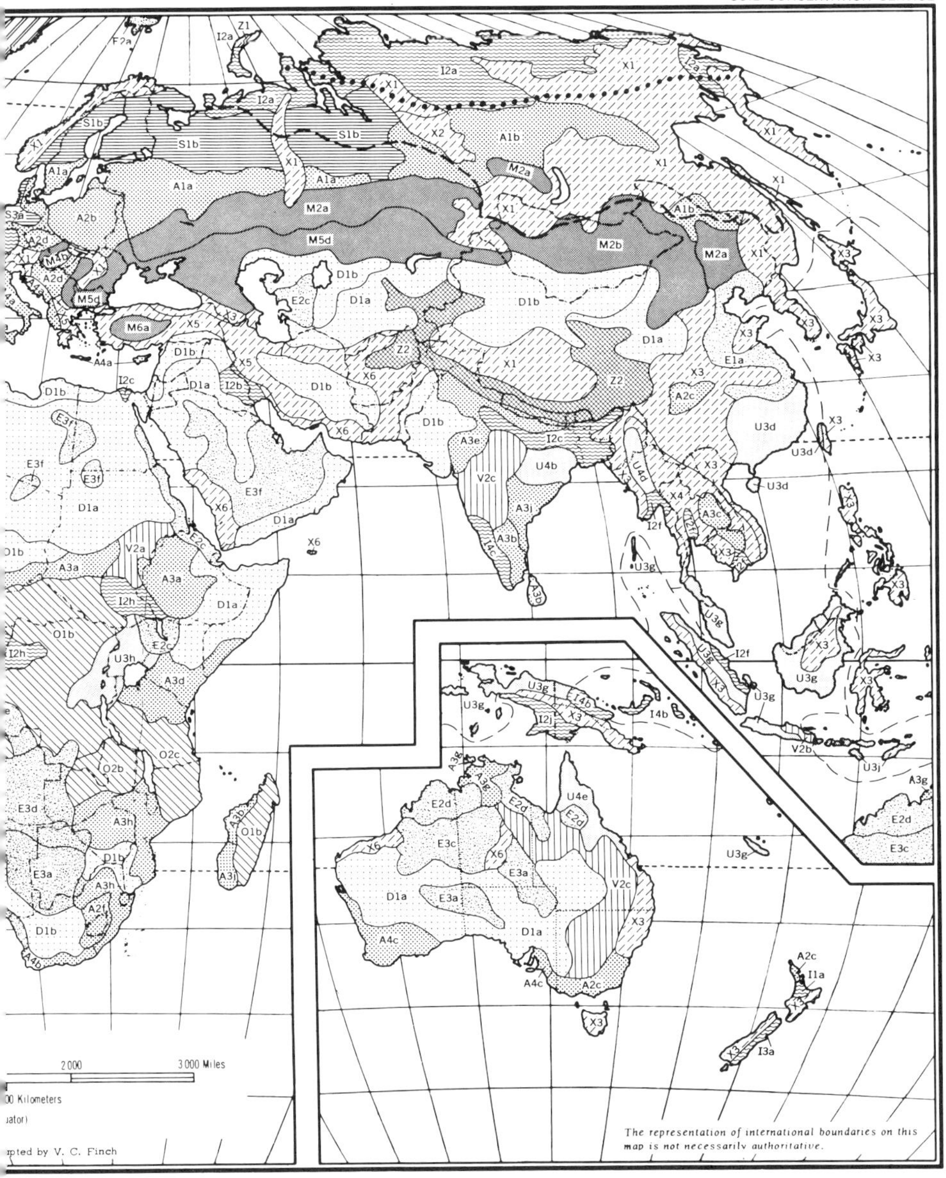

MAY 1972
USDA SCS HYATTSVILLE MD 1972

Desert, Sierozem, Reddish Brown, Solonchak, and Solonetz of the 1938 classification.

In the United States, Aridisols occur in large arid areas of the western states. They also are characteristic of the other great deserts of the world (fig. 8-9). Under irrigation, many Aridisols are very productive, but large areas without irrigation are either nonproductive or limited to extensive livestock ranching or nomadic herding.

Entisols

Entisols show little or no development of pedogenic horizons. They may occur anywhere under any vegetation cover or in any climatic regime; consequently, they match closely the definition of their counterparts in the 1939 system, the Azonal soils. Development of pedogenic horizons may have been retarded by (1) inert parent material, such as quartz sand, in which horizons do not form readily (Regosols of 1938); (2) formation from hard rock, such as limestone, which when weathered leaves little regolith (fig. 8-6); (3) lack of time for horizons to form in youthful materials, such as recent deposits of volcanic ash and alluvium; or (4) situation on steep slopes where rate of erosion exceeds rate of soil formation (Lithosols of 1938).

The following five suborders of Entisols are classified on the basis of moisture, materials, or erosion surfaces:

Aquents: Seasonally or permanently wet.
Arents: Horizons destroyed by plowing, spading, or earth moving by humans.
Fluvents: Well-drained, recent alluvial soils of flood plains, alluvial fans, and deltas (fig. 8-2).
Orthents: Slightly developed soils on recent erosional surfaces (figures 8-6 and 8-7).
Psamments: Slightly developed soils on dune and cover sands.

Except for Arents, each suborder divides into six or seven great groups on the basis of soil temperature, soil moisture regime, quartz content, etc.

Naturally, productivity varies greatly in Entisols. Many Fluvents are very productive, whereas soils of other suborders are too wet (Aquents) unless artificially drained, or steep and subject to accelerated erosion (Orthents). Psamments are droughty because of their high permeability, and often remain in natural pastures. Sandy soils, as those in the Nebraska Sandhills, often have excellent shallow groundwater supplies available for irrigation, and they can be quite productive. We covered additional characteristics of Entisols in the section on Azonal soils of the 1938 classification.

Histosols

Soils in which organic matter predominates are Histosols; formerly they were Bog soils, moor, peat, and muck. With the exception of the suborder of Florists which were classed as Lithosols, Histosols were Bog soils in the

TABLE 8-14 Global Extent of Soil Orders

Order	Percent of Land Area	Area (1,000 km^2)	(1,000 mi^2)
Alfisols	13.2	17,138	6,630
Aridisols	18.7	24,385	9,434
Entisols	8.3	10,759	4,162
Histosols	.9	1,170	453
Inceptisols	8.9	11,538	4,464
Mollisols	8.6	11,143	4,311
Oxisols	8.5	11,058	4,278
Spodosols	4.3	5,558	2,150
Ultisols	5.5	7,222	2,794
Vertisols	1.8	2,340	905
Mountain Soils	19.7	25,578	9,895
Rugged Mountains	1.6	2,072	802
Totals	100.0	129,960	50,278

Source: Estimates by the Soil Geography Unit, Soil Conservation Service, U.S. Department of Agriculture, January 1972.

1938 classification. Other great soil groups in the Hydromorphic suborder of 1938, however, are not classified as Histosols in the present system; for example, most Half-Bog soils belong to the histic subgroups of Aquepts and Aquolls; Planosols are included mostly in Aqualfs and Albolls; and Ground-Water Podzols are classed as Aquods.

Suborders of Histosols are defined by soil moisture regimes and by degree of decomposition of organic matter, whereas great groups are delimited in part, by soil temperature regimes. Suborders of Histosols are:

Fibrists: Relatively undecomposed organic matter.
Folists: Not saturated most of the time; less than 1 m to a lithic or paralithic contact.
Hemist: Moderately decomposed organic matter.
Saprist: Greatly decomposed organic matter.

Histosols occur over a wide range of climatic conditions, which creates difficulty in generalizing their suitability for crops. Where climate permits and artificial drainage is present, Histosols are very productive for certain vegetables. Poor drainage, however, may cause serious management problems, including land subsidence, danger of fires in dry organic matter, and oxidation of sulfur, which produces extremely acid and toxic conditions possibly eliminating all plant growth. Histosols are widely distributed but their total area is the smallest of the orders, occupying less than 1 percent of the world's land area (table 8-14).

Inceptisols

Inceptisols are young soils showing more profile development than Entisols but less than soils of other orders. They occur most often on geologically young materials or on recent (post-Pleistocene) erosion surfaces, in subhumid to humid climates from the tropics to subpolar regions. An exception is areas where precipitation exceeds evapotranspiration every month of the year, as in frigid but not pergelic soil temperature regimes. Here Inceptisols develop on more clayey parent materials, and Spodosols form in more sandy materials. Soils with an aridic (torric) moisture regime are excluded from the Inceptisol order.

Inceptisols may have any kind of epipedon, but the most common horizon sequence is an ochric or umbric epipedon over a cambic horizon. Duripans and fragipans may be present in some. All soils having a plaggen epipedon (table 8-5) are Inceptisols, and any underlying soil is considered to be "buried."

There are six suborders of Inceptisols, each having from four to ten great groups, except that Plaggets are all included under one great group of the same name. The suborders are:

Andepts:	From Japanese, *ando*: *an*, black; *do*, soil. Mostly volcanic ash soils; some great groups formed in parent materials derived from other volcanic or sedimentary rocks; allophane is an important constituent; high cation-exchange capacity.
Aquepts:	Wet moisture regimes; poor to very poor drainage.
Ochrepts:	Ochric epipedon; light colored, brownish, well-drained Inceptisols of mid and high latitudes.
Plaggets:	Plaggen epipedon formed by cultivation and manuring.
Tropepts:	Brownish to reddish, freely drained Inceptisols of the tropics; low in silicate clays or volcanic materials.
Umbrepts:	Umbric epipedon; acid, dark reddish or brownish, freely drained, rich in organic matter; mostly in hilly or mountainous regions of mid to high latitudes.

The 1938 system included Inceptisols in several great soil groups: Ando soils, Regosols, Humic Gley, Alluvial soils, Sols Bruns Acides, Hydrol Humic Latosols, Brown Forest soils, Calcisols, Latosols, and many others. Agricultural and nonfarming uses are diverse because of the wide range of climates in which Inceptisols occur, and because of the great variety of terrain and parent materials. Some areas of Aquepts, including the flood plain of the Mississippi River from Cairo, Ill., to the Gulf of Mexico, are highly productive if properly drained. Other Inceptisols, such as some of the Ochrepts of the Appalachians, were abandoned to forest after early attempts at cultivation failed.

Mollisols

Mollisols are mainly very dark colored, base-rich soils of temperate grasslands, although where parent materials are derived from chalk or marl, some are under forest with climates ranging from humid midlatitude to wet tropical. Most have mollic epipedons, with argillic, natric, or calcic horizons.

A few may have albic horizons (fig. 8-4), and some have duripans or petrocalcic horizons. Mollisols are not restricted to any moisture or temperature regime, but enough moisture to support perennial grasses seems to be essential. Surface horizons in most Mollisols have granular or crumb structure that does not harden when dry; thus the term *mollic* epipedon, meaning "soft."

Mollisols divide into seven suborders on the basis of soil moisture and soil temperature regimes, except in the order of Rendolls in which parent material is diagnostic. The suborders of Mollisols and their salient characteristics are:

Albolls: Albic horizon and fluctuating groundwater table (fig. 8-4).
Aquolls: Aquic soil moisture regime (wet) unless artificially drained; natural vegetation of grasses, sedges, and forbs.
Boralls: Cool to cold soil temperature regimes; well drained; under shortgrass steppes.
Rendolls: Highly calcareous parent material; under forests.
Udolls: Udic soil moisture regime (moist); warm soil temperature regimes; mostly tallgrass prairie vegetation.
Ustolls: Ustic soil moisture regime (winter dry); warm soil temperature regimes; under tall- to shortgrass prairies and steppes (fig. 7-1).
Xerolls: Xeric soil moisture regime (summer dry); warm soil temperature regimes; under bunchgrass and shrubs.

Mollisols are among the world's finest agricultural soils. Udolls are widespread in the western corn belt of the United States where adequate rainfall assures high yields of corn and soybeans, while farther west the drier, but nonetheless highly fertile, Ustolls form the basis for commercial wheat production. Almost a fourth of all soils in the United States belong to the order of Mollisols (table 8-15). Outside the United States and the prairie provinces of Canada, extensive areas of Mollisols occur in the Soviet Union, Mongolia, and North China, and in northern Argentina, Uruguay, Paraguay, and southernmost Brazil. Some occur within the humid tropics where parent materials derive from limestones and marls. The author has observed both Rendolls and Udolls on Malaita, in the Solomon Islands, under second-growth tropical rainforest in areas receiving more than 5,080 mm (200 in) of rainfall. The parent material for these tropical Mollisols derived from limestone bedrock in an area of karst topography.[3]

Oxisols

Oxisols of subtropical and tropical regions are the most highly weathered of all soil orders, occuring on gently sloping surfaces of great geological age. They are reddish, yellowish, or grayish mixtures of quartz, kaolin, sesqui-

[3] Karst topography has numerous sinkholes, caverns, and other features produced by the solution of limestone caused by percolating water. The name derives from the *Karst* area of Yugoslavia.

TABLE 8-15 Approximate Area of Kinds of Soil in the United States

Soil Order and Suborder	Area (mi²)	Extent (%)
Alfisols	483,450	13.4
Aqualfs	36,400	1.0
Boralfs	107,500	3.0
Udalfs	211,550	5.9
Ustalfs	94,650	2.6
Xeralfs	33,350	0.9
Aridisols	414,400	11.5
Argids	310,900	8.6
Orthids	106,500	2.9
Entisols	284,250	7.9
Aquents	8,200	0.2
Fluvents	10,750	0.3
Orthents	188,450	5.2
Histosols	18,750	0.5
Fibrists	8,300	0.2
Hemists	7,050	0.2
Saprists	3,400	0.1
Inceptisols	655,700	18.2
Andepts	66,950	1.9
Aquepts	411,100	11.4
Ochrepts	153,450	4.3
Umbrepts	24,100	0.7
Mollisols	896,000	24.6
Aquolls	46,100	1.3
Borolls	176,800	4.9
Udolls	170,450	4.7
Ustolls	318,400	8.8
Xerolls	184,250	4.8
Oxisols	500	<0.02
Spodosols	183,050	5.1
Aquods	25,900	0.7
Orthods	157,150	4.4
Ultisols	463,050	12.9
Aquults	41,250	1.1
Humults	27,550	0.8
Udults	357,650	10.0
Xerults	36,600	1.0
Vertisols	35,300	1.0
Uderts	13,500	0.4
Usterts	21,500	0.6

Source: Soil Survey Staff 1975, 429.

oxides of iron and aluminum, and organic matter, without clearly marked horizons. Because Oxisols form on ancient stable surfaces, it is common for weathering to extend to great depths, producing a thick regolith. Lack of expanding clay horizons and the presence of quartz sand makes them relatively permeable and, as a group, resistant to erosion when cultivated.

Oxisols occur where the present climate ranges from arid to perhumid, their presence in arid climates suggesting that there have been major climatic changes since the soils developed. Most Oxisols have low productivity for cultivated plants because of low cation-exchange capacity. Large areas have sparse natural vegetation cover even though moisture is ample during a long rainy season. Other areas support luxuriant rainforests and commonly are farmed under shifting cultivation (see chapter 6). Some Oxisols are in areas where savannas, probably of anthropic origin, replaced forest. Other Oxisols, formed from basic rocks, such as limestone and basalt, are quite productive under proper management. Commercial plantations often use these better quality Oxisols if transportation facilities are accessible.

Classification lags, primarily because they are not extensive in the United States, even in Puerto Rico and Hawaii. There are five suborders, based largely on soil moisture regimes:

Aquox: Aquic soil moisture regime (wet); plinthite within 30 cm (11.8 in) of surface.
Humox: High organic content (humus); humid, cool climates of tropical highlands or higher latitudes.
Orthox: Short dry season or none; equatorial; rainforest or anthropic savanna.
Torrox: Arid climate; relic of former pluvial period; relatively high base exchange; productive if irrigated.
Ustox: Ustic moisture regime; near Tropics of Cancer or Capricorn; savanna, semideciduous or deciduous forest.

Only Orthox and Ustox suborders are extensive enough to appear on world soil maps (fig. 8-9).

Oxisols include most, but not all, Laterites, Groundwater Laterites, and Latosols of the revised 1938 classification. Some formerly designated Latosols or Laterites are now Ultisols, Alfisols, and Inceptisols because they lack an oxic horizon as strictly defined in the present taxonomy. It is evident from world soil maps that all soils found between the Tropics of Cancer and Capricorn are not Oxisols—only about 35 percent. They are most extensive in tropical South America, making up an estimated 70 percent of the Amazon Basin (Falesi 1974), and in humid tropical Africa, making up about 40 percent. In other humid tropical areas, Oxisols are less common, accounting for only about 4 percent in humid tropical Asia (National Research Council 1982). A summary of the geographical distribution within the humid tropics is provided in table 8-16 and figure 8-10.

Table 8-16 Geographical Distribution of Soils of the Humid Tropics in Millions of Hectares and Percent Based on Dominant Soil in FAO Maps

Soil Order and Suborder	Humid Tropics Total		Humid Tropical America		Humid Tropical Africa		Humid Tropical Asia	
	Ha	(%)	Ha	(%)	Ha	(%)	Ha	(%)
Oxisols	525	(35)	332	(50)	179	(40)	14	(4)
Ultisols	413	(28)	213	(32)	69	(16)	131	(34)
Inceptisols	226	(15)	61	(9)	75	(17)	90	(24)
Entisols	212	(14)	31	(4)	91	(20)	90	(24)
Alfisols	53	(4)	18	(3)	20	(5)	15	(4)
Histosols	27	(2)	—		4	(1)	23	(6)
Spodosols	19	(1)	10	(1)	3	(<1)	6	(1)
Mollisols	7	(<1)	—		—		7	(2)
Vertisols	5	(<1)	1	(<1)	2	(<1)	2	(<1)
Aridosols	2	(<1)	—		1	(<1)	1	(<1)
Total	1,489	(100)	666	(100)	444	(100)	379	(100)

Source: Data derived from NAS, *Ecological Aspects of Development in the Humid Tropics,* Washington, D.C.: National Academy Press, 1982, table 2-3, p. 41.

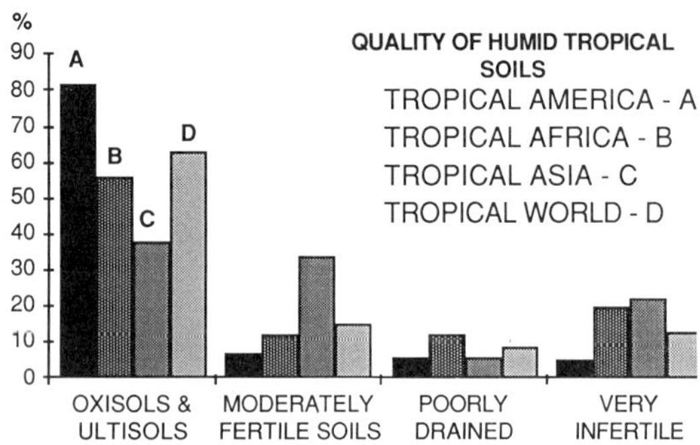

Fig. 8–10. General quality of soils of the humid tropics by percent of the total soils in each area. (Data from NAS, 1982.)

Spodosols

Spodosols include most, but not all, soils called Podzols, Ground-Water Podzols, and Brown Podzolic. Their main diagnostic feature is a spodic horizon, a subsurface horizon with accumulations of organic matter and sesquioxides of aluminum, with or without sesquioxides of iron (fig. 7-16). In undisturbed Spodosols one finds normally an overlying eluvial horizon consisting of a layer of gray to light gray quartz sand (albic horizon), which gave the name Podzol to these soils (fig. 7-17).

Spodosols form in sandy parent materials mostly under coniferous forests, although some occur under broadleaf deciduous trees or heaths. Some species of vegetation, such as pine trees and heather (*Calluna vulgaris*), produce particularly acid humus and encourage Spodosol development (see chapters 5 and 7). Some are in tropical areas where parent materials are quartz sands, and vegetation cover is rainforest, palm forest, or anthropic savanna.

Moisture regimes of Spodosols are primarily aquic or udic, although some have xeric regimes, and they may have any temperature regime. Spodosols are naturally acid infertile soils, but they may respond well to good management, particularly applications of lime and nitrogen. Under cultivation, the spodic horizon may be destroyed by biologic activity; then the soil no longer would be classed as a Spodosol. There are four suborders:

Aquods: Aquic (wet) soil moisture regime; shallow groundwater or extremely humid climate; natural vegetation ranges from Sphagnum in cool areas to palms in the tropics.

Ferrods: A high ratio of free iron to organic carbon in the spodic horizon (Bir); moisture regime not aquic.

Humods: A high ratio of organic carbon to free iron in some or all subhorizons of the spodic horizon; freely drained; vegetation ranges from *Calluna vulgaris* heaths to tropical rainforests (fig. 7-16).

Orthods: Well drained, with accumulation of aluminum, iron, or organic carbon in the spodic horizon in which no one dominates. They are the most common Spodosols in northern Europe and the United States.

Although Spodosols occur in widely divergent climates, mostly they develop in higher middle latitudes. We find large areas (Orthods and Aquods) from the Great Lakes region eastward to New England and in central and eastern Canada, and important areas of Aquods occur in Florida and on other parts of the coastal plain. They dominate in northern Europe in a belt from Scandinavia to western Siberia, and considerable areas of Humods developed under anthropic heaths on sandy glacial outwash plains of northern Germany and Denmark (see chapters 5 and 6). In the tropics, some so-called "giant podzols," described in chapter 7, are excluded arbitrarily from the Spodosol order by a limit of 2 m (6.5 ft) on the thickness of the albic horizon. Soils with an albic horizon greater than 2 m (6.5 ft) in thickness are grouped with Entisols (Soil Survey Staff 1975).

Ultisols

Ultisols are deeply weathered red and yellow soils containing considerable amounts of silicate clays, along with more highly weathered oxide clays. The argillic horizons have a base saturation of less than 35 percent. Ultisols generally developed beyond the limits of Pleistocene glaciation; thus, they formed on landscapes and from parent materials already old, geologically. Under more intense weathering of the humid tropics, however, Ultisols may occupy younger surfaces, whereas Oxisols occupy older, more stable landscape positions. Most soils of this order formerly were Red-Yellow Podzolic, Reddish-Brown Lateritic, Groundwater Laterites, and include some former Humic Gley and Low Humic Gley soils.

Ultisols may have mesic, thermic, or hyperthermic soil temperature regimes, and aquic, udic, xeric, or ustic soil moisture regimes. In the southeastern United States, most formed under coniferous or deciduous forests, whereas in south China, broadleaf evergreen forests occupied the land before agricultural settlement. In many other parts of the world, tropical savannas, probably anthropic, are present on Ultisols.

Ultisols are divided into five suborders, on the basis of soil moisture regimes, except for Humults:

Aquults: Aquic moisture regime (wet); gray or olive color; ochric or umbric epipedon; formed in alluvium and marine deposits.

Humults: Well drained; humus-rich, dark colored (umbric epipedon) in midlatitudes; lighter colored (ochric epipedon) but still humus-rich in low latitudes; formed under coniferous forests in midlatitudes and rainforests in low latitudes.

Udults: Udic soil moisture regime (well-distributed rainfall); humus-poor; light colored surface horizons (ochric epipedon) and yellowish brown to reddish argillic horizons; under forests or anthropic savannas.

Ustults: Ustic soil moisture regime (dry winter); little organic carbon; ochric epipedon; mostly reddish, but may be yellowish brown in regions of high rainfall; under deciduous forests or anthropic savannas.

Xerults: Mediterranean climates (dry summer); moderate to small amount of organic matter; ochric epipedon over brownish to reddish argillic horizon; coniferous forests.

Udults are the most extensive soil suborder in the southeastern United States, from New Jersey south and west to Texas. Humults are found from northern California to the Willamette–Puget Sound lowlands of Oregon and Washington, and limited areas of Xerults occur in northwestern California and southern Oregon. In South America there are Ultisols in central and southern Brazil, coastal Ecuador, and south-central Chile. Udults occur in West Africa and in East Africa near Lake Victoria, in northern Australia, on the islands and mainland of Southeast Asia, and in South China. Next to Oxisols, Ultisols are the most extensive of the humid tropics, occurring in about 27 percent of the area (table 8-16).

Vertisols

The central concept of Vertisols is the same as Grumusols in the revised 1938 classification, although some Rendzinas, Regosols, and Alluvial soils of the older system now are incorporated (fig. 8-5). Basically, Vertisols are clayey soils that develop deep, wide cracks sometime during the year in most years, and maintain a high bulk density between the cracks. Although Vertisols make up a relatively homogeneous order based on soil morphology, they show considerable variation in moisture regimes, soil colors, base saturation, percentage of carbonates, and depth to bedrock. Most have soil temperature regimes thermic or warmer, but some may have mesic regimes.

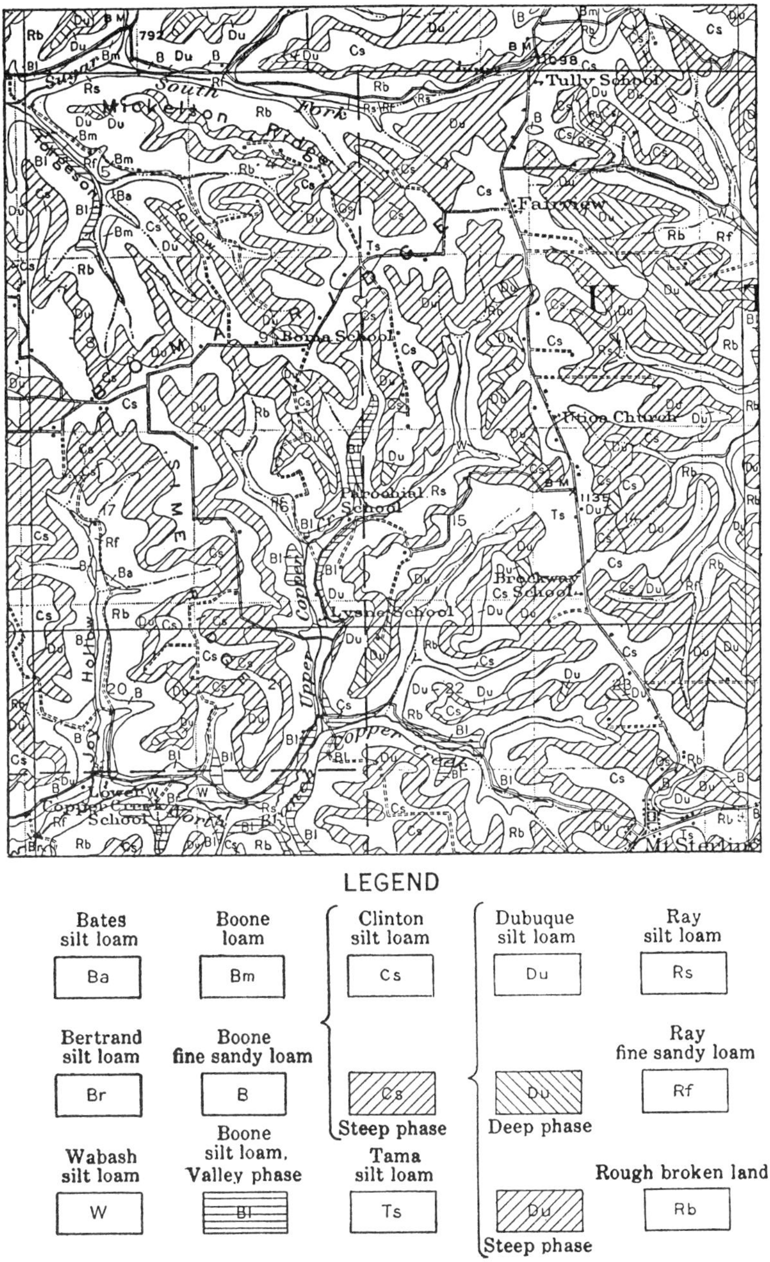

Fig. 8–11. Reproduction of a portion of a published soil map of Crawford County, Wis., illustrating the information carried on a large-scale soil map. The original map is in color. In addition to soil types, soil maps give important information, including drainage and cultural features, such as streams, roads, boundaries, and structures. (From USDA 1938, 1,010.)

The clay mineralogy of most in the United States is montmorillonitic, whereas some soils in other areas may be illitic.

Vertisols divide into four suborders on the basis of soil moisture regimes. Each suborder, except the Torrerts, subdivides into two great groups on the basis of color; high chroma soils receive the prefix *Chrom-* (Gr. *chroma*, color), and low chroma soils, *Pell-* (Gr. *pellos*, dusky). High chroma soils have a readily visible color other than black, gray, or white, whereas low chroma soils are dominantly white, gray, or black. Suborders are:

Torrerts: Arid climates (torric soil moisture regime); cracks may be open all year in most years.

Uderts: Humid climates (udic soil moisture regime); cracks open and close irregularly depending on weather.

Usterts: Wet-dry climates (*Am*, *Aw*, *Caw*); ustic soil moisture regime; cracks close once or twice each year.

Xererts: Mediterranean climates (*Cs*); xeric soil moisture regime; cracks closed in winter, open throughout the summer.

Natural vegetation on Vertisols ranges from forests, through grass, to desert shrubs. Large areas in the tropics and subtropics remain under savannas and open forests.

Vertisols often present serious problems when cultivated because of their high content of expanding clays. Because of high base-exchange capacity they often are very productive if properly managed, especially when compared to soils of other orders with which they often associate.

Soil Surveys and Local Soil Maps

Since soil survey work is localized, the concern of the field mapper is with soils at the series level. Although many soil maps were published before the adoption of the present soil taxonomy, the series category remains relatively unchanged from the older system.

Figure 8-11 is part of a published county soil map from Wisconsin; the original map is in color. Recent soil maps are reproductions of aerial photography on which soil symbols and boundaries are overprinted. A bulletin accompanying the county soil map contains information about physiography and climate, the soil mapping units, soil management and productivity, and engineering applications. These county soil maps and soil survey bulletins contain valuable information for farmers, engineers, and planners.

BIBLIOGRAPHY

Part 1

CLIMATE
(Chapters 1–4)

Ackerman, A. E. 1941. The Köppen Classification of Climates in North America. *Geographical Review* 31 (1941): 105–11.

Ackerman, E. A., and George O. G. Löf. 1959. *Technology in American Water Development.* Baltimore: The Johns Hopkins Press.

Ahrens, C. Donald. 1985. *Meteorology Today.* 2d ed. St. Paul, Minn.: West Publishing.

Alexander, Tom. 1974. Ominous Changes in the World's Weather. *Fortune* (February): 90.

Baes, C. F., Jr., H. E. Goeller, J. S. Olson, and R. M. Rotty. 1977. Carbon Dioxide and Climate: The Uncontrolled Experiment. *American Scientist* 65 (May–June): 310–20.

Barber, Richard T., and Francisco P. Chavez. 1983. Biological Consequences of El Niño. *Science* 222 (16 December): 1,203–10.

Barry, Roger G., and Richard J. Chorley. 1982. *Atmosphere, Weather, and Climate* 4th ed. London: Methuen.

Bearsdley, Tim. 1989. Greening the Summit: The Environment Arrives on the International Political Agenda. *Scientific American* (September): 17–20.

Blake, Donald R., and F. Sherwood Rowland. 1988. Continuing Worldwide Increase in Tropospheric Methane, 1978 to 1987. *Science* 239 (4 March): 1,129–34.

Blumenstock, D. I., and C. W. Thornthwaite, 1941. Climate and the World Pattern. In *Climate and Man, Yearbook of Agriculture, 1941,* 98–127. Washington, D.C.: Department of Agriculture.

Bolin, Bert, B. R. Döös, Jill Jäger, and Richard A. Warrick. 1986. *The Greenhouse Effect, Climate Change and Ecosystems.* New York: John Wiley & Sons.

Bolin, Bert, and Robert B. Cook, eds. 1983. *The Major Biogeochemical Cycles and Their Interactions.* New York: John Wiley & Sons.

Boucher, Keith. 1975. *Global Climate.* New York: John Wiley & Sons.

Brooks, C. E. P. 1970. *Climate Through the Ages.* New York: Dover (rpt. of 1949 ed.).

Bryson, R. A., and D. A. Baerreis. 1967. Possibilities of Major Climatic Modification and Their Implications: Northwest India, a Case for Study. *Bulletin of the American Meteorological Society* 48 (No. 3): 136–42.

Bryson, Reid A., and Thomas J. Murray. 1977. *Climates of Hunger*. Madison: University of Wisconsin Press.

Budyko, M. I. 1982. *The Earth's Climate: Past and Future*. New York: Academic Press.

Carter, Douglas B., and John R. Mather. 1966. *Climatic Classification for Environmental Biology*. Publications in Climatology vol. 19, no 4. Elmer, N.J.: C. W. Thornthwaite Associates Laboratory of Climatology.

Cess, R. D., et al., 1989. Interpretation of Cloud-Climate Feedback as Produced by 14 Atmospheric General Circulation Models. *Science* 245 (4 August): 513–16.

Chang, Jen-Hu. 1968. *Climate and Agriculture*. Chicago: Aldine.

Changnon, Stanley A., 1989. *The 1988 Drought: Papers Prepared by Regional Climatic Centers and Allied Agencies*. Rockville, Md.: National Climate Program Office, NOAA.

Chapman, Clark R. 1977. *The Inner Planets*. New York: Charles Scribner's Sons.

Clark, William C. 1989. Managing Planet Earth. *Scientific American* (September): 47–54.

Clark, William C., and R. E. Munn, eds. 1986. *Sustainable Development of the Biosphere*. London: Oxford University Press.

Critchfield, Howard J. 1983. *General Climatology*. 4th ed. Englewood Cliffs. N.J.: Prentice Hall.

Dort, Wakefield, Jr., and J. Knox Jones, Jr. 1970. *Pleistocene and Recent Environments of the Central Great Plains*. Lawrence: University Press of Kansas.

Dougal, Merwin D., E. Robert Baumann, and John F. Timmons. 1970. *Physical and Economic Factors Associated with the Establishment of Stream Water Quality Standards, Vol. 2*. Ames: Iowa State Engineering Research Institute.

Flora, Snowden D. 1953. *Tornadoes of the United States*. Norman: University of Oklahoma Press.

———. 1956. *Hailstorms of the United States*. Norman: University of Oklahoma Press.

Forrester, Frank H. 1984. Groundwater: Buried Treasure. *American Forest* 90 (August): 18–21, 62–63.

Gibbons, D. C. 1986. *The Economic Value of Water*. Washington, D.C.: Resources for the Future.

Girdler, R. W. 1964. The Palaeomagnetic Latitudes of Possible Ancient Glaciations. In A. E. M. Nairn, ed., *Problems in Palaeoclimatology*, 155–78. New York: Interscience Publishers.

Graedel, Thomas E., and Paul J. Crutzen. 1989. Changing the Atmosphere. *Scientific American* (September): 58–68.

Gribbin, John. 1976. *Forecasts, Famines and Freezes: Climate and Man's Future*. New York: Walker and Company.

———., ed. 1978. *Climatic Change*. London, Cambridge University Press.

Griffiths, John F., and Dennis M. Driscoll. 1982. *Survey of Climatology*. Columbus, Oh.: Charles E. Merrill.

Hansen, J., and Lebedeff. 1988. Global Air Temperatures: Update Through 1987. *Geophysical Research Letter* 15: 323–26.

Harrison, Lucia. 1935. *Daylight, Twilight, Darkness, and Time*. New York: Silver, Burdett.

Hoyer, B. E., J. E. Combs, R. O. Kelley, C. Cousins-Leatherman, and J. H. Seyb. 1987. *Iowa Groundwater Protection Strategy*. Des Moines: Iowa Department of Natural Resources.

Hoyt, John C. 1936. *Droughts of 1930–34*. Water-Supply Paper 680. Washington, D.C.: U.S. Department of Interior, Superintendent of Documents.

Hunten, D. M., L. Colin, T. M. Donahue, and V. I Moroz, eds. 1983. *Venus*. Tucson: University of Arizona Press.

Huntley, B., and H. J. Birks. 1983. *An Atlas of Present Pollen Maps for Europe: 0–13,000 Years Ago*. London: Cambridge University Press.

Huntley, Brian, and I. Colin Prentice. 1988. July Temperatures in Europe from Pollen Data, 6000 Years Before Present. *Science* 243 (24 February): 1,007–1,008.

Iowa National Resources Council. 1953. Water Resources and Water Problems, Des Moines River Basin. Des Moines: INRC.

———. 1958. Water Resources and Water Problems, Northeastern Iowa River Basins. Des Moines: INRC.

Johnson, Arthur H. 1986. Acid Deposition: Trends, Relationships, and Effects. *Environment* (May): 6–11.

Johnson, Wendell C. 1965. *Wind in the Southwestern Great Plains.* Conservation Research Report No. 6. Washington, D.C.: U.S. Government Printing Office.

Kates, Robert W., Jesse H. Ausubel, and Mimi Berberian, eds. 1985. *Climate Impact Assessment.* New York: John Wiley & Sons (for the Scientific Committee on Problems of the Environment [SCOPE]).

Kates, Robert W., and Ian Burton, eds. 1986. *Geography, Resources, and Environment.* Vol. 2, Themes from the Work of Gilbert F. White. Chicago: University of Chicago Press.

Kendrew, W. G. 1949. *Climatology.* London: Oxford University Press.

———. 1953. *The Climates of the Continents.* London: Oxford University Press.

Kerr, Richard A. 1988. Is the Greenhouse Here? *Science* 239(5 February): 283–84.

———. 1988. Ozone Hole Bodes Ill for the Globe. *Science* 241(12 August): 283–84.

———. 1989. How to Fix the Clouds in Greenhouse Models. *Science* 243(6 January): 28–29.

———. 1989. New Way to Switch Earth Between Hot and Cold. *Science* 243(27 January): 480.

———. 1989. Hansen vs. the World on the Greenhouse Threat. *Science* 244(2 June): 1,041–43.

———. 1989. Volcanoes Can Muddle the Greenhouse. *Science* 245(14 July): 127–28.

Kessler, Edwin, ed. 1983. *The Thunderstorm in Human Affairs.* Norman: University of Oklahoma Press.

Kivelson, Margaret G., ed. 1986. *The Solar System: Observations and Interpretations.* Englewood Cliffs, N.J.: Prentice Hall.

Krueger, Arlin J. 1983. Sighting of El Chicón Sulfur Dioxide Clouds with the Nimbus 7 Total Ozone Mapping Spectrometer. *Science* 220 (24 June): 1,377–79.

Lamb, H. H. 1966. *The Changing Climate.* London: Methuen.

———. 1985. *Climatic History and the Future.* Princeton, N.J.: Princeton University Press.

Landsberg, Helmut E. 1981. *The Urban Climate.* New York: Academic Press.

Laporte, Leo F. 1977. Paleoenvironments and Paleoecology. *American Scientist* 65 (November–December): 720–28.

Leovy, Conway B. 1977. The Atmosphere of Mars. *Scientific American* 237(July): 34–43.

Lindstrom, D. R., and D. R. MacAyeal. 1988. Scandinavian, Siberian, and Arctic Ocean Glaciation: Effect of Holocene Atmospheric Co_2 Variations. *Science* 239(5 February): 563.

Lockwood, John G. 1985. *World Climatic Systems.* London: Edward Arnold.

Lydolph, Paul E. 1985. *The Climate of the Earth.* Totowa, N.J.: Rowman & Allanheld.

McLaren, Digby J., and Brian J. Skinner. 1989. *Resources and World Development.* London: Oxford University Press.

Magaritz, Mordeckai, and Aaron Kaufman. 1983. Paleoclimate in Desert Regions. *American Scientist* 71 (September–October): 514–21.

Mather, John R. 1974. *Climatology, Fundamentals and Applications.* New York: McGraw-Hill.

Matthews, Paul O. 1984. *Water Resources, Geography and Law.* Washington, D.C.: Association of American Geographers.

Mehringer, Peter J., Jr., James E. King, and Everett H. Lindsay. 1970. A Record of Wisconsin-Age Vegetation and Fauna from the Ozarks of Western Missouri. In Wakefield Dort, Jr., and J. Knox Jones, Jr., eds., *Pleistocene and Recent Environments of the Central Great Plains.* Lawrence: University Press of Kansas.

Meinzer, O. E. 1923. *The Occurrence of Ground Water in the United States.* Water Supply Paper 489. Washington, D.C.: U.S. Geological Survey.

———. 1927. *Large Springs in the United States.* Water Supply Paper 557. Washington, D.C.: U. S. Geological Survey.
Meyers, J. S. 1962. *Evaporation from the 17 Western States.* Professional Paper 272-D. Washington, D.C.: U.S. Geological Survey.
National Aeronautics and Space Administration. 1982. Our Prodigal Sun. *NASA Facts* (April).
National Research Council. 1973. *Weather and Climate Modification: Problems and Progress.* Washington, D.C.: National Academy of Sciences.
———. 1975. *Understanding Climatic Change.* Washington, D.C.: National Academy of Sciences.
———. 1976. *Climate and Food: Climatic Fluctuation and U.S. Agricultural Production.* Washington, D.C.: National Academy of Sciences.
National Research Council. 1986. *Acid Deposition: Long-Term Trends.* Washington, D.C.: National Academy Press.
Neuberger, Morris, James Edinger, and William Bonner. 1982. *Understanding Our Atmospheric Environment.* 2d ed. San Francisco: Freeman.
Norwine, Jim. 1978. *Climate and Human Ecology.* Houston: D. Armstrong.
Oliver, John E. 1973. *Climate and Man's Environment.* New York: John Wiley & Sons.
Peltier, W. R., and A. M. Tushingham. 1989. Global Sea Level Rise and the Greenhouse Effect: Might They be Connected? *Science* 241(23 September): 806–10.
Pittock, A. B., L. A. Frakes, D. Jenssen, J. A. Peterson, and J. W. Zillman, eds. 1978. *Climatic Change and Variability: A Southern Perspective.* Cambridge: Cambridge University Press.
Ponte, Lowell. 1976. *The Cooling.* Englewood Cliffs, N.J.: Prentice Hall.
Rahn, Kenneth A., and Douglas H. Lowenthal. 1986. The Acid Rain Whodunit. *Natural History* 95(No. 7, July): 62–65.
Ramanathan, V., R. J. Cicerone, H. B. Singh, and J. T. Kiehl. 1985. Trace Gas Trends and Their Potential Role in Climate Change. *Journal of Geophysical Research* 90:5,547–66.
Ramanathan, V., et al. 1989. Cloud-Radiative Forcing and Climate: Results from the Earth Radiation Budget Experiment. *Science* 243(6 January): 57–63.
Roberts, Leslie. 1989. Does the Ozone Hole Threaten Antarctic Life? *Science* 244(21 April): 288–89.
Roberts, Walter Orr, and Henry Lansford. 1979. *The Climate Mandate.* San Francisco: W. H. Freeman.
Robinson, T. W. 1958. *Phreatophytes.* Water Supply Paper 1423. Washington, D.C.: U.S. Geological Survey.
Rosenberg, Norman J., Blaine L. Blad, and Shashi B. Verma. 1983. *Microclimate: The Biological Environment.* New York: John Wiley & Sons.
Rosenberg, Norman J., William E. Easterling III, Pierre R. Crosson, and Joel Darmstadter, eds. 1989. *Greenhouse Warming: Abatement and Adaptation.* Washington, D.C.: Resources for the Future.
Schneider, Stephen H. 1988. Doing Something About the Weather. *World Monitor* 1(No. 3, December): 28–37.
———. 1989. The Greenhouse Effect: Science and Policy. *Science* 243(10 February): 771–81.
———. 1989. The Changing Climate. *Scientific American* (September): 70–79.
———. 1989. *Global Warming: Are We Entering the Greenhouse Century?* San Francisco: Sierra Club.
Schneider, Stephen H., and Randi Londer. 1984. *The Coevolution of Climate and Life.* San Francisco: Sierra Club.
Sears, Paul B. 1980. *Deserts on the March.* 4th ed. Norman: University of Oklahoma Press.
Shaw, Robert A. 1977. Climate and Weather for Food Production. In E. R. Duncan, ed., *Dimensions of World Food Problems,* 105–17. Ames: Iowa State University Press.
Shaw, Robert H., and W. C. Burrows. 1966. Water Supply, Water Use, and Water Requirement. In W. H. Pierre, S. R. Aldrich, and W. P. Martin, eds., *Advances in Corn Production,* 121–42. Ames: Iowa State University Press.

Shear, J. A. 1964. The Polar Marine Climate. *Annals, Association of American Geographers* 56:310–17.
Sigurdsson, Haraldur, and Steven Carey. 1988. The Far Reach of Tambors. *Natural History* (No. 6, June): 66–73.
Smith, Nigel J. H. 1981. Colonization Lessons from a Tropical Forest. *Science* 214(13 November): 755–61.
Stommel, Henry, and Elizabeth Stommel. 1983. *Volcano Weather: The Story of 1816, The Year Without a Summer*. Newport, R.I.: Seven Seas Press.
Stothers, Richard B. 1984. The Great Tambora Eruption in 1815 and Its Aftermath. *Science* 224(15 June): 1,191–98.
Strahler, Arthur N. 1975. *Physical Geography*. 4th ed. New York: John Wiley & Sons.
Thomas, W. L., Jr., ed. 1956 [1984]. *Man's Role in Changing the Face of the Earth*. Chicago: University of Chicago Press.
Thompson, B. W. 1975. *Africa: The Climatic Background*. Ibadan, Nigeria: Oxford University Press, Nigeria.
Thompson, Louis M. 1984. *Effects of Weather Variability and Climate Change on Corn and Soybeans*. Indianola, Iowa: Simpson College.
Thornthwaite, C. W. 1948. An Approach Toward a Rational Classification of Climate. *Geographical Review* 38:55–94.
Thornthwaite, C. W., and Associates. 1962–65. *Average Climatic Water Balance Data for the Continents*. Publications in Climatology, pts. 1–8, vols. 15–18. Elmer, N.J.: Laboratory of Climatology.
Thornthwaite, C. W., and J. R. Mather. 1955. *The Water Balance*. Publications in Climatology, vol. 8, no. 1. Centerton, N.J.: Drexel Institute Laboratory of Climatology.
———. and ———. 1957. *Instructions and Tables for Computing Potential Evapotranspiration and the Water Balance*. Publications in Climatology, vol. 10, no. 3. Centerton, N.J.: Drexel Institute Laboratory of Climatology.
Toon, Owen B., and James B. Pollack. 1980. Atmospheric Aerosols and Climate. *American Scientist* 68(May–June): 268–78.
Tricart, J., and A. Cailleux. 1972. *Introduction to Climatic Geomorphology*. New York: St. Martin's.
Trolls-Smith, J. 1959. Fortidens Danmark ("Denmark of the Past"). In Stenalderen, Turistforeningen for Danmark Arbog 1959, 7–32. Copenhagen, Denmark: Det Berlingske Bogtrykkeri.
———. 1961. Jernalderens Naturforhold ("Natural Environment of the Iron Age"). In *Jernalderen, Turistforeningens Arbog 1961,* 7–20. Odense, Denmark: Fyens Stiftsbogtrykkeri.
Turner, B. L., ed. 1989. *The Earth as Transformed by Human Action*. London: Cambridge University Press.
U.S. Department of Agriculture. 1955. *Water: The Yearbook of Agriculture, 1955*. Washington, D.C.: USDA.
———. Soil Conservation Service. 1960. *Soil Classification: A Comprehensive System, in Approximation*. Washington, D.C.: Government Printing Office.
———. 1981. *1980 Appraisal*. Pt. 1, *Soil, Water, and Related Resources in the United States: Status, Condition, and Trends*; Pt. 2, *Analysis of Resource Trends*. Washington, D.C.: U.S. Department of Agriculture.
U.S. Department of Commerce, NOAA. 1988. *National Climate Program, 1987 Annual Report*. Rockville, Md.: National Climate Program Office.
———. 1989. *The Drought of 1988 and Beyond*. Rockville, Md.: National Climatic Program Office.
U. S. Senate, Select Committee on National Water Resources. 1960. *Evapotranspiration Reduction*. Washington, D.C.: Committee Print No. 21.
U.S. Water Resources Council. 1978. *The Nation's Water Resources*. Vol. 2, *Water Quality and Related Land Considerations, 1975–2000*. Washington, D.C.: U.S. Government Printing Office.

Voropayev, Grigori. 1985. Water and Man. *The Unesco Courier*. (January): 4–7.
Waggoner, Paul E. 1984. Agriculture and Carbon Dioxide. *American Scientist* 72(March–April): 179–84.
Waite, Paul J., and Mary M. Partington. 1986. *Iowa Precipitation Variations: Past, Present, and Future*. Des Moines: Iowa Department of Agriculture, State Climatology Office.
Wernstedt, Frederick L. 1972. *World Climatic Data*. Lemont, Pa.: Climatic Data Press.
White, Gilbert F. 1975. *Flood Hazard in the United States: A Research Assessment*. Boulder: University of Colorado Institute of Behavioral Science.
Wigley, T. M. L. 1989. Possible Climate Change Due to SO_2-Derived Cloud Condensation Nuclei. *Nature* 339(1 June): 365–67.
Wilson, E. O., and Frances M. Peter. 1988. *Biodiversity*. Washington, D.C.: National Academy Press.
World Commission on Environment and Development. 1987. *Our Common Future*. London: Oxford University Press.
Wright, H. E., Jr., and David G. Frey. 1965. *The Quaternary of the United States*. Princeton, N.J.: Princeton University Press.
Yoshino, Masatoshi M., ed. 1971. *Water Balance of Monsoon Asia: A Climatological Approach*. Honolulu: University of Hawaii Press.
Ziegler, Philip. 1969. *The Black Death*. New York: Harper & Row.
Zwally, H. J., A. C. Brenner, J. A. Major, R. A. Bindschalder, and J. G. Marsh. 1989. Growth of Greenland Ice Sheet: Measurement. *Science* 246:(22 December): 1,587–91.

Part 2

VEGETATION
(Chapters 5–6)

Aagesen, Aa. 1961. *The Population*. (Separate text and photographs to accompany Niels Nielsen, ed., *Atlas of Denmark*, vol. 2.) Copenhagen: C. A. Reitzels Forlag.

Akin, Wallace E. 1963. Reclamation of the Danish Heath. *Journal of Soil and Water Conservation* 18(3, May–June): 97–102.

———. 1968. *The North Central United States*. Princeton, N.J.: Van Nostrand.

Allaway, W. H. 1957. pH, Soil Activity, and Plant Growth. In USDA, *Soil, The 1957 Yearbook of Agriculture*, 67–71. Washington, D.C.: U.S. Government Printing Office.

Bartlett, H. H. 1956. Fire, Primitive Agriculture, and Grazing in the Tropics. In W. L. Thomas, Jr., ed., *Man's Role in Changing the Face of the Earth*, 692–720. Chicago: University of Chicago Press.

Buckman, Harry O., and Nyle C. Brady. 1969. *The Nature and Properties of Soils*. 7th ed. New York: Macmillan.

Bunting, B. T. 1959. Vallundgaard: A Study of Land Use in a Podsolized Area of Western Jutland. In *Transactions and Papers, 1959*. (Publication no. 26), 89–105. London: Institute of British Geographers.

Bure, Kristjan, ed. 1961. *Jernalderen [Iron Age]: Turistforeningen for Danmarks Aarbog 1961*. Odense, Denmark: Fyens Stiftsbogtrykkeri.

Butzer, Carl. 1964. *Environment and Archaeology: An Introduction to Pleistocene Geography*. Chicago: Aldine.

Caufield, Catherine. 1985. *In the Rainforest*. New York: Knopf.

Chang, Jen-Hu. 1968. *Climate and Agriculture: An Ecological Survey*. Chicago: Aldine.

———. 1970. Potential Photosynthesis and Crop Productivity. *Annals of the Association of American Geographers* 60:92–101.

Chapman, H. W., L. S. Gleason, and W. E. Loomis. 1954. The Carbon Dioxide Content of Field Air. *Plant Physiology* 29:500–503.

Clements, Harold M., Sr. 1969. *The Mechanization of Agriculture in Brazil*. Gainsville: University of Florida Press.

Coleman, V. T., and A. Mehlich. 1957. The Chemistry of Soil pH. In USDA, *Soil, The 1957 Yearbook of Agriculture*, 72–79. Washington, D.C.: U.S. Government Printing Office.

Collinson, A. A. 1977. *Introduction to World Vegetation*. London: George Allen & Unwin.

Colman, E. A. 1953. *Vegetation and Watershed Management*. New York: Ronald Press.

Cooper, J. P., ed. 1975. *Photosynthesis and Productivity in Different Environments*. International Biological Programme 3. Cambridge: Cambridge University Press.

Daubenmire, Rexford F. 1959. *Plants and Environment*. 2d ed. New York: John Wiley & Sons.

———. 1978. *Plant Geography, with Special Reference to North America*. New York: Academic Press.

DeLaubenfels, David J. 1970. *A Geography of Plants and Animals*. Dubuque, Ia.: William C. Brown.

Dort, Wakefield, Jr., and J. Knox Jones, Jr., eds. 1970. *Pleistocene and Recent Environments of the Central Great Plains*. Lawrence: University of Kansas Press.

Estes, James R., Ronald J. Tyrl, and Jere N. Brunken, eds. 1979. *Grasses and Grasslands*. Norman: University of Oklahoma Press.

Eyre, S. R. 1963. *Vegetation and Soils: A World Picture*. Chicago: Aldine.

———., ed. 1971. *World Vegetation Types*. New York: Columbia University Press.

Fleming, Peter. 1934. *Brazilian Adventure*. New York: Charles Scribner's Sons.

Fuller, H. J. 1948. Carbon Dioxide Concentrations of the Atmosphere Above Illinois Forest and Grassland. *American Midland Naturalist* 39:247–49.

Gerasimov, I. P. 1983. *Geography and Ecology*. Moscow: Progress Publishers.

Golly, Frank B., et al. 1975. *Mineral Cycling in a Tropical Moist Forest*. Athens: University of Georgia Press.

Good, Ronald. 1964. *The Geography of the Flowering Plants*. New York: John Wiley & Sons.

Graham, Edward N. 1956. The Re-creative Power of Plant Communities. In William L. Thomas, Jr., ed., *Man's Role in Changing the Face of the Earth*, 688. Chicago: University of Chicago Press.

Grigg, D. B. 1974. *The Agricultural Systems of the World: An Evolutionary Approach*. Cambridge: Cambridge University Press.

Hove, Th. 1962. *Hedeselskabets Hedeundersogelser, 1938–53 [The Heath Society's Heath Development, 1938--55]*. Beretning No. 7. Viborg, Denmark: Hedeselskabets Forskningsvirksomhed.

James, Preston E. 1959. *Latin America*. 3rd ed. New York: Odyssey Press.

Jensen, Einar. 1937. *Danish Agriculture: Its Economic Development*. Copenhagen: J. H. Schultz Forlag.

Jensen, Martin. 1954. *Shelter Effect*. Copenhagen: Danish Technical Press.

Jordan, Carl F., ed. 1981. *Tropical Ecology*. Stroudsburg, Pa.: Hutchinson Ross.

Jordan, Carl F. 1982. Amazon Rain Forests. *American Scientist* 70 (July–August): 394–401.

Kampp, Aa. H. 1959. *Landsbrugsgeografiske Studier over Danmark [Agricultural Geography of Denmark]*. Copenhagen, Denmark: C. A. Reitzels Forlag.

Knudsen, Kr. H. 1960. *Vanding i Landbruget [Irrigation in Agriculture]*. Copenhagen, Denmark: Landbrugets Informationskontor.

McGaughey, Stephen E., and Hans M. Gregersen, eds. 1983. *Forest Based Development in Latin America*. Washington, D.C.: Inter-American Development Bank.

McPherson, W. W., ed. 1968. *Economic Development of Tropical Agriculture*. Gainsville: University of Florida Press.

Matthei, Oscar R. 1963. *Manual Ilustrado de las Malezas de la Provincia de Nuble*. Chile: Universidad de Conceptión, Escuela de Agronomía.

Mitchell, Andrew W. 1986. *The Enchanted Canopy*. New York: Macmillan.

Moran, Emilio F. 1981. *Developing the Amazon*. Bloomington: Indiana University Press.

National Academy of Sciences. 1976. *Productivity of World Ecosystems*. Washington, D.C.: National Academy of Sciences.

National Research Council. 1980a. *Conversion of Tropical Moist Forests*. Washington, D.C.: National Academy of Sciences.

———. 1980b. *Research Priorities in Tropical Biology*. Washington, D.C.: National Academy of Sciences.

———. 1981. *Atmosphere-Biosphere Interactions: Toward a Better Understanding of the Ecological Consequences of Fossil Fuel Combustion*. Washington, D.C.: National Academy Press.

———. 1982. *Ecological Aspects of Development in the Humid Tropics*. Washington, D.C.: National Academy Press.

Nelson, J. G., and M. J. Chambers, eds. 1969. *Vegetation, Soils and Wildlife*. Toronto: Methuen.

Nelson, Michael. 1973. *The Development of Tropical Lands: Policy Issues in Latin America*. Baltimore: Johns Hopkins University Press (for Resources for the Future, Inc.).

Nielsen, N. C. 1953a. Arealets Benyttelse [Improvement of the Area]. In *Hedens Opdyrkning i Danmark ("Heath Reclamation in Denmark")*, 393–415. Viborg, Denmark: Det Danske Hedeselskab.

———. 1953b. Landejendomme og Befolkning [Land Division and Settlement]. In *Hedens Opdyrkning i Danmark ("Heath Reclamation in Denmark")*, 387–92. Viborg, Denmark: Det Danske Hedeselskab.

Oliver, John E. 1973. *Climate and Man's Environment*. New York: John Wiley & Sons.

Polunin, Nicholas. 1960. *Introduction to Plant Geography and Some Related Sciences*. New York: McGraw-Hill.

Prior, Jean Cutler. 1976. *A Regional Guide to Iowa Landforms*. Iowa City: Iowa Geological Survey.

Richards, P. W. 1966. *The Tropical Rain Forest*. Cambridge: Cambridge University Press.

Schou, Axel. 1940. *The Landscapes*. (Separate text and photographs to accompany Niels Nielsen, ed. *Atlas of Denmark*, vol. 1). Copenhagen, Denmark: H. Hagerup.

Scientific American Books. 1957. *Plant Life*. New York: Simon & Schuster.

Shelford, Victor E. 1963. *The Ecology of North America*. Urbana: University of Illinois Press.

Shimek, Bohumil. 1910. *Geology of Harrison and Monona Counties*. Quoted in Jean C. Prior, *A Regional Guide to Iowa Landforms*. Iowa City: Iowa Geological Survey, 1976.

Silverberg, Robert. 1967. *The World of the Rain Forest*. New York: Meredith Press.

Simmons, Ian. 1973. *Biogeographical Processes*. New York: John Wiley & Sons.

Thomas, William L., Jr., ed. 1956. *Man's Role in Changing the Face of the Earth*. Chicago: University of Chicago Press.

Thompson, Louis M., and Frederick R. Troeh. 1978. *Soils and Soil Fertility*. 4th ed. New York: McGraw-Hill.

Thorpe, H. 1957. A Special Case of Heath Reclamation in the Alheden District of Jutland, 1700–1955. In *Transactions and Papers, 1957 (Publication No. 17)*, 87–121. London: The Institute of British Geographers.

Trolls-Smith, J. 1961. Jernalderens Naturfolhold [The Natural Conditions of the Iron Age]. In K. Bure, ed., *Jernalderen [Iron Age]: Touristforeningen for Danmark, Aarbog 1961*, 7–20. Odense, Denmark: Fyens Stiftsbogtrykkeri.

Truog, E. 1946. Soil Reaction Influence on Availability of Plant Nutrients. *Soil Science Society of America Proceedings* 11:305–308.

U.S. Department of Agriculture. 1949. *Trees, Yearbook of Agriculture, 1949*. Washington, D.C.: U.S. Government Printing Office.

———. 1957. *Soil, the 1957 Yearbook of Agriculture*. Washington, D.C.: U.S. Government Printing Office.

Vale, Thomas R. 1982. *Plants and People: Vegetation Change in North America*. Washington, D.C.: Association of American Geographers.

Voight, G. K. 1962. The Role of Carbon Dioxide in Soil. In T. T. Kozlowski, ed., *Tree Growth*: New York: Ronald Press.

Walter, Heinrich. 1973. *Vegetation of the Earth in Relation to Climate and the Eco-physiological*

Conditions. Trans. from 2d German ed. by Joy Wieser. Heidelberg Science Library Vol. 15. New York: Springer-Verlag.

Ward, Jay. 1949. Naval Stores: The Industry. In USDA, *Trees, Yearbook of Agriculture, 1949*, 286–291. Washington, D.C.: U.S. Government Printing Office.

Wernstedt, Frederick L. 1972. *World Climatic Data*. Lemont, Pa.: Climatic Data Press.

Wright, H. E., Jr. 1970. Vegetational History of the Central Great Plains. In W. Dort, Jr., and J. K. Jones, Jr., eds., *Pleistocene and Recent Environments of the Central Great Plains*, 157–72. Lawrence: University of Kansas Press.

Part 3

SOILS
(Chapters 7–8)

Agricultural Extension Service. 1949. *Principal Upland Soils of Iowa, Agron. 49 (rev.)*. Ames: Iowa State College.

Akin, W. E., and R. I. Dideriksen. 1967. The Geology and Soils of Iowa. *Journal of Soil and Water Conservation 22* (No. 3, May–June): 110–14.

Baldwin, M., C. E. Kellog, and J. Thorp. 1938. Soil Classification. In *Soils and Men, 1938 Yearbook of Agriculture*, 979–1,001. Washington, D.C.: U.S. Department of Agriculture.

Batten, James W., and J. Sullivan Gibson. 1977. *Soils: Their Nature, Classes, Distribution, Uses, and Care*. Rev. ed. University, Ala.: University of Alabama Press.

Bear, Firman E., H. Wayne Pritchard, and Wallace E. Akin. 1986. *Earth: The Stuff of Life*. 2d ed. Norman: University of Oklahoma Press.

Berger, Kermit C. 1972. *Sun, Soil, and Survival: An Introduction to Soil*. Norman: University of Oklahoma Press.

Bridges, E. M. 1970. *World Soils*. London: Cambridge University Press.

Brown, Lester R. 1987. Sustaining World Agriculture. In L. R. Brown, ed., *State of the World, 1987*, 122–38. New York: W. W. Norton (for the Worldwatch Institute).

Buckman, H. O., and Brady, N. C. 1974. *The Nature and Properties of Soils*. 8th ed. New York: Macmillan.

Bunting, B. T. 1965. *The Geography of Soils*. Chicago: Aldine.

Buol, S. W., F. D. Hole, and R. J. McCracken. 1980. *Soil Genesis and Classification*. 2d ed. Ames: Iowa State University Press.

Cruickshank, J. G. 1972. *Soil Geography*. New York: John Wiley & Sons.

Donahue, Roy L., J. C. Shickluna, and L. S. Robertson. 1971. *Soils: An Introduction to Soils and Plant Growth*. 3d ed. Englewood Cliffs, N.J.: Prentice Hall.

Dudal, R. 1968. Definitions of Soil Units for the Soil Map of the World. In *World Soil Resources Report 32*, map. Rome, Italy: FAO.

Eckholm, Erik P. 1976. *Losing Ground: Environmental Stress and World Food Prospects*. New York: W. W. Norton (for the Worldwatch Institute).

Eyre, S. R. 1963. *Vegetation and Soils*. Chicago: Aldine.

Falesi, Italo Claudio. 1974. Soils of the Brazilian Amazon. In C. Wagley, ed., *Man in the Amazon*, 201–29. Gainesville: University of Florida Press.

Glinka, K. D. 1927. *The Great Soil Groups of the World and Their Development*. Trans. C. F. Marbut. Ann Arbor, Mich.: Edwards Brothers.

Golley, F. B., J. T. McGinnis, R. G. Clements, G. I. Child, and M. J. Duever. 1975. *Mineral Cycling in a Tropical Moist Forest Ecosystem*. Athens: University of Georgia Press.

Grim, Ralph E. 1968. *Clay Mineralogy*. 2d ed. New York: McGraw-Hill.

Hunt, C. B. 1968. *Geology of Soils*. San Francisco: W. H. Freeman.

Jordan, Carl F., ed. 1981. *Tropical Ecology*. Stroudsburg, Penn.: Hutchinson Ross.

Kellogg, C. E. 1941. Climate and Soil. In U.S. Department of Agriculture, *Climate and Man, 1941 Yearbook of Agriculture*, 265–91. Washington, D.C.: U.S. Government Printing Office.

———. 1963. Why a New System of Soil Classification? *Soil Science* 96:1–5.

Limbird, A. 1986. An Approach to Teaching Soils Geography in Canada: The Classification Dilemma. *The Operational Geographer/La Géographie Appliquée* 11:18–21.

Marbut, C. F. 1936. Soils of the United States. Pt. 3 of *Atlas of American Agriculture*. Washington, D.C.: U.S. Department of Agriculture.

Moran, Elilio F. 1981. *Developing the Amazon*. Bloomington: Indiana University Press.

National Research Council. 1982. *Ecological Aspects of Development in the Humid Tropics*. Washington, D.C.: National Academy Press.

Nelson, Michael. 1973. *The Development of Tropical Lands: Policy Issues in Latin America*. Baltimore: Johns Hopkins University Press.

Riecken, F.F., and G.D. Smith. 1949. Lower Categories of Soil Classification: Family, Series, Type, and Phase. *Soil Science* 67:107–15.

Shrader, William D. 1977. Soil Resources—Characteristics, Potentials, and Limitations. In E. R. Duncan, ed., *Dimensions of World Food Problems*, 118–35. Ames: Iowa State University Press.

Smith, G. D. 1963. Objectives and Basic Assumptions of the New Soil Classification System. *Soil Science* 96:6.

Soil Survey Staff. 1960. *Soil Classification: A Comprehensive System, 7th Approximation*. Washington, D.C.: Soil Conservation Service, USDA.

———. 1975. *Soil Taxonomy*. Agricultural Handbook No. 436. Washington, D.C.: Soil Conservation Service, USDA.

Thompson, Louis M., and Frederick R. Troeh. 1978. *Soils and Fertility*. 4th ed. New York: McGraw-Hill.

Thorp, J., and G. D. Smith. 1949. Higher Categories of Soil Classification: Order, Suborder, and Great Soil Groups. *Soil Science* 67:117–26.

U.S. Department of Agriculture. 1938. *Soils and Men, 1938 Yearbook of Agriculture*. Washington, D.C.: U.S. Government Printing Office.

———. 1957. *Soil, The 1957 Yearbook of Agriculture*. Washington, D.C.: U.S. Government Printing Office.

Wagley, Charles, ed. 1974. *Man in the Amazon*. Gainesville: University of Florida Press.

Wall, J.R.D., and J.R.F. Hansell. 1974. *Land Resources of the Solomon Islands*. Vol. 3: Malaita and Ulawa. Surbiton, U.K.: Land Resources Division, Ministry of Overseas Development.

Went, F. W., and N. Stark. 1968. Mycorrhiza. *BioScience* 18:1,033–39.

Wolf, Edward C. 1987. Raising Agricultural Productivity. In L. R. Brown, ed., *State of the World, 1987*, 138–56. New York: W. W. Norton (for the Worldwatch Institute).

Young, Anthony. 1976. *Tropical Soils and Soil Survey*. Cambridge Geographical Studies 9. Cambridge: Cambridge University Press.

INDEX

Page numbers in italics refer to figures or tables